E

H

n

n

DO NOT REMOVE
CARDS FROM POCKET

11/15/96

WILEY-PRAXIS SERIES IN SPACE SCIENCE AND TECHNOLOGY
Series Editor: **John Mason, B.Sc., Ph.D.**
Consultant Editor: **David Sloggett, M.Sc., Ph.D.**

This series reflects the significant advances being made in space science and technology, including developments in astronautics and space life sciences. It provides a forum for the publication of new ideas and results of current research in areas such as spacecraft materials, propulsion systems, space automation and robotics, spacecraft communications, mission planning and management, and satellite data processing and archiving.

Aspects of space policy and space industrialization, including the commercial, legal and political ramifications of such activities, and the physiological, sociological and psychological problems of living and working in space, and spaceflight risk management are also addressed.

These books are written for professional space scientists, technologists, physicists and materials scientists, aeronautical and astronautical engineers, and life scientists, together with managers, policy makers and those involved in the space business. They are also of value to postgraduate and undergraduate students of space science and technology, and those on space-related courses (including psychology, physiology, medicine and sociology) and areas of the social and behavioural sciences.

SATELLITE CONTROL: A Comprehensive Approach
John T. Garner, Aerospace Consultant, formerly Principal Ground Support Engineer, Communications Satellite Programmes, ESA-ESTEC, Noordwijk, The Netherlands

LIVING AND WORKING IN SPACE: Human Behavior, Culture and Organization, Second edition
Philip Robert Harris, Executive Editor, *Space Governance* Journal; Vice President, United Societies in Space, Inc.

THE NEW RUSSIAN SPACE PROGRAMME: From Competition to Collaboration
Brian Harvey, M.A., H.D.E., F.B.I.S.

Forthcoming Titles

METALLURGICAL ASSESSMENT OF SPACECRAFT PARTS AND MATERIALS, Second edition
Barry D. Dunn, Head of Metallic Materials and Processes Section, ESA-ESTEC, The Netherlands

SOLAR POWER SATELLITES: A Space Energy System for Earth
Peter E. Glaser, Vice President (retired), Arthur D. Little Inc, USA, Frank P. Davidson, Coordinator, Macro-Engineering Research Group, Massachusetts Institute of Technology, USA, Katinka I. Csigi, Principal Consultant, ERIC International, USA

SOLAR SAILING: Technology, Dynamics and Mission Applications
Colin R. McInnes Department of Aerospace Engineering University of Glasgow

LIVING AND WORKING IN SPACE

Human Behavior, Culture and Organization
Second edition

Philip Robert Harris, Ph.D., M.S., B.B.A.
President, Harris International Ltd
Executive Editor, *Space Governance* Journal
Executive Vice President, United Societies in Space, Inc.
Associate Fellow, American Institute of Aeronautics and Astronautics

JOHN WILEY & SONS
Chichester • New York • Brisbane • Toronto • Singapore

Published in association with
PRAXIS PUBLISHING
Chichester

Copyright © 1996 Praxis Publishing Ltd
The White House,
Eastergate, Chichester,
West Sussex, PO20 6UR, England

First published in 1992
This Second edition published in 1996 by
John Wiley & Sons Ltd
in association with Praxis Publishing Ltd

Wiley Editorial Offices

John Wiley & Sons Ltd, Baffins Lane,
Chichester, West Sussex PO19 1UD, England

John Wiley & Sons, Inc., 605 Third Avenue,
New York, NY 10158-0012, USA

Jacaranda Wiley Ltd, G.P.O. Box 859, Brisbane
Queensland 4001, Australia

John Wiley & Sons (Canada) Ltd, 22 Worcester Road,
Rexdale, Ontario M9W 1L1, Canada

John Wiley & Sons (Asia) Pte Ltd, 2 Clementi Loop #02-01,
Jin Xing Distripark, Singapore 0512

Library of Congress Cataloguing-in-Publication Data
Harris, Philip R. (Philip Robert), 1926–.
  Living and Working in Space : human behavior, culture,
  and organization / Philip Robert Harris.—2nd ed.
  p. cm. -- (Wiley-Praxis series in space science and technology)
  Includes bibliographical references and index.
  ISBN 0-471-96255-4 (cloth : alk. paper). — ISBN 0-471-96256-2 (pbk. : alk. paper)
  1.Space culture/management. 2. Space habitation/settlement. 3. Space
  industrialization/commerce. I. Title. II Series.
  TL795.7.H37 1996   95-48038
  338.0919—dc20   CIP

ISBN 0-471-96255-4 Cloth
ISBN 0-471-96256-2 Paperback

Printed and bound in Great Britain by Hartnolls Ltd, Bodmin

Dedicated to THE ENVOYS OF HUMANKIND
—the world's Astronauts and Cosmonauts, some of whom gave their lives to advance human migration into space, like the CHALLENGER SEVEN!

IN MEMORIAM:
THE CREW OF THE SPACE SHUTTLE *CHALLENGER*
Flight 51-L, 28 January 1986

[Back row, l to r: Ellison Onizuka, S. Christa McAuliffe, Greg Jarvis, Judy Resnik. Front row, l. to r.: Mike Smith, Dick Scobee, and Ron McNair. Photograph courtesy of National Aeronautics and Space Administration.]

"We'll continue our quest in space. There will be more shuttle flights ... . More teachers in space. Nothing ends here. Our hopes and journeys continue."

—U.S. President Ronald Reagan

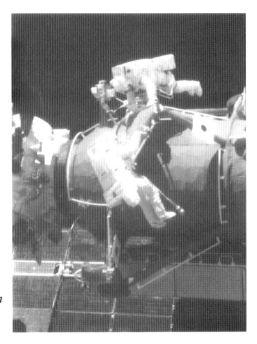

INTERPLANETARY WOMEN OF THE YEAR '96

Astronauts Linda M. Godwin and Shannon Lucid

Dr Linda M. Godwin, NASA mission specialist, engaged in a space walk from the Space Shuttle **Atlantis** cargo bay on March 27, 1996. In an era of international cooperation, this Shuttle was docked for the third time to the Russian **Mir (Peace)** Space Station which at nearly 240 tonnes is the largest spacecraft ever assembled in orbit. Pictured on our book cover, Godwin, a physicist, was teamed up with Army Lt. **Col. Michael Richard Clifford,** in the EVA installing of four panels used in **Mir** environmental experiments to trap cosmic space debris, while testing 1000 samples of coating material for a future **International Space Station**. It was the first time in 22 years since the demise of **Skylab** that Americans worked outside a space station!

On that same Mission (STS076-73-00G), astronaut **Shannon Lucid,** was dropped off on **Mir's** orbiting apartment and laboratory, to stay for almost five months with cosmonauts both named Yuri (**Onufrienko** and **Usachev**). **Lucid** will conduct scientific experiments with her Russian colleagues, and upon her return to Earth will set a new spaceflight endurance record, namely, the first American and first woman to be in orbit for that length of time. Proving that age and sex are no barriers to **living and working in space**, **Shannon**, 53, was born in Shanghai, China, where her American parents were missionaries; at six months after her birth, the family was interned in a Japanese prison camp for several months. Today **Lucid's** own family consists of husband, Michael, a petrochemist, and three adult children. With a laptop computer, she will communicate with them via e-mail, while working a busy five-day week!

Table of Contents

List of Exhibits

The author acknowledges with gratitude all who have contributed to these exhibits and who have been noted in the captions. * Asterisk designates an aerospace illustration by Dennis M. Davidson, Astronomical Artist for the Hayden Planetarium.

Cover—Shuttle Astronauts Teamwork at **Mir** Station

Dedication Page—Space Shuttle **Challenger** Crew, Flight 51-L

Chapter 3

Chapter 4

Chapter 5

Chapter 6

Notes on Contributors

Criswell, David R. (Appendix B). Director, Institute for Space Systems Operations at the University of Houston; associate administrator of the Texas Space Grant Consortium for universities throughout that state. With a Ph.D. from Rice University, he has been an astrophysicist for thirty years. His prolific research has resulted in many patents, including one as co-inventor of the Lunar Solar Power System. An AIAA Associate Fellow, Dr Criswell has published over 150 technical papers and conducted numerous studies for University Space Research Association and NASA, one of which resulted in *Automation & Robotics for the National Space Program* (NAGW629, 1985). His career has included positions with TRW Inc., Lunar and Planetary Institute, and California Space Institute. In l995 as a recognition of his accomplishments, the World Bar Association bestowed on him its **Space Humanitarian Award** and the National Space Society elected him to its Board of Directors. (Dr Criswell's address is ISSO, University of Houston, Houston, TX 77204, U.S.A.)

Davidson, Dennis M. (Art Director). Astronomical Artist, Hayden Planetarium, American Museum of Natural History, where he directs their computer graphics program; currently President, International Association of Astronomical Arts. Mr Davidson, a specialist in aerospace and astronomical art, contributed many original paintings and illustrations to this *second edition*. He is a Regent of United Societies in Space, Inc., and art editor of its journal, *Space Governance*. Formerly an illustrator with the California Space Institute and the School of Medicine at the University of California—San Diego, his clients range from NASA and aerospace companies to GTI Corporation and *Time-Life Books.* His commercial works appear in many space and medical publications, including the cover of the British journal, *Spaceflight*; *Air & Space*; *Popular Science*; *Discover*; and *Contact*, as well as covers for books by such publishers as Bantam, Random House, Silver-Burdett, Ellis Horwood, and others. A biology graduate from the University of California—San Diego, his artistic works have been exhibited in galleries from New York to Madrid. In 1992, he was selected to participate in NASA's Fine Arts Program. (Mr Davidson's address is Hayden Planetarium, American Museum of Natural History, Central Park West at 79th St., New York, NY 10024, U.S.A.)

Doré, Roland (Foreword). President, International Space University; former director, Canadian Space Agency. An engineering graduate of Ecole Polytechnique, Université de Montrèal, he received his M.S. and Ph.D. in Mechanical Engineering from Stanford University. At the Ecole Polytechnique, he held a number of administrative posts from dean to president and chairman of the board. His prestigious appointments in Canada have included membership in their National Research Council, as well as vice president of the Natural Sciences and Engineering Council. Recipient of numerous awards, Dr Doré has published over 70 scientific papers and lectured extensively at international conferences. When Canada's prime minister appointed him to head CSA, he promoted that space agency's role in the peaceful use and development of space, particularly through space science and technology, for the benefit of humankind. (Dr Doré's address in Strasbourg is ISU, Parc d'Innovation, Blvd. Gonthier d'Andernach, 67400 Illkirch, France.)

Goldman, Nathan C. (Chapter 9, Poems). Attorney specializing in space law in Houston, Texas; adjunct professor of space law in South Texas College of Law; research associate, Rice University. Author of numerous articles on space law, commerce, and policy, plus three books on *American Space Law* (1996); *Space Policy: An Introduction* (1992); and *Space Commerce: Free Enterprise on the High Frontier* (1985). A Regent of United Societies in Space, Inc., Dr Goldman holds a Ph.D. in political science from Johns Hopkins University and a J.D. from Duke University. Formerly, he was a NASA Faculty Fellow, an assistant professor of government at the University of Texas-Austin, and a Fellow there of the Institute of Constructive Capitalism. (Dr Goldman's address is 2328 Dryden Rd., Houston, TX 77254, U.S.A.)

Harris, Philip R. (Author). Management and Space Psychologist, President, Harris International; Executive Vice President, United Societies in Space, Inc.; Assoc. Fellow, American Institute of Aeronautics and Astronautics ... Previously, Dr Harris has been both a corporate and a college vice president, international management consultant for 160 human systems, and held numerous professional and research positions. The last-mentioned include Senior Scientist, NETROLOGIC Inc.; research associate, California Space Institute; education director, Air/Space America; research consultant, State of California Law Enforcement Command College; project director/contractor, Office of Naval Research ... "Phil" has published some 200 articles, and currently serves as Executive Editor of the journal, *Space Governance*, as well as co-editor of the **Managing Cultural Differences Series** of eight volumes. Author/editor of forty professional books, the most recent include: *Managing Cultural Differences,* fourth edition (1996), *New Work Culture and HRD Transformational Management* (1996); *Multicultural Law Enforcement (1995); High Performance Leadership,* second edition (1994) ... His honors include **Space Humanitarian Award**, World Bar Association; four **Journalistic Awards for Excellence**, Aviation/Space Writers Association; **Fulbright Professor to India**, U.S. State Department; **Torch Award** for outstanding human resource contributions, American Society for Training and Development ... He received his Ph.D. and M.S. in psychology from Fordham University, his B.B.A. in business from St. John's University, as well as certification as a psychologist from the University of the State of New York. (Dr Harris' address: Harris International Ltd., 2702 Costebelle Drive, LaJolla, CA 92038, U.S.A.)

O'Donnell, Declan J. (Appendix D). A tax attorney, Dr O'Donnell is president of the World Bar Association. He is the president and founder of United Societies in Space, Inc., a citizen organization promoting innovative strategies for developing outer space through alternative legal, governance, and financial means. Formerly a staff aide to the Governor, State of Colorado, he has served on the Colorado Local Government Study Commission staff. His treatises on space law have been published in the American Bar Association's *The Air and Space Lawyer*, *Space Policy, Behavioral Science, Space Power, Space Governance*, and in the 1995 proceedings of the International Lunar Exploration Conference. Declan graduated from the University of Notre Dame and Michigan State Law School from which he received a J.D. (Dr O'Donnell's address is Declan Joseph O'Donnell, P.C., 6841 South Yosemite, Englewood, Colorado 80112, U.S.A.)

Robinson, George S. (Appendix A). Attorney at Law, President, Ocean-Space Services, and Chairman, Council of Regents/United Societies in Space, Inc. Recently, he retired as associate general counsel of the Smithsonian Institution. His previous legal counsel service has included NASA, Federal Aviation Administration, and the U.S. Department of Transportation (Commercial Space Advisory Committee). Although his undergraduate studies were in biology, his LL.M. and Doctor of Civil Laws degrees were from McGill University's Institute of Space Law. A member of several professional and governmental advisory committees promoting space commerce, Dr Robinson is an adjunct professor at George Mason University and in 1965 lectured at Oxford University. A prolific writer in both space philosophy and governance, George was co-author of the classic book, *Envoys of Mankind—A Declaration of First Principles for the Governance of Space Societies* (1986) and *Space Law: A Case Study for the Practitioner* (1992). (Dr Robinson's address is 8458 Meadows Road, Warrenton, Virginia 22186, U.S.A.)

Acknowledgements

In writing *Living and Working in Space*, the author is indebted to Dennis M. Davidson, the talented astronomical artist of American Museum's Hayden Planetarium, whose illustrations in both editions help us all to visualize our human future in space. My appreciation is also extended to the other space artists whose work appears in our exhibits, as well as to National Aeronautics and Space Administration, and the European Space Agency for the many photographs which they supplied ... . Furthermore, I am grateful to Roland Doré, president of the International Space University, for writing the foreword of the second edition, just as I was to the NASA strategic planner and space philosopher, Jesco von Puttkamer, for providing the foreword in the original edition, some of whose exhibits we have retained in Chapter 1 of this 1996 edition.

This volume in the **Wiley–Praxis Series in Space Science & Technology** benefits from the collaboration of many scholars, but three have contributed significantly to our appendices—Dr George S. Robinson, formerly assistant general counsel of the Smithsonian Institution; Dr David R. Criswell, director of University of Houston's Institute for Space Systems Operations; and attorney, Declan J. O'Donnell, founder and president of United Societies in Space, Inc. (Refer to *Notes on Contributors*.)

I am indebted to many esteemed colleagues whom I first met and admired in 1984 at the California Space Institute on the campus of Scripps Institution of Oceanography in La-Jolla. There, during a NASA Summer Study, some of their information and insights contributed to the subject matter of this text. The author wishes to particularly acknowledge in that regard: Dr James Arnold, first director of the California Space Institute; Dr Stewart Nozette, now with the U.S. Air Force Phillips Laboratory; NASA representatives, Drs Mary Fae and David McKay, as well as Drs Michael Duke and Wendell Mendell, all of the Johnson Space Center. At that time, I was most appreciative of the opportunity to be a Faculty Fellow in that intensive workshop where I researched a novel on lunar industrializations, but actually ended publishing this professional book instead.

However, these pages that you read reflect the scholarship of other colleagues, some of whom cooperated in this venture when its publication was initially conceived as an anthology, especially: Roger M. Bonnet, director of ESA's Scientific Programmes; B. J. Bluth of NASA headquarters; Joseph V. Brady of The Johns Hopkins University School of Medicine; James D. Burke, California Institute of Technology, Jet Propulsion Laboratory;

Julian Christian of Universal Energy Systems; Jacques Collet, head of ESA's Long Term Programme Office; Angel and Patricia Colon of the Georgetown University Medical School; Mary M. Connors of NASA—Ames Research Center; Frank Davidson, director of MIT's Macroengineering Research Group; Ben R. Finney of the University of Hawaii; Nathan C. Goldman, Attorney at Law and adjunct professor at Rice University and the University of Houston; Albert A. Harrison of the University of California—Davis; H. H. Koelle, professor emeritus, Technische Universität Berlin; William E. MacDaniel, professor emeritus of Niagara University; James Grier Miller of the University of California—Los Angeles/San Diego and Mrs. Jessie L. Miller; Brian T. O'Leary, former Apollo astronaut, author/lecturer; Namika Raby of the California State University—Long Beach; John M. Talbot of the Federation of American Societies in Experimental Biology. Some of their ideas and findings have been synthesized and reported in this volume.

Without the unfailing support of my wife, the professor, Dr Dorothy L. Harris, during its twelve years of preparation and then revision, this work might not have reached you. In assembling the final manuscript, I owe special appreciation to my former co-workers at NETROLOGIC INC.—Dan Greenwood and Sarah Bode, as well as our friend, D. Eileen Bannard, who assisted with the original index. Finally, **profound thanks** to Clive Horwood, chairman of Praxis Publishing in Chichester, England, who had the foresight to bring you, the reader, not only the first printing but this second edition of *Living and Working in Space*.

Foreword

The space world is changing.

During the first decades of the space age, the main motivating factors behind space activities were national security and prestige. Space infrastructures (launchers, satellites, space probes) were developed under a technological thrust. Space applications were deemed valuable platforms for demonstrating national capabilities related to advanced technologies. Space was (and still is) a highly visible activity. That was the epoch of the space race.

The present decade has seen the advent of a new political equilibrium worldwide and with it a major, sustained, economic world crisis. Faced with these now conditions, it is the usefulness factor involved in space activities which is slowly becoming the sector's main driving force. We witness space activities as they bridge ideological and geographical frontiers. A vivid example of the internationalization of space is the **International Space Station**. The 14 countries involved in this project are prepared to invest tremendous effort in the task of demonstrating the feasibility of working and living together in space.

With the approach of the new millennium, we are entering a new space age in which space is put to the service of humanity. Space will be valued for its ability to serve humanity in telecommunications, earth observation, space science and human exploration of space. We will judge the value of space technologies by their ability to serve human needs such as food production, medical and health services, rural education, disaster warning and mitigation, environmental protection, navigation services, communication, understanding of the universe and extending humanity beyond the earth frontier. Fulfilling the user's needs will be of rime importance in the coming years. This will trigger a shift from government-led programmes to activities initiated and fully developed by the private sector. It will generate the creation of a multitude of applications on the ground and, consequently, a multitude of small enterprises to generate these applications. Launchers, satellites, space platforms and ground segments will be considered as tools to achieve other goals, and not as an end in themselves. The role of national and regional space agencies will change dramatically; instead, their role will be confined mainly to the realms of space science and manned space activities, and to support the development of long-leading technologies.

The dreams of Verne, Tsiolkovsky, Clarke and Asimov are becoming reality. As the students of the International Space University Summer Session '95 in Stockholm put it, they "envision worldwide benefits through space activities and their application; benefits based on the ideas of the common good of humanity and of shared global consciousness" (VISION 2020. An International View of the Future, Executive Summary). These students also maintained that the future success of world space activity would be closely linked to its capacity to serve humanity now. Further, the coming of the new millennium will see the establishment of the new paradigm: Space of Service to Humanity/SOS Humanity.

In making the case for "Human Enterprise on the High Frontier", Philip R. Harris has generated a book which places human considerations right back into the heart of space activity. In that sense, *Living and Working in Space* is a book for the new millennium. It contains a profusion of historical and prospective information. It describes today changes which will cause an impact on the future of space for many decades. It proposes a quantity of new ideas, either created or collected by the author, in fields ranging from techniques to management, from space business to political and legal considerations.

Dr Harris has drawn on his 30 years' experience in international research and consulting to produce a truly universal book which spans disciplines and national frontiers and extends beyond our present boundaries.

Those who are interested in the human side of manned space activities, in the international and interdisciplinary nature of past and future manned space programs, should read *Living and Working in Space*. I am convinced they will have as much pleasure in reading it as I have had.

January 1996

Roland Doré
President
International Space University

Prologue

Since the time our ancestors climbed down from trees and walked upright, the human species has always probed new frontiers! Over three million years, we manifested this characteristic for exploration repeatedly—as when early Man migrated from Africa across the planet, or when Europeans opened up the Pan American and African continents, or when Americans pushed across the Western frontier. With each terrestrial expansion, we developed more sophisticated mental models of our world, more complex social relations and increased control over matter.[1] Through these achievements, civilization has advanced, vast resources developed, and settlements established. Now in the last part of the 20th century, humans are taking faltering steps off their earthly homeland, moving beyond the atmosphere to another frontier which Isaac Asimov describes as much vaster and incomprehensibly richer. Going aloft to explore and develop these resources, he envisions as a great goal for Earth's peoples, one to fill our hearts and minds with glory and satisfaction, making narrow suspicions and hatreds seem small and unimportant.

But relatively few appreciate how these are truly "giant steps for mankind" as we fly into the real *new world*, one free from gravity and population limitations, as well as atmospheric impurity. Outer space is an ideal realm for experimentation and production that is impossible on Earth. Its endless assets can enrich the human family, possibly eliminating poverty on this planet. Permanent stations, outposts and bases in that orbital environment provide an unparalleled vantage point for scanning the cosmos and understanding the universe. Already both manned and unmanned spacecraft transmit information and images about other planets and galaxies within our Solar System. Space satellites have proven most persuasively their value for improving our global communication and agriculture, for predicting the weather and tracking human activities, for studying the Earth's topography and oceans, for understanding our own fragile biosphere, in terms of both problems and resources. In accomplishing such advances for the benefit of the global human family, we have taken high risks, investing large financial, material, and human resources. We have witnessed peak human performance among the thousands of space workers, both on the ground and in orbit, who make it all possible, but especially among the several hundred or so of our species who actually made it aloft.[2] We also have experienced failures and loss of life in the last forty, pioneering years which inaugurated the Space Age. But in the process, humankind has literally gone from the cradle of this Earth to its Moon and beyond. The impact of these initial space endeavors has tremendous implications and hope for our future, impelling us to change our collective image of our species. *We are no longer earthbound—maybe our real home is out there!*

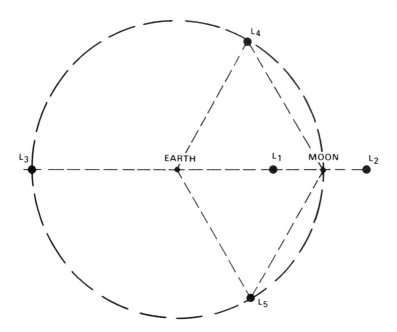

Exhibit 1—Libration Points/Zones in Outer Space. **The Lagrange or Libration Point One is 35 000 miles above the Moon's center, in direct line with Earth. The five L-points pictured are ideal positions to launch humanity to other worlds.** *Source:* National Commission on Space, Pioneering the Space Frontier, New York, NY: Bantam Books, 1986, P. 132.

So after millions of years as terrestrial beings, *Homo sapiens* in ever increasing numbers is beginning to migrate offworld. Human enterprise in space, so far, is manifested in both unmanned and manned activities. It is the creative human mind that is responsible for launching automated spacecraft to far-off planets, as much as for the building and staffing of a space shuttle or station. Extraterrestrial human activity increasingly involves artists, architects, film makers and entrepreneurs, as well as planetary scientists, engineers, technicians, and aerospace planners. The problems of transporting equipment and personnel safely into both lower and geosynchronous orbits necessitate the current emphasis in the space program upon the natural sciences, engineering, and technology. We need hardware and software solutions to get out of *Earth's gravity well*, especially inexpensive and safe launch systems.[3] However once we gain access to space, hopefully at lower costs, what do we do when we get there? With the publication of groundbreaking volumes like *Social Sciences and Space Exploration, Living Aloft* and *From Antarctica to Outer Space*, the National Aeronautics and Space Administration demonstrated greater awareness of the human requirements inherent in extended spaceflight.[4] The National Commission on Space further articulated the vision of *Pioneering the Space Frontier* through people.[5] Exhibit 2 from that momentous report illustrates the realm of space that is the focus of this book.

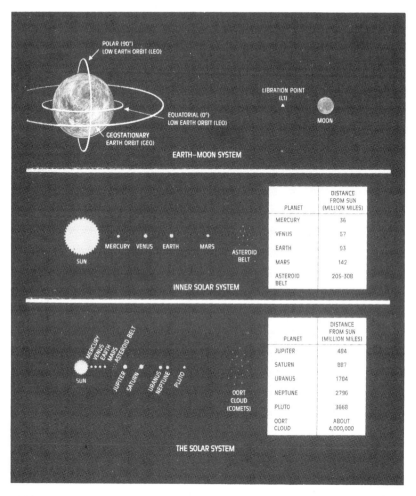

Exhibit 2—The Realm of Outer Space. **Exploration begins in the Earth–Moon System, extending out into our Solar System, and then out into the universe/s!** *Source*: National Commission on Space, *Pioneering the Space Frontier*. New York, NY: Bantam Books, 1986, p. 7.

Space developments for the remainder of this decade and the 21st century will demand greater contributions from the behavioral, biological, and information sciences. Expanding human presences throughout the Solar System will accelerate scientific investigations by psychologists, sociologists, anthropologists, and communication specialists, as well as by biologists, physicians, and experts in ergonomics and ecology. Construction of space stations and lunar bases lay the groundwork for a space infrastructure in the next century, leading to lunar industrialization and settlement, manned missions to Mars, mining of the asteroids, and eventually human colonies orbiting in space or established on other planets. Beginning with a handful of astronauts and cosmonauts, extending to space construction workers or *technauts*, human population up there is likely to escalate during the coming Millennia to thousands of spacefarers, whether as visiting scientists, tourists or colonists. The politician, teacher, journalist-observers who may fly the Shuttle or a rocket are forerunners of the millions of space travelers to come.

To study this emerging phenomenon of human behavior, culture, and potential aloft, this book was first compiled, largely from the perspective of behavioral scientists, in time for *International Space Year '92*. For this author, the possibilities for such a work became evident during the exhilarating Apollo missions of the 1960s when I served as a management and psychological consultant to NASA.[6] Involved in an executive and management development project which took me from NASA headquarters to many of its field centers, I began to sense that the organizational cultures of that space agency and its aerospace contractors were centered naturally upon the technical and engineering aspects of space undertakings, influencing broader decisions on human living and working aloft. Later in 1984 as a NASA Faculty Fellow during a summer study at the California Space Institute, I participated in analysis of space resources as the technological springboards for space developments in the next century, the proceedings of which were eventually published in five volumes.[7] Fortunately, during this strategic planning for a lunar base, one of the four workshops was devoted to examination of economic and system tradeoffs, as well as management, political and societal issues related to space activities. As team leader then for a group of diverse scholars considering the human dimensions of the lunar enterprise, I was exposed to interdisiciplinary input on the subject. It came from varied perspectives of psychiatrist and psychologist, political scientist and attorney, biologist and design engineer, anthropologist and architect, in addition to an economist and management specialist. Since NASA encouraged us to publish our insights, some of these findings were summarized and shared in the original edition of *Living and Working in Space,* supplemented by a synthesis from recent literature of other distinguished social scientists and writers on the topic of space.

Thus, after fourteen years of study, this *second edition* has evolved from my own research and that of esteemed colleagues, some of whom have been recognized in the acknowledgements section. Those who did peer review of the first edition published by Ellis Horwood observed that this is "the only comprehensive book on the subject." Now this text has emerged as part of the *Wiley–Praxis Series in Space Science & Technology.*

Living and Working in Space, second edition, is a completely revised and updated version of the original work. It has been carefully rewritten from a more international perspective, though the United States space program is used as a "case in point" relative to the world's space agencies. This expanded edition has two additional chapters, plus an extra appendix, as well as new mini case studies, experiential anecdotes, and the latest references on the subject matter. I am honored that Dr Roland Doré, president of the International Space University, consented to provide a new and insightful foreword. His phrase *"A Case for Human Enterprise on the High Frontier"* underscores the basic premise of this book, written for the reader with a general interest in the human side of living and working in a zero or low gravity environment. There are also numerous new exhibits in this edition, many of these illustrations by my former California Space Institute colleague, Dennis M. Davidson, now president of the International Association of Astronomical Arts. As a behavioral scientist, I present here tentative conclusions and prognosis regarding some of the human factors involved in mass migration aloft.

Living and Working in Space, second edition, consists of ten chapters, which begin with an examination of an emerging global space ethos, and the synergy so vital to our extraterrestrial expansion. We explore some of the human dimensions involved, apart from scientific and engineering considerations. Current and potential behavioral science contributions in the transition of *Homo sapiens* beyond our home planet are analyzed, especially from the perspective roles of anthropology, sociology, psychology, and living systems theory. Then reader attention is directed to the influence and the impact of space exploration on Earth's cultures, including the cultures of space organizations, as well as in creation of a new space culture. Next a chapter is devoted to a review of human performance aloft, and what can be done to ensure higher levels of orbital team behavior. Within the framework of a living systems approach, consideration is given to creation of a space personnel deployment systems for more effectively sending and returning greater numbers of people to and from orbit. A new chapter describes the kind of macrothinking necessary for global space planning, whether by the public or private sectors. Innovation in managing space enterprises follows to stimulate reader mind-stretching about adopting new strategies of macromanagement. A special case study is included on the development of the forthcoming International Space Station. In treating the diverse challenges in space industrialization and settlement, special emphasis is placed upon the prospects of space commerce, law, and policy. Finally, the last chapter closes by looking at alternative strategies for lunar economic development since the Moon has to be the first step toward exploring our universe. Each chapter has a bibliography of the latest references should the reader wish to probe further into the subject matter. These references are to be found in their own section, starting on p. 369. Before this current revision, all of the chapters in this exposition have previously been published in professional journals and subject to peer review. Four of them have received journalism awards for excellence from the Aviation/Space Writers Association.

The Epilogue proposes **action** plans for translating space visions into realities. Appendix A, thanks to Dr George Robinson, is a reprint of the 'Declaration of First Principles for the Governance of Outer Space Societies,' which should prove useful for student discussions. Appendix B, thanks to Dr David Criswell, is a case study on lunar industrialization related to beaming Solar Energy from the Moon for a clean, sustainable source of energy. In Appendix C, your author selectively compiled, a 'Directory of Space Resources and Publications.' An additional Appendix, D, thanks to attorney Declan O'Donnell, proposes a governance model through a future Treaty on Jurisdiction in Outer Space, which could facilitate development of a **spacefaring civilization.** Finally, we end with an author/subject index on the primary points covered in these pages.

Living and Working in Space, second edition, is intended primarily as a text for professionals and advanced students in the behavioral and space sciences. However, I hope it will appeal to other audiences, such as:

- thoughtful and intelligent people who are intrigued by the human adventure of advancing upward and outward into our universe. The latter includes not only the public who are generally supportive of space exploration, but the hundreds of thousands of space activists worldwide who accelerate "high ground" developments through their organizations, publications, and lobbying ...

- university students and researchers involved in academic courses related to space science, habitation and settlement ...
- social scientists and futurists wishing to advance their own knowledge and contribution to interstellar migration ...
- business leaders, managers, and entrepreneurs concerned about space commerce and its challenges ...
- planetary scientists and space engineers exercising their technical capabilities in space activities who wish to understand more about the human needs aloft of those who are to benefit from their expertise.

We remind all readers that this second edition is again dedicated to those **envoys of humankind**, the world's astronauts and cosmonauts, some of whom died to advance human migration into outer space!

January 1996

Philip R. Harris, Ph.D.
Management/Space Psychologist
Executive Vice President,
United Societies in Space, Inc.
Associate Fellow, American Institute of
Aeronautics and Astronautics
LaJolla, California, U.S.A.

1

Toward a Global Space Ethos

The year 2000 will inaugurate several technological ventures for space exploration that feature global cooperation. The 21st century may prove to be a period when space settlement and commercialization became feasible because of collaboration among competing nations and organizations. The very complexity and scope of space enterprises necessitate such an approach. The late U.S. Senator Matsunaga from Hawaii realized this when he observed:[1]

> **At a certain point, anything other than international exploration of the cosmos from our tiny planet will cease to make any sense at all ... we must develop policies that respond to the unfolding realities of the Space Age, that moves out to meet it on its own uniquely promising terms. Without such policies, earthbound civilization can only wind up recoiling upon itself.**

The International Space Year (ISY) of 1992 marked the publication of this book's first edition. With the sponsorship of many countries and space organizations, ISY encompassed worldwide space missions and projects, involving government officials, scientists, industrialists, students, and the public at large. The ISY premise necessitated unprecedented unity throughout the global space community and among national space agencies, becoming a harbinger of what could be done in the future. Because the task of exploring space has grown too large for one nation, given the boundless nature of the universe, the ISY activities became a milestone in the practice of space **synergy** (refer to Chapter 2). Some thirty-five years before that year the Space Age began during a Cold War between superpowers, so ISY organizers intended its undertakings to have a "transforming" effect on this new era in human development. [2] The United Nations (1990) endorsement of ISY emphasized programs that would improve:

- the management of the Earth and its resources;
- long-term education in space science, technology, and applications;
- public education on the role of space science and technology with reference the Earth's environment and the rest of the universe.

On the 500th anniversary of Christopher Columbus' discovery of the New World, ISY focused on the development of the high frontier through space policies and programs that reflected both the infinite nature of the universe, as well as the interconnectedness of all life on this planet. The celebrations inspired a reappraisal of the human species, and our role in colonizing our Solar System. At the ISY World Congress in Washington, D.C. (August 28–September 9, 1992), a climax was achieved around the conference theme of "Discovery, Exploration, and Cooperation—Hallmarks of this Planet's Increasingly International Advance into Space."[3] The World Space Congress presented an opportunity for 5000 thoughtful delegates, of which this author was one, to revitalize and revamp their psyches by considering our future in space. Among the thousands of insightful speakers there, perhaps a couple of selective quotations may underscore the significance of this landmark event:[4]

> **The most promising words ever written on the maps of knowledge are** *terra incognita*—**unknown territory ... . If the mere displacing of the Earth from the center of the solar system was so disturbing to thoughtful laymen then, what must be the consequences in our time of the discovery that our whole solar system, our whole Milky Way, our whole galaxy, our whole Universe is only a negligible peripheral one of countless billions ... . Perhaps we are no longer** *homo sapiens*, **but rather** *homo ludens*—**at play in the fields of stars.**
>
> —Daniel Boorstin, author of *The Discoverers* and Librarian of
> Congress Emeritus

> **The large scale of international cooperation required for building settlements on the Moon and Mars would inspire the peoples and governments of the entire planets to see what can be accomplished if they lay down their ancient fears and hatreds.**
>
> —Daniel S. Goldin, NASA Administrator

As the Renaissance defined European identity in the past by liberation of individual creative power, so the Space Age gives meaning to today's generations. As president of the National Space Society, Charles Walker, who has himself orbited the Earth, believes that **space is liberty**—it provides the liberty to inquire into unique places; to open our minds to new knowledge; to be ever thrilled by new discoveries; to experiment with new technologies on a scale impossible on the home planet; to voluntarily experience isolation in a way that creates new social structures and beliefs; to make more oases for life, more biospheres, more earths.[5] Outer space does offer unsurpassed freedoms to our species, but only if humans learn to cooperate in confronting challenges of living and working off the Earth.

Robert Reich, Harvard economist, wrote that the challenge for a country is not necessarily coming up with new policies, but with new visions.[6] The premise of this volume is that one such vision should include exploring the space frontier as a powerful way for developing human potential! The means for achieving this has been the process of discovery, ever dawning. "Humankind, throughout history, relentlessly challenges the borders of experience" (*Final Frontier*, Jan/Feb 1991). The nature of our species is to explore the unknown, to pursue the far horizon, to forge new frontiers. Pursuing the space vision is another manifestation of our humanity. While a solid case for going aloft can be made on the basis of resources that can be utilized for the benefit of our planet and its inhabitants, we are becoming extraterrestrials because we are human.[7] Through exploration and discovery, the quest broadens our perceptual vistas, as well as our knowledge base. Driven by destiny to extend human civilization beyond our own Solar System, the bold journey through space satisfies our spirit to know, while increasing our coping skills and expanding human culture.

A New Ethos
Perhaps the underlying need for humankind is to articulate and support this as a new ethos. **Ethos** is defined as the **fundamental character or spirit of a culture.** It is the underlying sentiment that informs the beliefs, customs, practices of a society. If **redefining** the American[*], Russian, European, or Asian ethos is taken as cases in point, one dimension or priority to be expressed is national purpose for development of outer space.[8] Currently, the majority of global inhabitants do not perceive space, its exploration and utilization, as central to our species' wellbeing. In general, humanity is still terrestrially oriented. People have yet to fully grasp the deep significance of migrating aloft, and its importance to this planet and its peoples! Four mini cases of different world regions are offered next to describe the emerging **space ethos** developing in the nations of each area.

1.1 NORTH AMERICAN SPACE CASE STUDY—U.S.A. AND CANADA

Over the past few hundred years, the New World experience has generated a unique national ethos—Attitudes and convictions inherent in the heterogeneous cultures of both the United States of America and Canada. As U.S. Secretary of Labor Robert Reich also noted in a recent interview, "One of the saving graces of Americans has been a willingness and ability to roll up their sleeves and get on with the task, putting ideology aside" (*Los Angeles Times*, May 3, 1987, II:1). This has been demonstrated in the way that country mobilized behind a national goal in the '60s to put a man on the Moon. In this decade, such national purpose can be manifested again to ensure permanent **human** presence on the space frontier. When concerns for space ventures truly become part of U.S.A.'s ethos, they will dominate the public's assumptions, values, and dispositions. If the United States had such an ethos, the issuance of the watershed report of the National Commission on Space (1986) would have been really big news; implementation today of its recommendations would now be a priority.[9] That work was followed by the National Aeronautics and Space Administration's (NASA) own "Ride Report" (1987), recommending strategies such as illustrated in Exhibit 3. Both reports are examples among many of major study findings on the nation's space future which never get translated into actual U.S. policy that is backed by both government and public commitment.[10] Similarly, if the country had a space ethos, the actions of the National Council on Space would have been watched as avidly as the

---

[*]Although **American** can be applied equally to all the peoples of Pan America, North and South, it is used here in the popular sense of U.S.A. citizens.

stockmarket; instead for a second time, a Presidential administration has currently disbanded that Council.

Were an ethos on space cultivated, then meaningful national space policies and strategies would be formulated that gain both citizen and investor support. Then, U.S.

To energize a discussion of long-range goals and strategies for the civilian space program, four bold initiatives were selected for definition study, and evaluation: *Mission to Planet Earth* proposes to estabish and maintain a global observational system in space ... *Exploration of the Solar System* ... . *Outposts on the Moon*—By 2010, up to 30 people would be productively living and working on the lunar surface for months at a time ... . *Humans to Mars*—committed to the human exploration, and eventual habitation of Mars.'

NASA Leadership and America's Future in Space, 1987. NASA Task Force Report under Dr Sally Ride, Astronaut. (Quoted in *Space World*, Sept. 1987, p. 25.)

Exhibit 3—Outposts on the Moon. *Selenians,* **Lunar Dwellers, living and working with automated helpers on the Moon!** *Source*: original painting by Dennis M. Davidson for *Space Resources* (NASA SP-509). Washington, DC: U.S. Government Printing Office, 1992.

leadership in space would be assured through the 21st century by adequate financing and implementation of bold plans. Although the American people have been sporadically enthusiastic about space endeavors, particularly as related to the astronauts, their interest has not been translated into sustained resolution behind long-term space exploration. Just prior to this decade, a Yankelovich poll commissioned by *Time* magazine (February 22, 1988) provided insight into the attitudes of citizenry in this regard:

- 48% thought it important that the U.S. be the leading space nation;
- 72% favored cooperative space efforts with the former Soviet Union;
- 51% wanted space spending kept where it was then.

Two years later, the Roper Organization reported their findings on American public opinion about space development (*Final Frontier*, December 1990):

- 51% considered themselves knowledgeable about the U.S. space program;
- 35% were satisfied with U.S. government spending on space;
- 66% thought the main goal of this space program should be the Earth's environ;
- 61% wanted construction of the space station accelerated;
- 59% suggested equal emphasis on both manned and unmanned space missions;
- 59% would book a vacation, if it were possible, to the space station, while 73% wanted it to be to another solar system!

The space supporters in the U.S.A. represent only a small fraction of the general population. Some 400 000 of these activists who possess a **space ethos** have joined about fifty space advocacy groups to influence policy and legislation, to inform themselves on space matters and participate in space progress.[11] In the nation's capital, *Spacecause*, affiliated with the National Space Society, is effectively representing and delivering the pro-space message through lobbying, conferences, and public affairs efforts at the grass-root level. Another two million people in the aerospace industry, government and military apparently may be counted among the space community (refer to Appendix C for a listing of their organizations). The largest of these space groups—The Planetary Society and the National Space Society (NSS)—conduct and publish an annual survey of their membership on space policy and goals. In 1995, for example, NSS members manifested their **space ethos** by overwhelmingly voting for cheaper access to space and greater involvement of private enterprise in space exploration. When asked **why** explore and develop outer space, their principal reasons were:

- furthering technological development ...
- expansion of settlement frontiers ...
- improving the Earth's environment.

In reply to the question on how the U.S. Congress should support commercial space activities, the vast majority urged:

- eliminating adverse government regulations of space industry ...
- increasing government R&D contracts to private enterprise ...
- expanding privatization of space activities ...

- eliminating direct competition between NASA and private firms ...
- providing tax exemptions on space developed products.

The NSS survey discovered that 73% of respondents thought that the U.S. Senate should continue not to sign the 1979 UN Moon Agreement, while 87% favored Russian participation in the proposed International Space Station (*Ad Astra*, May/June 1995, p. 59). In January 1996, a poll was conducted by *The Washington Post* and *ABC News* of 1005 randomly selected Americans. It indicated that a majority interviewed **favored** NASA's performance; construction of an **International Space Station** with Western Europe, Russia, Canada and Japan; containment of U.S. space effort and budget at its current level; the belief that the benefits justify the costs of space exploration.

Yet a groundswell of popular mass conviction for space enterprise has yet to energize the Presidential administration, Congress, and the space agency itself toward synergistic support for the Space Station, lunar base, and manned Mars mission. Almost a decade ago, Thomas Paine, former NASA chief and chairman of the National Space Commission, expressed the frustration of the scientific and commercial communities succinctly:

> **The biggest problem is the lack of direction of the U.S. space program which adds to the difficulty in mounting major international programs that could enhance the exploration of space.**
>
> (*Los Angeles Times*, May 25, 1987, I:3).

Since that astute observation, the situation remains the same in the mid '90s. There is no national unanimity as to what America should do next in space and how much should be invested. There is no national **space ethos**, and little evidence of **space synergy**. The space agency signs international agreements, and the U.S. Congress squabbles over the funding to implement such. The old arguments of manned or unmanned spaceflights continue, while the planetary scientists with vested interests support automated missions of exploration and do not cooperate with the advocates of human space discovery and settlement.

Emerging American Space Ethos
Seven decades ago, an American scientist, Robert Hutchings Goddard, fired that first successful liquid-fuel rocket from a Massachusetts pasture and provided his country with the key to enter the space frontier. Since then, Dale Meyers reminds us that the U.S.A. has used that key to good purpose—sending inquisitive extensions of human intelligence to eight planets in our Solar System; examining our own star, the Sun, and the cosmos from outside of the obscuring atmosphere of Earth; and placing a dozen Americans on the surface of the Moon. Meyers, a former NASA deputy administrator, noted that these accomplishments marked the first time in the four-and-a-half-billion-year history of our planet that creatures, evolved from the chemicals in its crust, had achieved the ability to leave its surface and visit other celestial bodies in the system.

The American **space ethos** first began to emerge in response to the former Soviet Union's challenge with the launching of its first **Sputnik** in 1957. Several years later, on May 25, 1961, President John F. Kennedy called for national leadership in space achievement, and committed the country to a lunar conquest within a decade. Subsequent presidents gave mediocre support and declining resources to the space program until President George Bush early in the 1990s. One way to evaluate a nation's **space ethos** is what it

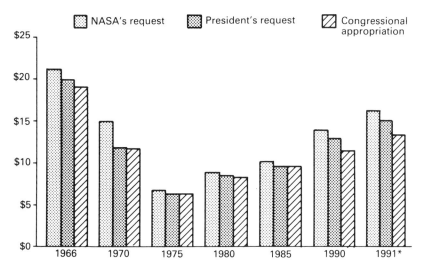

*Approved NASA budget for 1992 = $14.3 billion; SDI approved funding = $4.15 billion

Exhibit 4—The Rise and Fall of NASA's Fortunes. **NASA budget comparisons from 1966 to 1992 in terms of billions of '90 dollars.** *Source*: Andrew Lawler, "US Space Plans in Fiscal Gridlock," *Space News*, Nov, 5–11, 1990, Vol. 1:43, p. 1.

invests in space development—the U.S., for example, on average has been spending as much on space as the rest of the world put together. The annual battle of the Federal budget allocations to the space program is graphically illustrated in Exhibit 4 on "The Rise and Fall of NASA's Fortunes". By the middle of this '90s decade, a new Republican-led Congress seeks to reduce expenditures and cut the national debt, so the NASA FY 1996 budget proposal again diminishes. Compare the 1992 approved budget of $14.3 billion in that exhibit with the 1996 proposed budget of $14.2 billion illustrated in Exhibit 5. (Yet the latter represents an increase of 1991 space expenditures of $12 billion.) This agency budget breakdown in the second exhibit also indicates NASA mission priorities.

Now after forty years of remarkable and comparative progress in space technology and operations, the U.S.A. seemingly stumbles in successful efforts to establish itself permanently on the high ground. Even before the **Challenger** and other launch disasters, the Office of Technology Assessment (OTA) back in 1984 reported that the U.S. government had not adapted to the changing situation regarding its space program, and that there was a lack of national consensus about long-term goals and objectives for this country in space. To remain competitive, even then the OTA urged provisions for greater private sector investment in civilian space activities, as well as more international cooperation in space endeavors. The OTA study advocated involvement with Japanese and European partners for this purpose, calling for a more synergistic national policy and direction for space development.

Normally, we expect a head of state to articulate a country's ethos. Among the Presidents of the United States, only John Kennedy has succeeded in leadership with regard to space. From the beginning of Ronald Reagan's administration, that President appeared consistently supportive of space enterprise. This is a sample of his vision communicated to the citizenry over the course of two terms:

NASA FY 1995 Budget Summary* (in millions)		
	FY 1995 Enacted	FY 1996 Proposed
Human Space Flight	$ 5514.9	$ 5509.6
Science, Aeronautics & Technology	5943.6	6,006.9
Mission Support	2589.2	2726.2
Inspector General	16.0	17.3
National Aeronautical Facilities	400.0	– –
Total Budget Authority	14463.7	14260.0

Exhibit 5—NASA Budget Update for FY '96. **Update on NASA authorization legislation in the U.S. Congress, with breakdown into five major areas of expenditures.** *Source*: Legislative Research, The Office of Legislative Affairs, NASA Headquarters, published by the National Space Society (922 Pennsylvania Ave. SE, Washington, DC 2003, U.S.A.) in *The Month in Review: a Compendium of NASA Congressional Activities, October 1995.*

- July 4, 1982, Independence Day Address:

  Today we celebrate the 206th anniversary of our independence ... . The fourth landing of the *Columbia* is the historical equivalent of driving the golden spike which completed the transcontinental railroad. It marks our entrance into a new era ... . And now we must move forward to capitalize upon the tremendous potential offered by the ultimate frontier of space ... . Simultaneously, we must look aggressively to the future by demonstrating the potential of the Shuttle and establishing a more permanent presence in space.

With that speech, the President issued a directive on the same day detailing the basic goals of his space policy. It was to ensure the country's security and space leadership, obtain economic and scientific benefits through space exploitation, expand private sector space investments and involvement, promote cooperation with other nations in space for the enhancement of humankind.

- March 23, 1983, State of the Union Address:

  Our second great goal is to build on America's pioneer spirit and develop our next frontier: space. Nowhere do we so effectively demonstrate our technological leadership and ability to make life better on Earth ... . America has always been greatest when we dared to be great. We can reach for greatness again. We can follow our dreams to distant stars, living and working in space.

This address marked an endorsement of a free enterprise philosophy in space. With its analogies of clipper ships and Yankee traders, "The Great Communicator" outlined the prospects for space markets and commerce. Then Reagan instructed the National

Aeronautics and Space Administration to develop a permanently manned space station within a decade, and to invite other countries to participate as a means of strengthening peace. To ensure that companies interested in putting payloads into space have access, the President also directed the Department of Transportation through its new Office of Commercial Space Transportation to assist in developing a private sector launch service industry, and to remove regulatory barriers to space industrialization. Within six months, the President was to approve a National Space Strategy promoting initiatives for long-range studies of new launch vehicles, military operations in space, identification of major space goals, shuttle cost recovery, and reaffirmation of an $8 billion space station. {**Author's Note: in 1996, over a decade later and with expenditures of some $12 billion, that space station is still not in orbit, though it has now become truly international in planning. Refer to the station case study in Chapter 8.**}

As a national leader, President Ronald Reagan, as suggested in the above quotations, apparently had been in the forefront of expressing a tentative **space ethos**. In 1984, for example, on the occasion of the 25th anniversary of the space agency's founding, he again spoke of space as a vision of "limitless hope and opportunity." Addressing NASA employees at the National Air and Space Museum, the then Chief Executive observed that the space agency "has done so much to galvanize our spirit as a people, to reassure us of our greatness and of our potential." In 1988, toward the end of his Presidency, Reagan issued a definitive National Space Policy that endorsed commercial projects and development of Mars technology. But many citizens perceived a gap between Presidential rhetoric and decisions that advance national space consensus and policy. The President's words never reached down and moved the masses within the country to get behind space undertakings. Furthermore, most of the nation's financial resources devoted to space during the administration of that 40th President did not go to civilian space programs, but to military space programs such as "Star Wars" or the Strategic Defense Initiative (SDI).

A people's ethos is forged through both success and tragedy. The Apollo Moon landing in 1969 represented a peak experience for America, as well as for humanity in general. But in this nascent period of the American space program, the death of ten astronauts on the ground and in space tested the public willingness to accept the high risk inherent in space exploration. Until 1986, the shuttle transportation system seemed to be another model of achievement until both technological and human systems failure doomed the **Challenger** and its crew of seven. But even investigations of that failure contributed to the formulation of a **space ethos** and led to new advancements, as the report of the Presidential Commission on the Space Shuttle **Challenger** indicated.[12] The setbacks have also resulted in Presidential directives giving mission priorities on the shuttle to defense, not commercial, uses. While praising NASA reforms in implementing the recommendations of that "Rogers Commissions," U.S. Senator Donald Riegle reminded us that NASA cannot craft space policy; that must be done by the Executive Branch with the ratification of Congress and the people (*AAS Space Times*, May–June 1987). When he was chairman of the Senate Subcommittee on Science, Technology, and Space, Riegle urged a national debate on how much of the American space effort should be military in scope in contrast to commercial emphasis. Such widespread public discussion has yet to occur.

In the first half of the 1980s, Congress chided NASA administrators, scientists, and engineers for being too narrow in their space planning and lacking vision. Despite the constraints of budgetary cuts and investigations, agency personnel have combined with aerospace scholars and contractors to produce some impressive and futuristic schemes, as witnessed in the 1986 NASA publication, *Planetary Exploration through the Year 2000.* While the space agency has been a great technology driver, it has not aroused the spirit of the American people behind its plans. To its credit, NASA has engaged in extensive self-study and promulgated, as indicated, the 1987 Ride Report on America's future in space. Sometimes these internal studies are on NASA's initiative, but more often as a result of external pressure.

One such force was the National Commission on Space (NCOS) established by the Presidential authority. It consisted of fifteen space experts under the leadership of Dr Thomas O. Paine, former NASA administrator, who were directed to provide a fifty-year space program forecast. The task of these Commissioners was to synthesize national input and propose civilian space goals for 21st century America. Their farsighted report in 1986 entitled, *Pioneering the Space Frontier,* was the closest expression to date of developing American **space ethos**. It resulted from fifteen public forums held across the country, as well as communications received by mail, computer network, or testimony at Commission meetings. Representing a cross-section of American thought on space planning to the year 2035, the very readable report gave some indication of national concordance toward the construction of a space infrastructure, including the development of interplanetary space-craft and ports, space factories and settlements. The Commission recommended a triple program to:

(1) advance understanding of our Earth, its solar system, and the universe;
(2) explore, prospect, and settle the inner solar systems;
(3) stimulate space enterprise for the direct benefit of Earth's people.

To accomplish these goals over the next half century, the Commission estimated an investment of $700 billion by the American taxpayer and its international partners. How-ever, the Commissioners also envisioned big returns on that investment, as the first forty years of space enterprise has already proven. They were aware of a Government Account-ing Office estimate in 1982 that the payback is approximately $7 for every $1 invested in space. Therefore, NCOS expected returns on space investment which would be primarily in:

- advancement in national science and technology capabilities so critical to economic strength and security ...
- economic returns from space-based enterprises that capitalize on broad, low-cost space access ...
- opening up the vast resources of the space frontier to supplement the limitations of our own planet.

Assuming that the **International Space Station** will be erected before the turn of this century, the NCOS report projected that human habitation on the Moon and Mars would occur by year 2020.

This National Commission not only set forth a vigorous vision for the American civil-ian space program, but the rationale, methods and aspirations for its implementation. Its

"Declaration for Space" communicates a possible **space ethos** with statements such as this:

> With America's pioneer heritage, technological pre-eminence, and economic strength, it is fitting we should lead the people of this planet into space. Our leadership role should challenge the vision, talents, and energies of young and old alike, and inspire other nations to contribute their best talents to expand humanity's frontier ... . As we move outward into the Solar System, we must remain true to our values as Americans: to go forward peacefully and to respect the integrity of planetary bodies and alien life forms, with equality of opportunity for all.
>
> (National Commission on Space, 1986, p. 4)

The National Space Society with some 50 000 members endorsed this vision, adding:

> We believe the technologies and industries born on space frontier in the next few decades will drive the world's leading economies in the next century. Our role is to educate the public on the benefits of space developments and work with allied organizations to create the cultural and political context for an open frontier in space. We believe the United States must be a leader on that frontier, or it will cease to be the greatest hope for human liberty and freedom.
>
> (*Space World*, January 1988, p. 32)

Unfortunately, because of political conditions at the time, such as an Administration crisis over the "Iran Contra Affair," the excellent report and recommendations of the National Commission on Space were largely ignored at the White House. It was not until September 29, 1988, that the U.S.A. manned space program was reborn! This watershed event may have contributed to the re-emergence of a **space ethos**. After three years of launch failures and frustrations as a result of the **Challenger** accident, the NASA spokesman announced, "Americans have returned to space." The successful launch and subsequent return of the Shuttle orbiter, **Discovery**, brought over a million wellwishers to the Kennedy and Vandenberg space centers. Excitement spread across the nation—the masses were again inspired by the feats made possible through the high performance of space scientists, engineers, and the astronaut team. The world again watched and rooted, while the average citizen cheered. The media joined in celebrating the new beginning, but commentators cautioned that there is a need for the country to set new directions in its space endeavors. A tentative ethos was manifested after being in a hiatus since the Apollo Moon mission days. Broadcasters described the **Discovery** launch, "As the day Americans paused to dream again" (Dan Rather, CBS Evening News, September 29, 1988). They joined their countrymen in the urge to forge ahead aloft, but wisely asked, "What does this nation want to do in space?"[13] Response to that key question requires public participation if a critical mass of citizen support is to result—that has not yet happened.

The next Administration of President George Bush attempted to set that new direction for the nation's future in space. This was first done by reinstalling the National Space Council under Vice President Dan Quayle, appointing a former astronaut as the new NASA Administrator (Admiral Richard Truly), and designating a blue ribbon panel to untangle the fiscal and managerial knots that bind the space agency. Called the Advisory

Committee on the Future of the U.S. Space Program, it produced in 1990 another significant report on American space agency problems, expressing these nine concerns:

(1) lack of a national consensus as to what should be the goals of the civil space program and how they should be accomplished ...

(2) NASA's over-commitment in terms of program obligations relative to too little resources and allowance for the unexpected ...

(3) management inefficiencies caused by constant changes in project budgets of the space agency, exacerbated by actions sometimes necessary to extricate projects from technical difficulties ...

(4) institutional aging and insufficient planned organizational change within NASA ...

(5) incompatible personnel policies between civil service requirement and space agency needs for leading-edge technical specialists and managers ...

(6) natural tendency for space projects to grow in scope, complexity and cost, and the need to curb such unplanned expansion ...

(7) deterioration of NASA technology base which must be quickly rebuilt ...

(8) need for engineering systems to continuously monitor flaws in spacecraft technology and correct problems before they escalate into disasters ...

(9) overdependence of the civil space program on the Space Shuttle, and the need to use other heavy-lift launch vehicles for for all but missions requiring human presence.

Under its chairman, Norman R. Augustine of Martin Marietta Corporation, this prestigious expert panel made four principal recommendations concerning space goals, programs, affordability, and management. Relative to the themes of this book, these included:

• not only "Mission to the Planet Earth" (MTPE) focusing on environmental measuring from space, but a "Mission from Planet Earth" (MFPE) with a long-term goal of human exploration of Mars, preceded by a modified Space Station which emphasizes life sciences, an exploration base on the Moon, and robotics. MFPE would operate on a "go as you pay" basis with missions scheduled as adequate funds are available ...

• major reforms in management and personnel operations of NASA, including in the restructuring a post for an Associate Administrator of Human Resources responsible for acquiring and retaining the highest quality employees; an Associate Administrator for Exploration responsible for robotic and manned missions to the Moon and Mars; establishment of a systems concept and analysis group in a Federal Research and Development Center.

This report did profoundly impact NASA, and the resulting reorganization still goes on under its present Administrator, Dan Goldin.

In the long term, President Bush may have contributed much to articulating the U.S. **space ethos** principally through his Space Exploration Initiative (SEI)—see Exhibit 6). In a speech on July 20, 1989, commemorating then the 20th anniversary of the **Apollo 11** Moon landing, George Bush tried to lay out a framework for American space developments for the decade of the '90s and 21st century. "SEI" was to be a peaceful civilian endeavor in contrast with Reagan's "SDI", the Space Defense Initiative. The Space Exploration Initiative expressed noble objectives, including completion of Space Station **Freedom** in the 1990s; a return to the Moon to stay at the beginning of the new century; and a

Exhibit 6—Launch Out—Space Exploration: Cape Canaveral. **The Shuttle Space Transportation System Lifts Off from the Kennedy Space Center in Florida.** *Source*: National Aeronautics and Space Administration (NASA).

journey into tomorrow with a manned mission to Mars. The President called this vast endeavor a "New Age of Exploration" that would result in a lunar outpost by 2004, landing a human on Mars by 2019, and significant exploration and scientific investigation of the Moon–Mars system. In the 21st century, permanent settlements would be established on both, with an aim to exploiting their resources for strategic and commercial purposes. Initially NASA estimated SEI would cost $400 billion and the Congress was reluctant to back the initiative beyond preliminary studies of this ambitious program.

To help in analysis of the SEI prospects and priorities, the National Space Council also tried to reach beyond the space agency for innovative input about what directions and actions should take place in the near future. Thus, the Council encouraged NASA to create the "Outreach Program," a nationwide search for creative ideas and technologies to further the exploration of the Moon and Mars. To move toward new levels of ingenuity, the effort begun in 1989 tapped diverse sources from the National Research Council, and national laboratories to the strategic defense specialists and even the public at large by means of a "800" free telephone connection. To assist in the surveying and reporting the solicited input throughout the country, NASA contracted with two organizations—the Rand Corporation, a non-profit research group and think tank in Santa Monica, California; and the American Institute of Aeronautics and Astronautics, a professional society of the aerospace industry in Washington, D.C. In addition, the various federal agencies were invited to participate, and the many respondents were U.S. Departments of Energy and Defense, as well as NASA personnel. The Rand "toll-free" number produced 19 048 calls, while the deadline resulted in 1697 ideas to that collector and another 530 ideas through the AIAA conduit (1991). After a preliminary screening by Rand and AIAA, the best and the brightest ideas went to an independent advisory panel in Crystal City, Virginia, headed by ex-astronaut and former air force lieutenant general, Thomas Stafford. Called the "Synthesis Group," it reported the results to the National Space Council in Spring 1991.[14] The group organized into teams to review the ideas in terms of how a mission would leave the Earth, then travel to and land on the surface of the Moon and Mars. Both Rand and AIAA held public conferences around their findings. While limited in its outreach, the whole undertaking provided more than a coherent architecture for the Space Exploration Initiative. Indeed, it was a democratic and synergistic process that broadened the base of public involvement in the U.S. space program. As the director of program architecture for this Synthesis Group, Lt. Gen. Sam Armstrong (USAF Ret.) observed:

> A pioneering society is a growing society. We've always been a nation of pioneers. But the people have to want to do it.
>
> *Final Frontier*, Jan./Feb. 1991, p. 41

In a sense, this SEI report to the President compiled by the Stafford Synthesis Group may be the second best expression to date of the emerging American **space ethos**. The well-illustrated document effectively communicates what should be the country's space visions, and provides again a convincing rationale for returning permanently to the Moon and exploring Mars, both for scientific and commercial reasons, as this excerpt confirms:

NATIONAL SPACE VISIONS

- **Increase our knowledge of our Solar System and beyond**
- **Rejuvenate interest in science and engineering**
- **Refocus U.S. position in world leadership**
- **Develop technology with terrestrial applications**
- **Facilitate further space exploration and commercialization**
- **Boost the U.S. Economy**

Unfortunately, this SEI vision statement has yet to be implemented by the U.S.A.! But there was another initiative by the Bush Administration in the Fall of 1990 that may yet prove productive—namely, synergistic talks were begun not only with Canada, Japan, and the European community, but also with the then Soviet Union on the possibilities for an international lunar base and human expedition to Mars. American leadership in the past has inspired international cooperation in space which is increasingly followed by the world community. Globally, space agencies now have agreements with one another, not only in developed free enterprise nations of the First World, and with the former or present socialist economies of the Second World, but in many developing countries of the Third World.

U.S.A. Space Ethos Conclusions

Since President Kennedy in the '60s, no action to date by a White House Administration or the U.S. Congress has succeeded in creating a **space ethos** within American culture, though their past efforts as described above may contribute eventually to its development. For over two decades the government has promoted prestigious studies and reports as documented in this case, but failed to act on most of their recommendations. Having again disbanded the National Space Council, a significant action taken in the mid '90s by the new Democratic President, Bill Clinton, was to sign cooperative agreements with the Russian Federation to engage in a number of joint space ventures. When the Republican Party took over the Congressional leadership, the new Speaker of the House, Newt Gingrich, addressing an NSS Space Policy Seminar on Capitol Hill, asked:[15]

> **What we need to know from you is, if we want to maximize the human race's opportunity to move out into space ... then what should our policies be?**

Given that Dr Gingrich is a historian, the author would have hoped that he was familiar with the excellent policies proposed to, but largely ignored by, the national government for the last twenty years. However, he is also a professional futurist, so it was heartening to note his other remarks predicting the growth of space tourism, and calling for NASA to become a customer to private enterprise (particularly in space transportation), as well as for legislative enactments for temporary tax-free earnings on profits in space. Another hopeful sign on the same occasion came from the new chairman of the House Committee on Science, Rep. Bob Walker (R-PA), himself a committed space activist, who also opted for tax incentives to promote commercial space activities. Perhaps the most insightful remark at this Congressional seminar came from Tom Rogers, president of the Space Transportation Association, who reminded the audience that frontiers are opened by people with the support of their government!

Before the year 2000, the U.S.A. needs to engage in serious national discussions on the implications of **recommendations** from the National Commission on Space and the numerous national reports previously cited. One forum that might act as a catalyst for public support with a view to **decisions** might be a **White House Conference on Space Enterprise** (see Epilogue) [16] That event could markedly contribute to the creation of a **space ethos** which moves the nation forward vigorously on the high frontier!

As space policy expert, John Logsdon, ruefully observed: "the pathway to the Moon and Mars leads straight through Capitol Hill," meaning the U.S. Congress.[17] In the current climate of budget reductions and deficits, exacerbated by serious domestic problems and involvement twice in Middle Eastern or Gulf Wars, the Electronic Industries Association of Washington, D.C., forecast that the U.S. space agency budget will reach $20.1 billion by the year 2000 is very questionable (*Space News*, October 22–25, 1990, Vol. 1:41, p. 6). The EIA study of NASA programs found more immediate support for an Earth Observation System, which is what actually happened.

To underwrite human enterprise in space involves hard national choices on allocation of scarce resources to planetary science and astrobusiness enterprise, and/or to military and defensive space priorities. The decision-making process for long-term, large-scale space projects requires a sociopolitical environment that will ensure public support and participation in constructing a space infrastructure. It also demands the type of **macrothinking** discussed in Chapter 7. Translating today's space goals and policies into realities is much more demanding than the government's original decision to go to the Moon or to build a Space Shuttle. [18] Recently, the Congress had difficulty agreeing to pass a bill (H.R. 504) to fund an American Space Memorial in the nation's capitol to honor the deceased astronauts, or to concur on the Space Settlements Act (H. R. 4218) that would authorize NASA to assume leadership in the establishment of settlements in space! Current domestic controversy over downsizing and internationalizing the Space Station and its funding further undermines the collaboration the U.S. seeks from allied nations for this joint orbiting facility.

While mass media productions like *2001*, *Star Trek*, and *Star Wars* create a subculture in favor of space, the mainstream culture in the U.S.A. has yet to be "turned on" to the opportunities ahead on the final frontier aloft. Educational institutions have yet to prepare American youth for the New Millennial prospects of colonizing the galaxy!

Canadian Space Ethos
In the neighbor to the north, Canada, an ethos toward space is gradually arising. It was born with the launching of **Alouette 1** in September 1962 on the U.S. Thor–Agena vehicle, a cooperative program among Canada, the U.K., and the U.S.A. In 1972, Canada launched its first domestic geostationary communications satellite system. Under the leadership of the Canadian Space Agency, created in 1989 and headquartered on a new campus in St. Hubert, near Montreal, the nation's diversified space efforts coalesced. Until recently that organization and the indigenous aerospace industry were largely aligned with NASA and its space projects, having developed key components in both hardware and software for spacecrafts and satellites. Its principal activity has been creating a Mobile Servicing System (MSS), a **Canadarm** for use on the proposed space station, based upon Canadian astronauts' experience with a Shuttle robotic arm. With a budget projected through year 2000 of 2.4 billion Canadian dollars, space expenditures from April 1994 through March 1995 were $333 million for both public and private sectors. There is a growing involvement of CSA and Canadian researchers not just with NASA, but with the space agencies of Europe, Russia, and China. After 20 years of cooperation as an observer, on March 21, 1991, five agreements were signed between CSA and ESA for Canada's participation in these European programs—**Hermes** spaceplane development, Earth observation, and telecommunications. Two other ethos indicators are evident in Canada:-

* a growing space community with many publications and conferences;
* a steady increase in the allocation of the national budget to space.

Since CSA's former director, Dr Roland Doré, became president of the International Space University with its main campus located now in France, Canadians are more apt to identify with the global space community. The signs of this first became evident in the ISY '92 7th Conference on Astronautics of the Canadian Aeronautics and Space Institute with its theme "Canada in Space: a Coming of Age and New Horizons."[19] The discussions covered a wide range of **space** topics from research, technology and solar power to astronauts, education, and business prospects. However, space is still a small subculture in Canada, and has yet to penetrate mainstream thinking.

Similarly, to the south in Latin America, both students and scholars seek to meet the quintessential challenges in space science, technology, and settlement. Brazil seems to lead in the search for a **space ethos**, expending $143 million in '94 and developing its Alcantra Launch Center.

1.2 RUSSIAN SPACE CASE STUDY

The International Space Year proved to be a turning point in another way—in 1992 the Union of Soviet Socialist Republics imploded and collapsed as a political entity. It ended the Cold War, and the space race between superpowers, namely, the U.S.A. and the U.S.S.R. Certainly, both shared the position of space leadership, for the Soviets put the first satellite in orbit, as well as first man and woman in space, amassing during their regime three times the "manned" time in space in contrast to Americans aloft. They not only were far ahead in the number of rocket launches, but have had a functioning space station **Mir** in orbit for a decade of extended missions.[20] Premier Nikita Khrushchev once boasted, "the launch pad for our cosmonautics is socialism." Under the Soviets, it was

Exhibit 7—**Energia** Space Launchpad: Kazakhstan. **The most powerful rocket in the world today is Energia, which is launched outside of Russia in Tyuratam/Baikonour in the CIS republic of Kazakhstan.** *Source*: C. M. van den Berg, Earth Space Review, Vol. 1:4, 1992, p. 14 (Gordon and Breach Publishers, 820 Town Center Dr., Langhorne, PA 29047, U.S.A.).

always difficult to estimate how much was expended on both military and civilian space endeavors, but in 1989 the leadership stated that $4.8 billion was being spent on civil space, though Western analysts considered the figure somewhat high. When the first edition of this book was published (1992), it appeared the Russians were well ahead in creating their own unique **space ethos**, but now their commitment to space exploration is currently undermined by internal chaos, principally socio-political and economic in nature. Hapless cosmonaut Sergei Krikalev had to stay in orbit an extra five months because the government that sent him was no longer in power and the new one did not have money allocated for his retrieval rocket! The Soviet system has been replaced by a shaky political entity known as the Commonwealth of Independent States, which in 1992 formed its own CIS Space Agency (*Space News*, Nov. 23–29, 1992, p. 9). This new authority is supposed to carry out an inter-republic space agreement such as indicated in Exhibit 7. The major spacefaring republics are Russia, Kazakhstan, and Ukraine. The reality is competing power centers with questionable goals!

The Way It Was
But let us examine the "way it was" in the former Soviet Union with its centralized, totalitarian control and closed system. Back then, an interview with Dr John Logsdon, a leading space policy analyst, brought forth this interesting comment:

> What is striking about the Soviet Union is the country's commitment to space, which enables the government to carry out the program. That commitment comes from the Soviet leadership—particularly Khrushchev, but also Brezhnev—defining space in a way that the society internalized the belief that this was a part of their future.
>
> *Space World*, August 1986, p.16)

That internalized belief is an expression of the Russian ethos, which some observers maintain was a substitute for the religious practices that officially had been frowned upon for almost seventy years of Communism's reign.

Yuri Gagarin, the first human in space, and Alexei Leonov, the first person to walk in space, were revered like demigods, and still today are held in high regard by the masses. Then the public adulation of Soviet spacefarers was evident in the busts of gallant space giants which adorn buildings and parks. Museums, monuments, and murals pay homage to the men, women, and machines of the Space Age (*Air & Space*, August/September 1987, p. 115). In their pantheon of heroes, the cosmonauts were of mythical proportions—cosmodromes and space cities were built for their care and activity, space missions were eagerly followed by the masses through the media. Soviet space museums abound and were then well attended by the populace. The extent of this prior "worship" can be gleaned from these extracts of another Smithsonian article, "Commemorating Cosmonautics" by J. Kelly Beatty (*Air & Space*, June/July 1988, pp. 96–101):

• Crops may fail and policies flounder, but the Soviet Union's ongoing space triumphs remain an obvious source of pride among its people ... . The space heroes are not merely respected, they are adored ... . Atop a 25-meter spire besides Lenin Avenue in Moscow stands a 6-meter-tall likeness of Yuri Alekseyevich Gagarin, well muscled and angular ... . The heroic stature attributed to Soviet space explorers is reflected in the frieze on the

Cosmonaut Memorial, another gleaming spire that arches 90 meters into the sky. A Museum is situated beneath the memorial ... . No word describes it better than "shrine".
• Near the Cosmonaut Memorial, a collection of rockets and space artifacts serve as a large "theme park" among the Exhibition of Achievement of the National Economy ... . A space exhibit, the Cosmos Pavilion was added in 1966 and today it boasts more than 9500 m^2 of floor space and draws eight to nine million visitors a year ... . A huge inscription, "Cosmonautics, The Way to Peace" wraps around the wall. The theme is repeated often: in a huge portrait that dominates the rotunda, Gagarin holds a white dove ... . The first human in space serves as an enduring icon of the achievements reflected in Moscow's space museums.

Until recently, the former U.S.S.R. commitment could be perceived in the size of its space budget and personnel, supposedly double that of the U.S.A. (their ruble expenditures were said to be around $12–13 billion annually, a questionable figure). The extent of that nation's unflagging pursuit of the space was also evident in the then world's busiest spaceport, Plesetsk—north-east of St. Petersburg near the Arctic Circle; it has orbited more than a thousand payloads and tested 40 times more missiles annually than the U.S. In a *National Geographic* review (Oct. 1986, p. 455) of "Soviets in Space," its editor, Thomas Canby, concluded that space exploits stir the Soviet soul like religion and these stirrings are fanned by the government. Some believe that space successes helped the nation to overcome its inferiority complex.

The Russians, both past and present, also sought international participation in space. They have actively promoted scientific exchange among space scientists worldwide, and gradually opened their space facilities and programs to public view. They have included representatives on their spaceflights from both the former Eastern bloc and Western nations. In their 1988 flight plan for unmanned probes of the moons of Mars, they announced with pride that instruments from scientists of fourteen nations, including the U.S.A., were on their spacecraft. At the onset of President Mikhail Gorbachev's administration, policies were implemented promoting more openness and economic restructuring—welcome forces of change that were eventually to destabilize not only Soviet society, but its space program and ethos.[21] In the mid '90s, the inhabitants of the former U.S.S.R. are more concerned about survival amidst profound political and economic reform. This vast empire has been broken apart and its peoples in the midst of profound transition, contending with opposing forces who favor a centralized, socialist planned economy on the one hand, and democratic reformers pushing for a market-oriented, free enterprise system. Like its American counterpart, space planning in the nascent Commonwealth of Independent States is at a crossroads, grappling with rising costs, shrinking budgets, and weakening government support. Space programs and hardware that were once the showcase of the old Communist system are now being offered to Western customers for sale. The pride of Soviet space technology has been "mothballed"—the shuttle **Buran** and the heavy-lift expendable rocket, **Energia**. Conditions at the beginning of the decade were well described by two former directors of the Moscow Institute of Space Research:

—Roald Sagdeev, now a professor in the University of Maryland, addressed the George Washington University Space Policy Institute, and confessed that the Russian space program was caught up in the general collapse of the Soviet economy and central authority, being subject to claims of governments in the union's 15 ethnic republics (*Space Business News*, December 24, 1990, p. 3) ...

—Albert Galeev, speaking at the Institute on Global Conflict and Cooperation in the University of California–San Diego, admitted a year before the downfall of the U.S.S.R. that dramatic scaling back of their space missions was under way. He revealed that economic and political turmoil had brought an end to discussion of manned expeditions to Mars, and the new focus instead would be on unmanned missions (*Los Angeles Times,* January 13, 1991, p. B1).

The Way It Is

And what is the situation today in the space program of the newly constituted Russian Federation? Media reports were that at the beginning of his Presidency, Boris Yeltsin was unconvinced that space exploration could immediately benefit society and remedy its ills, so he had curtailed space spending. Yet to obtain foreign currency, he has been open to international space ventures if they utilize Russian space equipment and facilities, and provide jobs to their scientists and engineers. By 1993, Yeltsin's advisors were realizing that space was a good way to further Russia's international prestige and business. Although the bulk of spaceports are within Russian territory, some key launch pads, such as Baikonour, are in other CIS republics. Within the Russian Federation, control of the space program fragmented as ministries and the Academy of Science were reorganized, and the Russian Space Agency (RKA) came into being. In 1995, its budget is down 40% when set at 1.187 trillion rubles, about $270 million in foreign exchange markets. The RKA civilian agency shares facilities and spaceflight with the Military Space Forces (VKS). By Federation presidential decree, a Russian Center for the Conversion of the Aerospace Sector was set up to unify efforts both within Russia and the CIS. That Center not only engages directly in several large aerospace programs, but is intent on improving telecommunications, meteorological, ecological, and other space-based systems. Furthermore, the Russian government also agreed to increase financial support to the country's aerospace industry via tax breaks and grants (*Space News*, Nov. 8, 1993). With a severe funding crunch and increased privatization, space organizations pursue innovative strategies to survive and become more entrepreneurial, such as:

• selling or leasing space hardware and software, as well as facilities to other nations and foreign corporations ...
• selling satellite data, especially about weather and resources ...
• selling or exchanging Russian space expertise, particularly by consulting ...
• converting defense and space activities to civil or commercial operations (e.g. part of the Plesetsk Cosmodrome to the local Mirny community) ...
• training personnel from other countries in the Yuri Gagarin Cosmonaut Training Center and Star City, and even sending them in orbit to **Mir** space station for a fee ...
• transforming the Gagarin Center, in part, into a research institute which rents facilities and services to international customers ...
• increasing space tourism under **Cosmos Tours** with fees to visit former Soviet space facilities and museums, attend aerospace shows, as well as to experience cosmonaut training simulations ...

- founding of a space consulting firm, **The Space and We Co. Ltd.**, by ten cosmonauts to facilitate business relations and ventures in their country ...
- contracting for services of space and nuclear scientists/engineers who are underfunded or unemployed.

With regard to the latter need, Nobel laureate Leon Lederman commented:

> **Civilization would be much poorer without Russian Science. Apart from their outstanding discoveries in disciplines ranging from space to biology to particle physics to mathematics, Russian scientists provide the kind of intellectual stimulation and fertilization that makes science a global community.**
>
> *Popular Science*, August 1994, pp. 24–27.

The painful transformation of Russian space science and technology is under way, as these developments in the post-Soviet era indicate:

- Publication in English of *Space Bulletin*, a magazine by Moscow's Association for the Advances of Space Science and Technology; *Russian Aerospace Market*, an annual by St. Petersburg's Technoex-Conversion; *Samolynot*, an Russian/English-language aerospace magazine published in conjunction with the Aviation Maintenance Foundation International in the U.S.A. [22]
- International Conferences as the Russian Academy of National Economy's "Russian and the United States Stimulating Economic Progress through Cooperation" at the University of Maryland ... and Moscow Space Club's "Altai: the Outer Space Microcosm" in Russia's Altai Mountains, Bernaul. [23]
- Cooperation Agreements with other national space agencies, such as between RKA and NASA or ESA, on manned space infrastructure construction and other activities.
- Tashkent, Uzbekistan Agreements with ten former Soviet states, now CIS republics, provide for joint funding with Russia of civilian space activities, plus a separate accord with Kazakhstan on the future of the Baikonour Cosmodrome, located in that country—the Russian space program will have full access without restrictions.
- Entrepreneurial efforts to privatize state industry, such as the founding of the Defense Industrial Investment Company of Kaliningrad as an investment bank, using funds to underwrite promising ventures within Russia that produce for the civilian market (e.g., commercial applications of space technology); company is made up of industrial activities that were formerly government agencies.
- Formation of joint business ventures between Russian space organizations and foreign business corporations, such as between the Russian Academy of Sciences and the Russian–American Science Inc. of Falls, Church, Virginia (RAS); the Krunichev Strategic Missile Plant's agreement with Motorola Inc. to launch communication satellites for the American firm; and the same factory's venture with both Motorola and Lockheed Corporation on the Iridium project.
- Establishing Western marketing outlets for Russian space talent and material, such as Energia, U.S.A. to represent NPO Energia, the organization that built the **Mir** space station and other critical space hardware and now controls CIS manned space programs.
- International Maritime Satellite Organization (Inmarsat), a London-based consortium

of 64 nations, plans to use a Russian **Proton** rocket to launch one or more **Inmarsat 3** communications satellites.

Although its space program is severely strained, in 1994 Russia conducted more launches than any other one country or all the other spacefaring nations combined (48 successful launches and 1 failure). In that one year, it introduced four new types of satellite; kept its **Mir** station continuously occupied with cosmonauts, many from other nations; besides launching 18 **Proton** rockets, it added a new one—**Rokot**. In 1995, its plans call for three American Shuttle flights to **Mir**.

Russian Space Ethos Conclusion

Seemingly, the old Soviet ethos regarding space is being eroded in the '90s. In the new market-oriented Russia, if the price is right, everything is up for sale. Mementos of space heroes are posthumously pressed into duty to sell cars—the famous Cosmos Pavilion, a shrine of Soviet space feats, had been turned into a used-car lot while the exhibit is closed for repairs (*Los Angeles Times*, Nov. 23, 1992, p. H/5). Yet great leaders of the Soviet–military space complex had motivations other than pleasing the state, as General Grerman Titov explained in a 1992 California lecture to The Planetary Society—the first man to orbit Earth for more than 24 hours and once chief of the Soviet Defense Command explained that he and many of his colleagues were and are concerned not only with humanity's exploration of the solar system, but with gaining a deeper understanding of our own planet! Again the Russian ethos was revealed that same year when Vladmir L. Isvanov, flight testing director for the **Mir** space station, described plans for converting the Plesetsk Cosmodrome into a fully fledged space center to launch a variety of rockets, even into geostationary orbit. Then he asserted:

> **Russia is and will be a space power. Breathtaking ambitions are necessary to reach the spiritual heights, the heights of the Russian spirit. We began with the first *Sputnik* and we will go far beyond.**
>
> *Space Fax Daily*, October 6, 1992.

Despite the current chaos which has taken a toll on the Russian space program and caused numerous setbacks, Nicholas Johnson, an American who publishes annual reports on their space activities, believes the Russian space program has not collapsed and still operates over 150 satellites, though they are having to scale back on their space plans.[24] For Russians, space exploration has always had a deeper **cosmic** meaning. **Russian Cosmism,** best expressed by the visionary Konstantin E. Tsiolkovsky is a social movement dealing with the history and philosophy of the origin, evolution, future of the universe and humankind in their genetic unity and mutual influence. This "father of space travel," philosophized before his death in 1935:[25]

> **The biggest step in the history of mankind will be to leave the atmosphere of Earth to join the other planets.**

The future of the troubled Russian and CIS space programs seem to be in joint international ventures with other nations. These developments within the global space community will be discussed in the next chapter.

Exhibit 8—Europe's Spaceport for **Ariane**: Kourou. **Ariane-4 on the launchpad in French Guiana.** *Source*: European Space Agency (ESA).

1.3 EUROPEAN SPACE CASE STUDY

The multicultures of modern Europe have been developing a **space ethos** for centuries in their art, science, and literature. The early dreamers of space travel came from Europe—e.g., Italy's inventive artist, Michelangelo; France's science fiction writer, Jules Verne, and Romania's mathematician, Hermann Oberth. The drawings and writings of such visionaries inspired generations of future space scientists and engineers worldwide, but especially in Germany where rocket pioneering abounded in the first half of the 20th century. Europe's entry into the Space Age occurred in World War II, with the negative impact of missiles from Peenemunde upon a devastated England. However, with their minds on the Moon, many of the German rocket scientists there under Wernher von Braun, were transformed from war to peace by being absorbed into either the American or

Russian space programs.[26] In post-war Europe different countries generated a **space ethos** in unique ways and at different times. The major players in today's **European Space Agency (ESA)** are listed alphabetically below with a historical sketch:

• *France*—the Committee for Space Research was organized in 1959, followed by the formation of the National Center for Space Studies (CNES) in 1962; 23 years later the nation's space agency was to get a modern headquarters in the Ranguil suburbs of the medieval city of Toulouse where French **aerospace** industry and schools were also located. Finally, a Directorate General for Space was established in 1988 within the Ministry for Post, Telecommunications and Space ... . In 1965, France became the third country to orbit the rocket **Diamant** with its **Asterix 1** satellite; by 1990, it launched a series of satellites beginning **Spot 2** for remote sensing. The country's principal launch vehicle is now **Ariane 5** (it manages the ESA team for the minishuttle **Hermes** which has yet to be built). Arianespace became the private sector enterprise incorporated under French law to manage and market the Ariane launch services. In 1964, CNES chose Kourou near the equator to take advantage of the Earth's rotation for a geostationary orbit. There evolved the Guiana Space Center (CSG) which was to become Europe's official spaceport (see Exhibit 8). French technology has been an outstanding contributor to space progress, including on-going research for a cryogenic engine for powering **Ariane 5** and an imaging system for the U.S.'s **Clementine** lunar survey. In '95, it launched satellites **Helios-1** and **Telcom 2C**. Today CNES relies on a dozen major research laboratories as space contractors. Again as one ethos measure, in 1994 France spent $2.1 billion on its civil space programs and over $725 million on military space. Frenchmen have flown on both the American shuttle and the Russian space station. France is a prime mover in pushing Europe toward space independence, but as the country's immigrant population increases, one wonders if the fledgling **space ethos** has moved down from the intellectuals and scientists to the masses. The catalyst for this ethos may come from many prestigious global space organizations headquartered in France, such as the European Space Agency and the International Astronautical Federation in Paris, as well as the International Space University in Strasbourg.

• *Germany*—The German Aerospace Research Establishment was formed in 1969 within the Federal Republic by a merger of three organizations, one of which dated from 1907. By 1974, responsibilities for atmospheric research were absorbed. Under the Ministry for Research and Technology, the German Space Agency was founded in 1989 as a limited liability company. While Germany has participated in both NASA and ESA space projects, its own unique efforts have been directed toward the **Rosat** X-ray astronomical research satellite, **DFS** and **TUBsat** telecommunications satellites, the Sanger aerospace plane, and the **Columbus** module for the **International Space Station**. Germans and their experiments have flown on both the American shuttle and the Russian space station. With a 1994 civil space budget of $993 million, Germany is the second largest contributor to ESA funding.

• *Great Britain*—at the start of the Space Age, the U.K. showed great promise with its talented space scientists and engineers, vigorous activities by its space activists and their publications, and national launch vehicle program. The British Interplanetary Society and

BBC space specials have done much to raise citizen awareness. One positive development was the creation in 1985 of the British National Space Centre under the U.K. Department of Trade and Industry. A second was participation in 1990 by the U.K. Science and Engineering Research Council to fly the Along-track Scanning Radiometer as payload on the European Remote Sensing Satellite (ERS-1). Another is the publication by Butterworth-Heinemann in Oxford of the prestigious international journal, Space Policy. In the mid '90s, the U.K. principal space contributions are in instrumentations for Earth observation, and ESA's **Infrared Space Observatory**. But the nation's ethos for space seems to be diminishing temporarily as the Tory government cuts back its support for the aerospace industry in general, and space research in particular. Britain takes a token role in ESA projects such as **Ariane** and **Columbus**. In 1994–5 FY, its space budget was reduced to $290 million, well below its European counterparts. Britain's ambivalence about fully participating in the European Union somewhat undermines its activities within ESA—this problem with its continental neighbors goes back to the 1970s when its advocacy of **Europa** as a civil launch vehicle failed to gain their support. The average Briton has yet to realize that the way back to the glory days of "empire" lies on the high frontier.

- *Italy*—a country that moved early and actively by establishing a National Commission on Space Research in 1959, followed by spacecraft development. By 1964, it was cooperating with NASA to launch the **San Marco-1** satellite from a mobile platform off the coast of Kenya; this has evolved today into the **San Marco Scout** spacecraft. By 1979, Italy had a National Space Plan that led to the founding in 1988 of the Italian Space Agency (ASI) under its Ministry of Science and Technology. The nation has been a key participant in ESA, providing the use of the Italian Processing and Archiving Facility (I-PAF) located at ASI/Centro di Geodesia Spaziale in Matera. In 1994, Italy contributed $567 million to its space investment, less than previous years. Innovative Italian space research has ranged from **Italsat II** to programs with the U.S. on **Tethered Satellite Systems** and **Lageos II** to study tectonic motions with lasers, as well as **SAX X-Ray** satellite and **SAR-X**, a synthetic aperture radar satellite. At present, the Italian ethos for space is enthusiastic and very active, but delimited to a subculture of scientists, engineers, and students.

ESA Emergence

After decades the efforts by Western Europe to coordinate its space activities climaxed in 1975 with the formation of the **European Space Agency**, a merger of two prior organizations. The growth in the space science and launch program within the original thirteen member states parallels the progress of the larger entity, the European Common Market. By ESA's 20th anniversary, the latter has evolved into the expanding European Union, which is providing space with an ESA budget of $3.4 billion annually. Under its current **Horizon 2000**, ESA's long-term science program, the Council of member states has laid a realistic foundation for European space activities related to environmental sciences, telecommunications, launch systems, and manned spaceflight. For ten years now, these synergistic strategies have been implemented on the basis of worldwide partnerships with U.S.A., Russia, Japan, and Canada, as will be explained in the chapter to follow.

Headquartered in Paris, ESA facilities include the European Space and Research Center in The Netherlands (ESTEC, Noordwijk); the European Space Operations in Germany (ESOC, Darmstadt); and the European Space Research Institute in Italy (ESRIN, Frascati). Its official launch center is Kourou, French Guiana, with management delegated to Arianespace, and its European Astronaut Centre is located in Cologne. The Agency has done much to encourage EU member states enterprise in space science, Earth observation, and satellite communications. For example, in 1995, Finland became the 14th member state of ESA to participate in its programs, expending $4 million in space R&D. ESA also has an active program with the Swedish Space Corporation (SSC) at its **Estrange** center to track and control ESA satellites (Sweden space spending is around $102 million annually); with The Netherland's **ESTEC** Test Centre in Noordwijk (Dutch space spending is in '94 was $120 million and its largest ESA program is building a robotic arm for the Russian part of the International Space Station. The 1994 space investment of other ESA members not mentioned above are: Austria ($53 million); Belgium ($182); Denmark ($33 million); Ireland ($5.2 million); Norway ($33.7 million); Spain ($117 million, plus $30 million private space R&D); and Switzerland ($87.7 million public and $58 million private spending). After the collapse of the U.S.S.R., former eastern bloc countries wanted association with the agency, so cooperation agreements were signed with the Republic of Hungary in 1991, with the Government of Romania in 1992, and with the Republic of Poland in 1994—all indications that the **space ethos** is increasing in Central and Eastern Europe through information exchanges and Pan-European conferences. To assist Africa with remote sensing, mapping, and surveying, an ESA workstation was set up in Kenya (Nairobi). ESA has a long history of mission cooperation with both the American and Russian space programs, both manned and unmanned.

Despite some setbacks and launch failures, the European Space Agency has manifested steady successes, such as:

- the **Ulysses** spacecraft which rendezvoused with the planet Jupiter in 1992 ...
- the **ERS-1**and **2** satellites, establishing an ESA information network called **Earthnet**
- the mission extension of **Giotto (GEM)** instruments from the comet **Halley** encounter to comet **Grigg–Skjellerup** encounter ...
- the mapping of 120 000 stars by the satellite **Hipparcos** and the scientific experiments and data gathering of **Eureca** ...
- the participation in the **Hubble Space Telescope** under NASA's leadership ...
- the **Human Behavior in Space Simulation Studies** and the **Isolation Study for European Manned Space Infrastructure**, which will be discussed in the next chapter ...
- the **Spacelab** manned orbiting laboratory developed for use on the American space shuttle by ESA astronauts ...
- the free-flying laboratory called **Columbus** which is in preparation as an orbiting module with the forthcoming International Space Station (discussed in Chapter 8).

The Agency does not invent missions, but asks for proposals from scientists, both in Europe and elsewhere, whose peers then select the missions to be pursued. New programs include **Solar Heliospheric Observatory (SOHO)**; **Cluster**, four identical spacecraft to be launched on **Ariane 5** flying in tetrahedron configuration for polar orbits around Earth, observing the Sun's interaction with its magnetic field; the **Infrared Space Observatory**

(ISO) launched in 1995 in cooperation with both NASA and the Japanese Institute of Space and Astronautical Science. By 1996, the ESA/NASA joint venture of SOHO was sending back results of its sungazing.

European Space Ethos Conclusion

In the late 90's, the European Space Agency is moving into a global leadership position on lunar development (explained in a Chapter 7.3 case study). Advance planning under ESA's Office of Scientific Programmes produced a second major report in its *International Lunar Workshop—Towards a World Strategy for the Exploration and Utilisation of our Natural Satellite*. This offers rationales for permanently returning to the Moon that encompass philosophical, scientific, political, and socio-economic reasons.[27] At the 1994 Switzerland workshop which produced that report, Director R.M. Bonnet explained that ESA studies envision the Moon as "a vast space station offering many areas of cultural, as well as industrial resources—the closest space harbor beyond Low Earth Orbit." Therefore, ESA proposes a four-phase approach of exploration and development to be inaugurated at the turn of this century, open to international participation. If funded and implemented, it would become the most significant expression of the **European space ethos** to date. Some indication of this is evident in this quotation from the ESA Science Programme Committee on **The European Moon Programme** (ESA/SPC 94 43, Annex 1):

> **At the turn of the second millennium, our planet is at a major turning point. Politically, economically, and socially, humanity is in search of a new equilibrium. Such an equilibrium is necessary if we are to stop the continuous increase in unemployment which affects several hundreds of millions of the Earth's inhabitants. The need for large and global programmes ... is necessary for a new start in global economic development ... through peaceful initiatives which would offer a means to occupy the talents of everyone for the common good and the future of civilization ...**
>
> **Space programmes are among such large programmes, encompassing science and high-technology activities on a planetary scale, serving mankind, offering a mirror with which it can survey its home planet, while exploring further the limits of habitability and its visibility of the Universe.**

1.4 ASIAN SPACE CASE STUDY

Since the dawn of the Space Age many countries in the Middle and Far East have demonstrated interest in space, particularly with reference to communication satellites. The United Nations, particularly through its Office of Outer Space Affairs,[28] has sponsored conferences for participants from developing countries. These have centered on how space technology and resources might contribute to sustainable development and food production, communications for development and safety, modernization and infrastructure. If we take Asia as a case in point, Japan, with its First World economy, is obviously in the leadership position, followed by China and India, so these will become the subject of our review here. However, a **space ethos** seems also to be emerging in smaller Pacific Rim countries, such as Pakistan, Taiwan, and Korea. Their space programs and budgets are increasing, and almost 100% of every space dollar goes into productive, non-military use, instead of supporting a huge space bureaucracy.[29] Again, one ethos manifestation is what a country invests in space technology. Australia, for instance, spent about $6.7 million in 1995, plus $520 million for commercial purchases of space goods. The Australian Space Council's five-year plan includes development of a small satellite manufacturing industry; promoting Australian launch sites, such as the Woomera rocket range and proposed Cape York Spaceport; accessing remote sensing data, then extending its use within the Asia–Pacific market; performing space-related research through universities and high tech aerospace corporations; and participating in international space projects ... . Despite internal socioeconomic problems, Pakistan expends about $7 million annually on space. It has a national space agency, SUPARCO, which launched in 1995, **Badr-B**, its second LEO satellite. At its Islamabad receiving station, remote sensing imaging data are obtained and processed from many orbiting satellites of other countries. Perhaps the most unique of Pacific space endeavors is on the small island of Tanegashima with its $1 billion, modern, advanced Japanese launch center ... .

In 1994, ministers for space programs in the area gathered in Beijing to formulate a Regional Space Technology Program under the aegis of the UN's Economic and Social Commission for Asia and the Pacific. Regional cooperation—synergy—may be the best way for these diverse peoples and countries to cultivate the **space ethos** until they truly become spacefaring nations!

• *Japan*—in essence, this fourth nation to go into orbit has two space agencies, each funded separately. The oldest and smallest is The Institute of Space and Astronautical Science (ISAS), founded in the mid '50s, which launches scientific satellites from its Kagoshima Space Center on the southernmost main island of Kyushi. In collaboration with the University of Tokyo and Japanese electronic/motor companies, ISAS develops large solid rocket launchers, like the **Mu-Series**, which carries science satellites, like **Muses-A**, **Solar-A** and **Astro-D**, an X-ray satellite, its most sophisticated spacecraft involving extensive international cooperation. Now a multi-university institute, professors from many institutions come to conduct research, such as on **HIMES** (highly maneuverable engineering space vehicle) ... .

The other, founded in 1964, is the National Space Development Agency (NASDA) with the big budget, projects, and facilities. In 1972, NASDA established its Tsukuba Space Center, and has been launching diverse satellites ever since in both LEO and GEO; their names varied from **ETV-1** to series called **KIKU, UME, AYAME, HIMAWARI, YURI, SAKURA**, and **FUJI**. Their principal launch vehicles have been N-I/N-II, H-I/H-II; the latter is a Japanese design (49 m in length, 4 m in diameter, 256 tonnes in weight, with a payload capability of 2000 kg). The **H** series is launched from the Tanegashima Space Center in south-eastern Tanegashima Islands, Kagoshima prefecture. Their Earth Observation Center is in Hatoyama-machi, Saitama prefecture, while their Kakuda Propulsion Center is located in northern Kakuda City, Miyagi prefecture. Currently, the Japanese are focused on the Orbital Reentry Experiment Vehicle, the Vehicle Evaluation Payload and Space Flyer Unit for launch on **H-II** in geostationary orbits. For more than 30 years, NASDA has been expanding its international space role, and is now into material processing, life science experiments, preparation of payload specialists for the U.S.

Shuttle and developing **JEM** (Japanese Experimental Module) for use on the International Space Station before the year 2000.

Japan's Space Activities Committee coordinates the nation's space program among four government departments, facilitating joint ventures with the above agencies, universities, and private sector aerospace companies, as well as with other international agencies. Such cooperative projects include the 87 tonne J-1 rocket launched in 1995, the joint studies with the National Aerospace Laboratory (NAL), and vigorous lunar development research (refer to Chapter 10.l(B)). Within the latter, for example, a study group has proposed a **Lunar Energy Park**, and its plans extend to 2030.[30] There seems to be a strong and growing **space ethos** in Japan among the educated elite, especially in academic, scientific, and corporate communities. Involvement in the nation's space efforts is quite diversified, and in FY 94 over $2 billion was expended on space activities. Since the '80s Japan has held some 25 official meetings with the European Space Agency that has resulted in various cooperation agreements for sharing data and systems.

• *China*[31]—an ancient nation whose lore of Moon travel and experiments with rockets go back many centuries, China a likely to become a major spacefarer in the 21st century. Exhibit 9 is one indication of its current sophistication in space technology.

In this century, despite a civil war, the country's space activities began in the late 1950s when a development plan was formulated. In 1964 China developed and launched its first space vehicle. In 1970, it sent its first satellite **Mao-1** into orbit via the **Chang-Zheng-1 (CZ-1)** launch vehicle, thus inaugurating a series of recoverable communications satellites. By 1987, the **CZ-3** was engaged in the first commercial launch for foreign

Exhibit 9—A Chinese Control Room. *Source*: China Great Wall Industry Corporation (GWIC) and published in *Space Governance*, June 1995, Vol. 2:1, p. 20—"A Chinese Perspective on Space Development" by Deyong Kong.

countries, this time for the French company, Metra. By 1995, China had successfully launched more than 40 satellites with self-developed launch vehicles, particularly the **Chang-Zheng**, or **Long March** series. The country now has the Beijing Rocket Test Center and two launch facilities (Xichang Space Center and Jiuquan Space Center), as well as a TT&C satellite tracking network. At the time of writing, the Beijing Space Technology and Test Center was being constructed to provide clean operational environment from delivery of dispatched products of spacecraft to assembly and testing. There, the National Spacecraft Trial Base will have 100 000 m^2 floor area for space-related laboratories. In China, the Ministry of Astronautics (MOA) oversees space research and industry. Its technical consulting unit is the Science and Technology Committee, while its R&D entity is the Chinese Academy of Space Technology (CAST).

China and its universities attach great importance to space technologies and their applications for improving economic and social conditions within this huge, impoverished country. Its satellites are being utilized for television communication, disaster monitoring and forecasting, identifying natural resources, and even for mass education via the Satellite Distance Education Network which links 53 000 learning centers. Because the People's Republic of China is a closed system and there is an emphasis on military satellites, it is difficult to obtain information on how much the country is spending on space. Since 1985 the state has been moving into the commercial market by the decision to make its three-stage, **CZ-3** and **CZ-4** rockets available to foreign clients. The China Great Wall Industry Corporation is leading in the establishment of an independent space industrial system with international marketing capability. Under MOA, CGWIC is the sole organization for launch services and coordinates subcontractor partners. It has agents throughout the world—in the U.S.A., for instance, it is represented by Becker and Associates of McLean, Virginia. Also under MOA, China has two other companies in space-related business—Wanyuan and Lishen Microelectronics. In addition to launch and research services, China is selling reentry satellites to customers (**SETE** and **FSW-1/2**). China is also slowly joining the international space community—for example, in 1990 it began exchanges with the European Space Agency on launch services, and in May 1995, it signed a bilateral cooperation agreement with the U.S.A. In 1997, the Asia Telecommunications Co., partially owned by the Chinese CITIC Group, will fly the **ASIASAT-3** on a Russian **Proton** rocket. The government has always been actively engaged in UN programs on space applications, particularly with reference to environmental concerns, remote sensing, and space transportation systems. It is difficult for outsiders to estimate the extent of a **space ethos** within China, but it would appear to be centered among the educated elite. They have increasing interest in human spaceflight, life sciences and space medicine, planetary exploration and utilization of space resources.

• *India*—this people has been fascinated with spaceflight for centuries, for it is part of Hindu mythology. By the 18th century reality emerged when Tippu Sultan bombarded the British with rockets! By 1957, a corps of Indian scientists under Vikram Sarabhai were engaged in studies of cosmic ray physics at Ahmedabad's Physical Research Laboratory (PRI). In 1972, Prime Minister Jawaharlal Nehru created a National Committee for Space Research (NCSR) with Sarabhai as chairman. Although India was a Third World nation, Nehru encouraged space investment to solve socio-economic problems, and to make this country self-sufficient. For four decades, India has been seeking answers to its development dilemmas through space science and technology.[32]

Its space program began with a launching range for the United Nations which started at Thumba. Nearby, the Vikram Sarabhai Space Center (VSSC) was to evolve under the aegis of the Indian Space Research Organization (ISRO), created in 1969 and now part of the government's Department of Space (DOS). India has a history of joint ventures with the major spacefaring leaders—in 1972, the former Soviet Union launched India's first satellite, **Aryabhata**; in 1984, India's first cosmonaut, Rakesh Sharma, flew aboard **Salyut** 7 for a week; in 1988, the U.S.S.R. entered into the commercial satellite market by launching India's **IRS-IA** for remote sensing of its natural resources ... . In 1975, ISRO undertook the **SITE** satellite television project via NASA's **ATS6** satellite, thus demonstrating the social benefit of space-based broadcasting (farmers and students in 2400 villages received the communications in six states for a year). Because of Indian expertise in cosmic particle research, its payload experiments were part of an ESA ll-nation consortium which flew on NASA's **Spacelab** in the 1980s. At Cape Canaveral in 1982, **INSAT-IA** was lifted atop a **Delta** rocket, thus launching a communications revolution for the huge country and its teeming population. Built under contract by Ford Aerospace for the Indian Space Department, this was the world's first geostationary civilian satellite to combine telecommunications, TV broadcasting, and weather forecasting. **INSAT-IB** was launched from a U.S. space shuttle, helping to establish an **INSAT** National Satellite System for the disaster warnings, as well as broadcast communications—this spacecraft extended TV coverage to 70% of India's population.

Despite poverty and inadequate infrastructure, India has invested over $1 billion in space research and facilities that is having a payoff today. In 1995–96, the nation spent $330 million on space programs which employ over 15 000 citizens. Its return has been largely in multipurpose domestic satellites remote sensing, and launch vehicles. The heart of the Indian space program is in Bangalore, where both DOS and ISRO have their headquarters and the latter's Satellite Centre for design and fabrication ... . At the southern tip of India in Trivandrum, VSSC has become the main place for space R&D technology, especially the **SLV** rocket (used to launch India's first space satellite, **Rohini**), and the later **ASLV** and **PSLV** series of launch vehicles, as well as operation of the UN's TERLS station ... . On Sriharikota Island near Madras is the SHAR facilities for integration, testing, and launching of space vehicles, such as **SLV-3**, in addition to its ground tracking system for orbiting spacecraft ... . In the northeast at Ahmenabad, the Space Applications Centre (SAC) engages in space science and technology applications, such as telecommunications, space-based surveys, space meteorology and satellite geodesy ... . The Auxiliary Propulsion System Unit (APSU) has laboratories at both Bangalore and Trivandrum which design, develop and supply propulsion control packages. Apart from entities already cited, the government of India also sponsors the Development and Educational Communications Unit (DECU) and the National Remote Sensing Agency (NRSA). In 1996, India, now one of the six leading spacefaring powers, successfully deployed the Polar Satellite Launch Vehicle, placing **IRS-P3** into orbit.

Because of both heritage and the benefits that satellite communications has brought to the masses, it would seem that the **space ethos** in India is more widely diversified. It is evident in the fierce protection of village television by peasants, and thousands of Indians who flock to the Thumba space museum when it is opened. The people of India seem proud of their country's space achievement, and now accept and even enthusiastically support space spending. In the long run, that ethos will spread only if the late Vikram Sarabhai vision is fulfilled—namely, using India's rockets and satellites to deliver greater literacy, better living conditions, and less poverty!

Asian Space Ethos Conclusions
The scope of interest in space technology has risen dramatically among Oriental peoples in the last decade. Given Japan's economic and technical leadership, one would expect it to be pre-eminent in this regard within the region. But as far back as 1976, Indonesia became the first Asian country to enter the domestic satellite era when it acquired the U.S. built **Palapa** system—today with over 40 ground stations, it provides communication links for 3000 islands over 5000 km of ocean. The current generation of this Indonesian satellite uses surplus capacity on behalf of Malaysia, Singapore, Thailand, and the Philippines. What is remarkable is the vigor and progress in developing economies, such as China, India, and even Pakistan. Through its Islamic Space Institute, the last-mentioned republic has formed partnerships with Bangladesh, Egypt, Indonesia, Iran, Saudi Arabia, and Turkey to pool their resources in satellite construction, instrumentation, control systems, launch vehicles and facilities. Pakistan is also an example of how a Third World country can benefit from international cooperation—the hundreds of sounding rockets launched from Sonmiani beach near Karachi have carried payloads from the U.S., France, and the U.K. Regional satellite systems are another example of space **synergy**, the subject matter of our next chapter—**Arabsat** was financed by 22 countries in West Asia. Obviously, the **space ethos** is growing throughout the Middle and Far East, but the Indians seem to have the lead in terms of its impact upon its inhabitants.

1.5 CONCLUSIONS ON GLOBAL SPACE ETHOS

Humankind, whether in the East or in the West, senses that our species seems to be in a epochal transition to space-based living and in the process will create an entirely new space culture. Whether by manned or unmanned spacecraft, current generations are pushing back the high frontier by expanding into outer space which our descendants will inhabit. In the foreword for the first edition of *Living and Working in Space*, Jesco von Puttkamer, a NASA senior strategic planner, provided a unique illustration of human progress and future in space. It has been reproduced in Exhibit 10, and offers rare insights.

If ethos defines the character of a people, then it is evident that there is a strong microculture throughout the world which already possesses a **space ethos**! This subculture exchanges on the Internet, or through professional conferences and papers, as well through joint space ventures. Even within countries that lack a strong space program, one will find space advocacy movements. As more spacefarers go aloft and experience the "overview effect," they gain a new perspective from achieving that vantage point in orbit, so that even our planet is viewed as an interconnected ecosystem.[33] With this enhancement of our collective consciousness, "Earthbound Everyman" may be ready to cultivate not just a **space ethos**, but more, as this quotation from attorneys Robinson and White so aptly implies:[34]

Thus we stand in the late twentieth century on the threshold of extending old civilizations into space, perhaps even creating new ones in which our own sons and daughters may be extraterrestrials from every point of view.

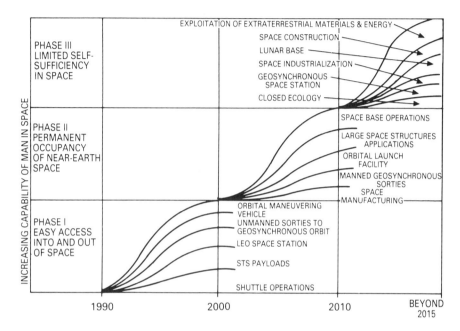

Exhibit 10—Human Progress in Space. *Source*: Jesco von Puttkamer from his "Foreword" to the first edition, Living and Working In Space. Chichester, Ellis Horwood, 1992, p. 15.

Mankind will not remain forever on Earth. In pursuit of light and space he will timidly at first probe the limits of the atmosphere and later extend his control throughout the Solar System.
—Konstantin Tsiolkovsky, Russian space philosopher, 1911

There is a clarity, a brilliance to space that simply does not exist on Earth, even on a cloudless summer's day in the Rockies. Nowhere else can you realize so fully the majesty of our Earth and be so awed at the thought that it's only one of untold thousands of planets.
—Gus Grissom, Second American Astronaut in Space, 1965

One of the major reasons the spirit of adventure has gone out of space exploration is that we have allowed bureaucracies to dominate too many of our scientific endeavors. Bureaucracies are designed to minimize risk and to create systematic procedures ... . The challenge for us is to get government and bureaucracy out of the way and put scientists, engineers, entrepreneurs, and adventurers back into the business of exploration and discovery. The 21st century should be a great century of exploration for humanity ...
—Newt Gingrich, Speaker, U.S. House of Representative; 1995

2

Human Development and Synergy in Space

The human species has demonstrated the capacity to dream and to envision the future, as well as to contemplate its past. Through the power of imagination, humans can conceive and often execute grand plans. We create images in our heads and sometimes act upon them. This conceptual ability powerfully influences human behavior, whether in terms of individuals or institutions, nations or humankind. It results in the setting of goals which energize people to impressive achievements. And so it is with outer space and the human expansion into the universe.

2.1 SPACE VISIONS OF HUMANKIND

For eons, humans thought they were Earthbound, but dreamed of leaving the planet and even of going to the Moon. The Man in or on the Moon was a recurrent theme in our mythology, so philosophers, poets, and novelists wrote about the possibility of our species moving beyond this terrestrial home of origin. Reality began when a 27-year-old Russian cosmonaut, Yuri Gagarin, orbited the Earth in the "cosmic ship" **Vostok** on April 12, 1961. Then on July 21, l969, the Apollo 11 LEM landed on the lunar surface, allowing Neil Armstrong and Buzz Aldrin actually to walk on the Moon![1] (See Exhibit 11). As longshoreman–philosopher Eric Hoffer reminded us, we are now forced to change our image of the humankind—we are *no longer Earthbound*; maybe we are only transients on this planet, perhaps *our real home is out there*! Space achievements challenge us to re-move the psychological binders and blinders on our collective self-concept, fostering the actualization of human potential.

As humans change mindsets about who and what we are, so will we alter our vision of outer space and our purposes in its exploration, development, and settlement. In a Los Alamos conference and proceedings on interstellar migration and the human experience, the assembled scientists set forth this vision: [2]

- For the past five million years of human physical and cultural evolution, the learned use of technology—both material and social—has enabled this exploring species to expand and cope with all aspects of this planet's environment and beyond ... .

Exhibit 11—First Lunar Landing of Humans. **Apollo 11** astronaut "Buzz" Aldrin is photographed on the Moon, July 21, 1969, by crewmate Neil Armstrong reflected in the visor—the one who made the first giant step for mankind, and said "That's one small step for a man, one giant leap for mankind." *Source:* NASA Headquarters.

- Technological prowess, revolutions, and contributions have not only driven human culture and procreation for the four billion inhabitants of this Earth, but are creating an entirely new space culture and being ... .
- Just as the ancient Polynesians navigated through images in their heads to create a new culture by colonization of the Pacific islands, so our descendants with emerging space technologies and explorers' bent may use the asteroids, comets, and planets to populate, settle, and civilize our Solar System ... .
- Since *Homo sapiens* adapts and evolves both biologically and culturally, the conquest and colonizing of space will facilitate a quantum leap in human evolution and diversity, eventually leading to crossing the gulf of light years to other star systems.

In an anatomy of new realities, the late Jonas Salk brilliantly philosophized on man unfolding and the "survival of the wisest."[3] With these ponderings, the polio pioneer Salk speculated on a "metabiological evolution" that results in an increase of consciousness, including of our evolution. Just as the biological evolution depends upon genes in a cell, the metabiological equivalent is *ideas* generated in the human mind. Such ideas affect the nature, characteristics, and behavior of the metabiological organism, whether as an individual or collectively in society. My friend, Salk the scientist, was intrigued by the way the human mind deals with the challenges posed by our environment, developing *metabiological* traits that are transmitted, like genes, to succeeding generations. Dr Salk envisioned the human family as a single organism requiring new responses and relationships, especially between our intuition and reason. He described evolution as a process of changing relationships. If Salk's thesis is correct, that human thought and creativity develop in response to our environment, then it is my contention that the space environment will produce the real metabiological evolution and immense cultural diversity. There—*beyond the Earth's atmosphere*—humankind may develop the wisdom to survive and diversify as part of the cosmos, thereby ensuring our continuance as a species!

When our primordial ancestors made the transition from ape to human, they moved from jungle to savanna, employing tools in their quest for survival. The initial stages of human development were first tribal, centered around hunting and food gathering; followed next by the agricultural period focused on farming, yet stimulating the beginnings of market places or cities. In both previous stages, our ancestors increasingly utilized technology that produced personalistic, flexible, and small-scale organization, as well as conflict. Subsequently, the third industrial stage employed mechanized technology and systems that laid the groundwork for going offworld. In our present fourth developmental stage, the *metaindustrial*, our energies are directed toward using information and communication technology, leading to the phenomenon of globalization.[4] One outcome today is that the management of complex, large-scale systems requires cooperation—all factors contributing to the movement from terrestrial to extraterrestrial living. To succeed aloft requires innovative research and synthesis in both space engineering and behavioral science technologies. As humans expand outward into the universe, anthropologist N. P. Tanner from the University of California—Santa Cruz comments that this migration into solar and star systems will necessitate cross-cultural skills and genetic diversity, whether we meet extraterrestrials out there or whether *we become* the differing extraterrestrials![5]

As intimated in our opening chapter, one is not likely to have a **space ethos** without a **space vision.** Multiple visions have been put forth about humankind's Diaspora in space by cosmic philosophers and prophets. The visionaries range from 19th century French science-fiction writer, Jules Vernes, and Russian science-teacher, Konstantin Tsiolkovski, to 20th century American engineering professor, Robert Goddard, and Romanian mathematician, Hermann Oberth. Their utopian dreams were translated to realities by teams of German, Russian, or American rocket engineers led by men like Sergei Korolev, Wernher von Braun, and Krafft Ehricke.[6] Their forecasts of yesterday are today's realities, such as satellite telecommunications, remote sensing, Earth observation and more. During the past several decades, visions of orbital space colonies and artificial worlds have been set forth in classic books in many languages; some of these authors have even suggested planetary engineering or *terraforming* for the restructuring of other planets into New Earths.[7] Since the 16th century, artists and now a variety of media specialists have illustrated and interpreted such space visions for the masses—from artists such as Michelangelo, Lucien

Rudaux, Chesley Bonestell, R. A. Smith, and David A. Hartley to film producers such as George Pal, Arthur C. Clarke, Gene Roddenberry, Steven Spielberg, and David Lucas. Under the leadership of its president, Dennis M. Davidson, the International Association of Astronomical Arts (IAAA) has drawn together artists from 14 countries to visualize our future in space. Using all medium and even computer graphics, these creative professionals help us to travel where we cannot yet go, and even to envision space settlement. Further, Association members also include those who have been in orbit, such as cosmonaut Alexei Leonov and astronaut Alan Bean.[8]

Currently, the differences in visions of the human futures in space center around whether this frontier is to be used to promote peaceful scientific exploration, commerce, and settlement,[9] or for military defense and eventually war.[10] Fortunately, with the demise in the '90s of the former Soviet Union and the Cold War, the former view is in ascendance, while the approach of "Star Wars" and the Strategic Defense Initiative diminishes. Going back to the U.S.A. as a *case in point*, the previous chapter described what the National Commission on Space (1986) set forth as an American vision of civilian space goals for the next fifty years. In *Pioneering the Space Frontier*, that Presidentially appointed group provided a scenario for a multitrillion-dollar, space-based economy dependent upon the support of the public with Administration and Congressional leadership. These distinguished proponents of space commercialization called for private–public sector partnerships in a threefold increase in space investments, research, technologies and transportation. The Commission advocated living and working in space on a permanent basis, and offered a road map for goal-achievement. This American dream of a space renaissance, unfortunately, was overshadowed by other realities, such as a failing economy, so that nation's plans for space exploration have been curtailed.[11]

The space activist movement naturally endorsed the direction of that Commission vision. Among them, one of the largest organizations—the National Space Society—articulated specific goals and a timetable for creating a **spacefaring civilization**:

- immediate inauguration of a lunar polar orbiter, lunar transport system, and Mars geothermal orbiter ...
- completion of a Space Station design aimed at a manned return to the Moon ...
- provision of adequate Earth-launch capability to support buildup of a permanent presence on the Moon by and processing/utilization of lunar resources by 2010 ...
- development of nuclear space power systems both for transportation and for use on the lunar surface ...
- research for second-generation launch and transportation systems to support larger space settlements while reducing launch payload costs and limitations ...
- public policies and subsidies for a U.S.A. commercial space transportation industry.[12]

In an attempt to sharpen its own planning and goal setting, as well as to enlist public support, the National Aeronautics and Space Administration has itself promoted a number of in-house studies to present its near-term views of the U.S.A.'s space future. Some of these reports and special publications are through the NASA Advisory Council or in conjunction with the National Academy of Science or with the National Academy of Public Administration (see Appendix C).[13] Under the aegis of astronaut Sally Ride, the agency expressed its most focused vision in a report on *Leadership and America's Future in Space*. As previously indicated, this discussed the immediate future in terms of proposals for:

- examination of global phenomena on this planet from the international perspective of geostationary and polar-orbiting platforms—the "Mission to Planet Earth";
- facilitation of a planetary science program of the outer planets and smaller bodies (e.g., asteroids), possibly including an unmanned Mars sample return—the "Exploration of the Solar System;"
- establishment of a scientific outpost on the Moon, beginning with robotic exploration and Lunar Geoscience Observer until several operational phases result in a base for 30 people engaged in scientific and industrial pursuits—"Outpost on the Moon;"
- inauguration of a Mars manned mission program culminating in a human landing on that planet or its moons, and eventual establishment of a permanent base—"Humans to Mars."

Alas, the country's socio-political and economic situation has limited the implementation of most of these visionary plans.

The development of a space ethos within a culture is dependent on the articulation of a space vision which, in turn, impacts the national image of a people. The opening chapter has described the leadership role of the American President in this process, particularly through White House interventions, such as the National Space Council. All this official, forward movement occurred within the past ten years despite national setbacks and spending curtailments caused by severe budget deficits, recession, and even a war in the Persian/Arabian Gulf.

However, government actions alone are insufficient if a nation's space vision is to be communicated to the masses. Too often technologists drive a space program, neglecting to involve the public. Alcestis Oberg, space writer, summarized the problem:

> **NASA and the aerospace community consistently have snubbed ordinary citizens from direct space participation. The Moon has been reduced to a NASA testbed, not a Plymouth Plantation, and Mars is another rock-collecting expedition, not a brave new world. The space program no longer serves the world, the people serve it.**
>
> *(Space News, Feb. 4–10, 1991, p. 15)*

She believes that would-be leaders have yet to demonstrate the space program's relevancy to national life.

To ensure success in the early attempts at permanent space habitation also requires broad interdisciplinary and institutional participation, as well as interagency and international collaboration. Professional associations in the space field have long promoted conferences and publications to express their vision of the future (such as, International Astronautical Federation, the British Interplanetary Society, the American Institute of Aeronautics and Astronautics, the American Astronautical Society, and other organizations described in Appendix C). Invariably, their published proceedings contribute to a body of literature making the case for space investment (such as, *The Human Quest in Space*, the AAS 24th Goddard Memorial Symposium, edited by Burdett and Soffen,

1987). Other clarifications of space vision may come from the published reports of re-search organizations, such as the National Academy of Public Administration, California Space Institute, and The Brookings Institution (the latter, for example, published in 1987, *Space and National Security* by research associate, Paul B. Stares). Non-profit organiza-tions may also issue their forecasts, as did the United Nations Association of the USA in their 1986 report (*The Next Giant Lead in Space—an Agenda for International Coopera-tion*). Annually, for instance, voluntary organizations gather to discuss our space futures, and again publish the proceedings which express their vision—two examples, are the United States Space Foundation's yearly **Symposium**, and the National Space Society's **International Space Development Conference.**[14] Sometimes informal networks, often gathered through the Internet, best express the space aspirations of inhabitants. For exam-ple, The National Space Society has a **Homepage** on the World Wide Web, where those plugged into the "information superhighway" can exchange their ideas of where we should be going in space. Another loosely formed alliance of professionals calls itself the **Mars Underground**, convening periodically in Boulder, Colorado. Its published papers provide a blueprint for manned missions to Mars, including precursor unmanned missions, re-quired experiments, and the technology necessary to maintain a permanent human presence on the Red Planet. So far four volumes of proceedings have been published un-der the title *The Case for Mars*, and others are under way.[15]

The space vision of the diverse peoples in the former U.S.S.R. has been somewhat obscured in the West until recently. In the past, it was enunciated by scholars such as Konstantin Tsiolkovsky, who wrote from 1883 to 1935; Yuri Kondratyuk, who described multistage rockets, solar energy power utilization, and orbiting space stations between 1916 and 1929; and Frederich Tsander, who designed forerunners of a space shuttle in the 1930s. Today, with more open communication and exchange, the world is beginning to perceive the Russian's farsightedness about space exploration, and to appreciate their accomplishments as humankind's largest space enterprise. With over a 100 launches annually to Earth orbit or beyond, their program generally represents 80% of the world's space missions!

In writing on the future manned spaceflight from the old Soviet perspective, Clark and Johnson analyzed their shift from lunar landings to space stations, like the now functioning **Mir**, and their current focus upon Mars.[16] During the mid 1970s, Soviet policy makers decided to exploit space for the benefit of their country's economy. With the interest in the late '80s of moving toward a market economy, a new national space organization, *Glavkosmos*, sought to market its space program and hardware internationally through. joint ventures. The next stage of space projects envisioned for the 1990s by the Russian Federation concentrates on cooperation in international projects. With a combination of heavy-lift vehicles such as **Energia** (SL-17), **Proton** (SL-13), and the **Buran** space shuttle, the Russians had hoped to lift both cargo and cosmonauts to an orbiting "Cosmograd" ("Spacetown") where 20 humans could live and work, principally in scientific and micro-gravity research. Because of severe funding shortages which have "mothballed" both **En-ergia** and **Buran**, the Russians have had to scale back that plan for partnership in the **International Space Station.** Interestingly, Russian scientists for the last few decades have been examining using space solar power for home consumption by means of orbiting power stations and satellite.[17] (Refer to Appendix B.)

Unfortunately, today internal socio-economic and political crises within the former U.S.S.R. are forcing the postponement of the implementation of many Russian space visions. Thus, long-range manned missions to Mars were sidetracked, while more limited, unmanned scientific probes of Mars and its moons are scheduled for the rest of this decade. During a 1991 lecture at the California Space Institute, Dr Albert Galeev, then director of their Institute for Space Research, revealed that within the Commonwealth of Independent States (CIS), only the fusion power community is still interested in the Moon, while the Institute of Geochemistry is focused on utilization of asteroids ... . Apparently, the Russian space ethos and vision are largely based on input from the scholarly and scientific communities, especially their prestigious Academy of Sciences.

What is needed before the turn of the century is global expression of how other space powers—such as the People's Republic of China, Japan, India, and the European Space Agency—envision humanity's future in outer space. We indicated some of their aspirations in the previous chapter. While the World Space Congress of ISY '92 did contribute to this understanding, there have also been useful regional conferences which have done the same at the local level, such as the Pacific Basin International Symposium on Advances in Space Science Technology and its Applications.[18] Obviously, the United Nation's Outer Space Affairs Office is best positioned to provide the forums for such sharing, but they have done this for developing countries.

In formulating a space vision, the basic question to be answered is **why go into space at all?** That was the theme of space essay contests sponsored by **United Societies in Space, Inc.** [19] In 1992, the winner was Jesco von Puttkamer, a NASA strategic planner, who reminded us that almost 500 humans, men and women from both East and West, have already been aloft. He envisions this evolutionary opening of cosmic space as a **cultural process.** For him, going into orbit is a social phenomenon, a force for change and growth in the human family. Having worked with von Braun on the **Saturn/Apollo** and **Skylab** programs, this aerospace engineer views spaceflight as a source of innovation bringing immense returns. von Puttkamer believes that we have an ethical obligation to invest in space on behalf of future generations. Accessing space requires a consciousness change, so we see humans and Earth–Space as one creative system, as illustrated in Exhibit 12. We go into space for the same reasons that prehistoric ancestors left the seas to breathe on land ... . The 1993 winner was Roland A. Foulkes, a University of Florida anthropologist, who analyzed space in various dimensions, including extraterrestrial. Like Victor Truner's view of anthropology, he contends we go to a far place (space) in order to appreciate a familiar place (Earth) better. So he confirms Gerard K. O'Neill's vision: the opening of the high frontier represents opportunity that challenges the best in us, and as we build there, we are given new freedoms to search for better governments, social systems, and ways of life.

In summary, the various visions of humankind's future on the high frontier generally conclude that SPACE IS A PLACE for:

- advancing human evolution and culture through space technology and the practice of synergy in the application of information and knowledge;
- improving the quality of human life both on Earth and offworld by utilizing the unlimited space resources;

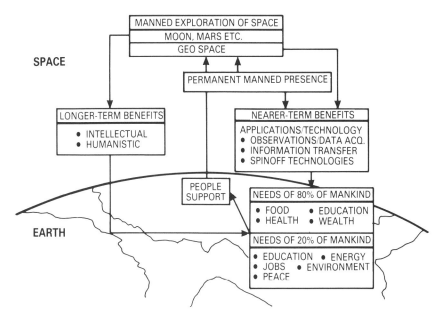

Exhibit 12—Development of Human–Earth–Space, One Creative System. *Source*: Jesco von Puttkamer, from his "Foreword," to the first edition, *Living and Working in Space*. Chichester: Ellis Horwood, 1992, p. 18.

- engaging in peaceful human pursuits, such as scientific exploration/experimentation, industrialization/settlement.

When viewed from 500 to a 1000 years from now, the renowned futurist and author, Alvin Toffler, believes that in terms of human evolution, today's primitive, faltering steps beyond Earth will likely be recognized then as the most important human project of our era, matched only by what is happening in computers and biology. He concluded:[20]

> **Space will not only shape our descendants' view of our time; it also shapes our view of the future ... . There is at the same time a growing post-national consciousness around the world, a globalist ideology with a planetary view, and space activity has clearly been a major contributor to its spread. The very image of "Spaceship Earth" has been a potent diffuser of post-national globalism.**

Yet, Toffler bewailed that no such positive vision of the future exists with reference to the 21st century, only confusion, blur, uncertainty, fear. Again, it is my contention that such a vision will emerge if global leaders utilize the high frontier as their common focus, thus energizing the planet's inhabitants outward and upward!

Perhaps humankind's vision for space was best expressed by J. D. Bernal, British physicist, in his 1929 classic book, *The World, the Flesh, and the Devil*:

> However, once acclimatized to space living, it is unlikely that Man will stop until he has roamed over and colonized most of the sidereal Universe, or that this will be the end. Man will not ultimately be content to be parasitic on the stars, but will invade and organize them for his own purposes.

2.2 LEAVING EARTH'S CRADLE—JOINT VENTURING

The artist's rendition in Exhibit 13 captures for me all the challenge in leaving our home planet to explore and settle the universe. **Spacekind** are multicultural beings of both genders who reach out to **Earthkind** to support them in the quest be it on orbiting vehicles and platforms, or on new worlds, such as Mars.

Exhibit 13—Leaving the Home Planet. *Source*: NASA's Lyndon B. Johnson Space Center, Houston, Texas 77058, U.S.A.

Since 1957 when *Sputnik* (meaning "satellite" in the Russian language) was launched, the central thrust of space endeavors for the past forty years has been upon technical not social development in space.[21] It is understandable why priority would have to be given to hardware and software which makes feasible both unmanned and manned operations in space.[22] Success in this regard enables our species to put humans on the lunar surface, to produce orbiters and space laboratories, to probe the far corners of the universe with automated spacecraft.[23] This first stage of space development was a time when rocket specialists, engineers, and technicians dominated planning and management. In this period of short-duration manned flights, the consistent, underrated elements in much of space research and operations, conferences and publications has been the **broader** human factors and life sciences. The Russian space program has shown more concern for these dimensions, while NASA's attention has been directed somewhat toward physiological and medical studies of the astronauts, and to a lesser extent the psychological qualities for effective living in outer space. It is no wonder, then, that veteran astronaut Norman Thagard, himself a physician, experienced such acculturation problems when he became the first American to stay aloft 115 days. Relative to his adjustment difficulties aboard the Russian **Mir** station, the NASA administrator publicly apologized for the Agency's lapse in Thagard's psychological preparation and support services, promising improvements before the second U.S. astronaut flew to **Mir** in March 1996 (*San Diego Union Tribune*, July 8, 1995). [See the front cover photograph and commentary on Shannon Lucid on page vi.]

The initial decades of space exploration and exploitation have been also been marked by debate over the advantages of unmanned versus manned flight. Planetary scientists make the case for less cost and hazards by the use of automation and robotics in spacecraft without crews. To illustrate this continuing debate within the U.S.A., the Space Science Board of the National Academy of Sciences expressed its vision that the main goal of the nation's civilian space program should be the advancement of scientific knowledge and its application to human welfare (*Commercial Space*, Winter 1987, p. 13). Furthermore, the Board recommended the use of automated spacecraft for all missions beyond lower Earth orbit from the years 1995 through 2015, a position at odds with the National Commission on Space's advocacy of permanent colonization of the Moon and Mars in the first quarter of the 21st century. The Board's argument was that the focus should be upon studies in astronomy and astrophysics, microgravity, solar/space physics, planetary exploration, and space medicine. The last mentioned point is that more research is needed on the effects of prolonged weightlessness before long-duration missions can be undertaken. One of the nation's pre-eminent space scientists, James Van Allen, articulated this viewpoint in a lecture at the University of California—San Diego. He maintained that the U.S.A. spent $150 billion on manned flight to get 133 different individuals aloft, and contrasts the accomplishments with magnificent scientific results obtained from unmanned flights at one-fifth the cost (*Los Angeles Times*, November 20, 1986, II/4). Professor James A. Arnold, a lunar chemist and first director of the California Space Institute, countered on the same occasion that longer manned flights go beyond science and the gathering of knowledge. He pointed to the inherent human drive to explore, to transform dreams of living and working outside the Earth's atmosphere into tomorrow's realities. Both in the U.S.A. and in the former U.S.S.R., this divergence in viewpoints is again manifest relative to missions to Mars.

The consensus emerging among space planners and policy makers is that both types of mission are required and are complementary. Whether manned or unmanned, both represent human extension into space. Experience has demonstrated the value of human inventiveness in space when computers and mechanical equipment fail and need to be repaired or replaced. An interview with former **Skylab** astronaut, Owen Garriott, among those few Americans with long spaceflight experience (two months), revealed the livability within that prototype U.S.A. space "station". While discussing the planning for the "international station" of the 1990s, he observed that NASA has a Congressional mandate to provide a "man-tended capability" in the research to be done there, as well as in "tele-science"—interactive research combining a groundbased team, a spaceborne team and automated instruments (*Space World*, Dec. 1986, 27–31). In the current era of tight budgets, it is shortsighted for planetary scientists to battle against human spaceflight as they struggle for more funding of robotic spacecraft.

The ultimate argument is that humankind goes into space because it is our nature and destiny, and other rationalizations are subservient to that overall purpose. Furthermore, public investment in the space program demands a human component.

The joint venture of multiple national space agencies to erect an **International Space Station** before year 2000 with "permanent manned capability" is discussed in a case study in Chapter 8 on macromanagement. Meanwhile, the Russians continue to extend the stay of multinational participants at **Mir** (Peace), their original orbiting space laboratory which now may be reached by rocket launchings of both manned (**Soyuz** capsule) and unmanned (**Progress** cargo carrier) space vehicles, as well as by docking of the American shuttle. One reason for extended flights in orbit and life science research aloft is to prepare humans for a 30-month round-trip to Mars. In the near-term, space scientists worldwide are planning before 2025 to establish manned bases at the Moon and/or Mars as a prelude toward colonization of the Solar System in the centuries to come. Whether this will be accomplished by the fiftieth anniversary of the Apollo Moon landings is more a matter of vision, leadership, and funding; we have the technical "know-how" but presently lack the international will and consensus.

Yet, this global thrust toward an enduring human presence in orbit marks the *next stage in space development*, a time when the biological and behavioral sciences must cooperate together in major contributions. Dr Elie Shneour, president of Biosystems Research Institute, in a lecture at the California Space Institute (March 4, 1987), indicated that closed ecosystems are possible in space. But for long missions beyond 100 days, much more research is required to create and maintain balanced ecological systems to sustain the quality of human life in a hostile environment. The perceptions of engineering planners tend to ignore or undervalue many factors that biologists consider critical to life and the reduction of stress on the spacefarer.[24] These factors—such as oxygen, food, water, circadian rhythms, or personal pollutants—tend to magnify on longer space flights. We have only begun to comprehend the near-Earth environment, no less ponder the human challenges beyond.[25] Since there is so little human experience for such extended spaceflight, beyond those of nuclear submariners, behavioral and biological scientists should join in combined research with space scientists and engineers regarding humans **living and working in microgravity** for a year or more!

The social science contribution to this research arena has barely begun.[26] In addition to the analysis of terrestrial analogs for offterrestrial living, there is much need for increased studies of:

- group behavior and dynamics in spaceflight, especially with heterogeneous, multidisciplinary, and multicultural crews in isolated, confined environments;
- culture and quality of life, including safety and recreation, to be created at space stations, lunar installations, and a Mars/Deimos outpost;
- recruitment, selection, training, support, and evaluation of long-duration spacefarers and their families;
- on-site expert systems and simulations for education and entertainment in orbit, as well as computerized systems for monitoring and diagnosing physical and psychological health;
- art and science of space management on Earth and in orbit;
- effective operation aloft of space commerce, manufacturing,
- and mining, as well as solar power systems;
- planning, establishing, sustaining, and governing space settlements.

The next chapter will explore these matters further.

2.3 EMERGING SPACE SETTLEMENT ISSUES

The above listing gives only some of the challenges that lie ahead with reference to permanent space habitation. To elaborate on a few of these **issues** for continuing research, consider:

(a) *Relations among crews* of diverse composition on long-term flights. Although there has been much study to date in the field of group dynamics, some of it directed toward life in confined quarters of submarines, there has been little transfer of this information and insight to the study of team behavior among spacefarers, and its implications for future space colonists. There is opportunity for immediate application of such behavioral science research to the preparation of crews for the Shuttle and **Mir.** However, the real challenge is in training multicultural crews for the **International Space Station**, so that they may function more effectively as synergistic teams with their ground support staff, as depicted in Exhibit 14.

There has also been significant investigations in the field of interpersonal and intercultural communication which is transferable to space habitation. During the December 1983 **Columbia** flight of six astronauts, an incident was recorded on television for all the world to observe. Those in flight were overwhelmed by the communications and requests from ground control, particularly from project managers of experiments. There should be increasing communication challenges as larger crews go into space for longer periods, not only among those in flight, but with those related to the mission back on Earth. In the next stage of flight, ground control will not only have to give up some of its domination, but more autonomy will have to be given to those in orbit. This is especially true of those at a

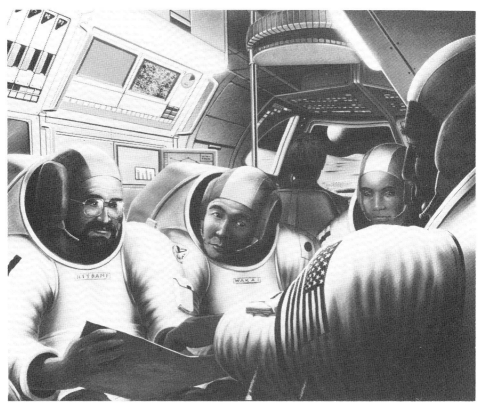

Exhibit 14—Transnational, Multicultural Crew Teams. Source: NASA's Lyndon B. Johnson Space Center, Houston, Texas 77058, U.S.A.

lunar outpost and base. Improved information technology alone is not sufficient; communication specialists must begin to apply their research to space living, starting with the prototype stations and bases in the immediate decades ahead.

At the 37th International Astronautical Federation Congress, Russian scientists presented the "Ethical Problems of Interaction Between Ground-based Personnel and Orbital Station Crewmembers."[27] They viewed long-term space missions as complex socio-man–machine systems whose effectiveness largely depends upon the quality of interaction between the subsystems. For the Russians, the ethical aspects concern the human relations which permeate every component of spaceflight and determine mission efficiency. Therefore, in addition to psychological and medical examinations before/during/after manned missions, they utilize socio- psychological and organizational–technical approaches to facilitate interaction, work and leisure, for improved interpersonal and intergroup relations. In the U.S.A., the leading researcher on spaceflight communication issues has been Dr Mary Connors at NASA Ames. She concludes that these issues range from the spacecraft environment to the psychodynamics of communication aloft, and recommends that a broad behavioral science perspective be used in investigations.

(b)   *Increasing research by Earth-bound scholars on space habitation*, or matters of permanent settlement in the microgravity environment is another issue to address. The human advances in space already alter our terrestrial culture. The more people reside aloft for longer periods, the likelihood is that they will cope and adapt in innovative ways to their low or gravity-free environment. [28] Many anthropologists, for example, seem fixated on the past within terrestrial communities, being reluctant to deal with the future space communities. As we will see in the next chapter, cultural anthropologists could make enormous contribution if they would direct the tools and the skills of their discipline toward the emerging space culture. One facet of culture is how people organize and manage themselves, say at a lunar base or a Martian settlement. Management scientists also could assist with the enormous task of organizing, transporting, supplying, and administering the material and human resources required for a permanent human community in space. Space management opens new vistas for research and inventiveness, particularly the macromanagement of large-scale enterprises (refer to Chapter 8). Initially, the operational administration of an **International Space Station** represents a unique opportunity for management scholars to inaugurate such studies.

(c)   *Personnel to be transferred to and from orbit* for colonization purposes is another challenging issue to be investigated by social and life scientists. Our experience in manned flights has been so very limited, in terms of both numbers and type of elite astronauts and cosmonauts. Most of these have been disciplined military males, with only a few scientists and women making it to the "high ground" with them. An occasional politician, even a Saudi prince, and one teacher have flown on the Shuttle, while the Russians have had more diverse spacefarers on their orbiting station, allowing guests ranging from French spacionautes to a Japanese journalist. In the next fifty years, diversity will be the norm for masses of people migrating aloft, starting with contract construction workers and ending with tourists. In *Voyagers to the West*, Harvard historian Ballyn reminds us that the colonists to the New World during the 18th century were largely poor, ill-used white artisans and indentured servants, as well as African slaves.[29] The prospects are that space colonists of the 21st century will be more affluent and self-directed, better educated and selected. Expertise is required of specialists in cross-cultural relocation and living in exotic environments to design systems for deployment and support of spacefarers (refer to Chapter 6).

(d)   *Global leadership in space settlement and colonization* will become a critical issue, beginning in the New Millennium. At present, no national government or space agency is seriously addressing the matter, and the leadership is likely to come from the private sector. A *case in point* is the 100th United States Congress, which passed legislation attached to NASA budget authorization to promote space settlement studies (refer to Chapter 3.1). Provisions in this foresighted document included a report every two years to the President and Congress by NASA reviewing its progress toward this end. Mark Hopkins, president of *Spacecause*, thinks that such legislation is of strong symbolic importance, and has documented how that bill was not being enforced and the reports delayed. In a conversation with the NASA official charged with justifying a lunar/Mars program, the author discovered from Carl Pilcher that two reports were filed but not publicized and that current changes within both the Administration and Congress had de-emphasized the concern.

However in 1995, the NASA Office of Spaceflight's Policy and Plans will issue a strategic report on *Human Exploration and Development of Space.*

The National Space Society's position paper on *Settling Space: The Main Objective* states that this is a unifying goal that can focus human efforts to create a **spacefaring civilization with communities beyond Earth.** Its rationale is that permanent space settlements requires a space-based economy to sustain self-sufficient growth aloft. To this end, NSS advocates:

(a) reduction in the cost of access by new space transportation systems;
(b) exploiting of nonterrestrial materials (NTM) and resources;
(c) standardization of vehicles and systems among nations so that they are compatible and less expensive;
(d) expansion of microgravity research **by robotic and human activities**.

Space settlement is an intellectual arena where sociologists, political scientists, architects, and urban planners may come into their own, especially in terms of technology transfer offworld. These and similar habitation issues will be considered in other chapters here on human enterprise in space. For the U.S.A., along with its Russian, Canadian, Japanese, and European partners, the initial experimental laboratory for such matters will be the forthcoming **International Space Station**. NASA proposes to start building it by 1997 as part of a growing space infrastructure, if funding is appropriated.

Space habitation is a world issue requiring a systematic mechanism for international exchange of information learned about living and working in an orbital environment. Perhaps the time has arrived for consideration of a **Global Space Administration** to coordinate exchanges not only among national space entities and private enterprise, but to promote a more comprehensive strategy for extraterrestrial living by our species.[30] Having studied the subject of **governance in space settlements** for some years, the author concludes that space agencies alone will never provide the solution, and that a new structure or authority will have to be created for this purpose in the near future.[31]

2.4 INTERDISCIPLINARY CONTRIBUTIONS TO SPACE HABITATION

If the next stage in exploration and utilization of the space frontier is to focus upon human migration and habitation, then technical considerations should be amplified to include contributions of knowledge from a broader array of human sciences. Peter Drucker, management guru, observes that technology can only be improved so much and that financial resources are limited; to expand quality control, productivity, and profitability, he proposes concentration on the unlimited potential of human resources. Until now aerospace planners and managers have had a narrow view of human factors, not giving these priority in their research and development. During a 1984 NASA summer study on strategic planning for a lunar outpost, as a Faculty Fellow this writer had an opportunity to consider some of the *non-technical aspects* involved in putting a human back permanently on the Moon.[32]

In Exhibit 15, artist Dennis Davidson illustrates my findings in a modular form, similar to the possible appearance of a future lunar installation. In a sense, it is a conceptual model of some **other** human dimensions that must be brought into space habitation, whether in

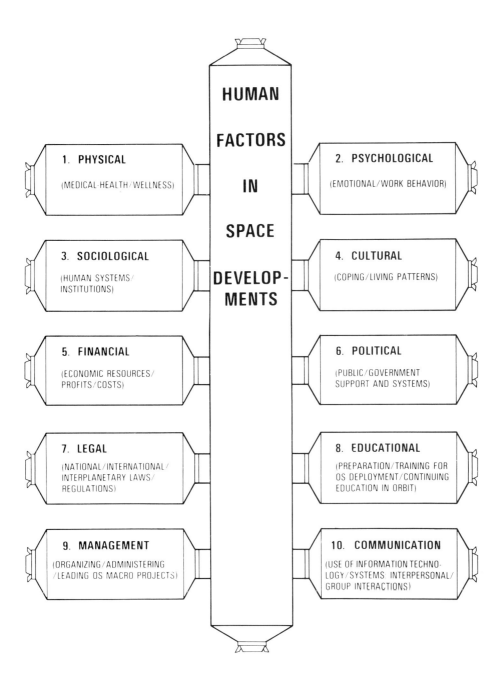

Exhibit 15—Human Factors in Space Development Beyond the Technical*. *Source*: Illustrated in form of lunar habitat modules by Dennis M. Davidson, astronomical artist, Hayden Planetarium, American Museum of Natural History, New York City.

an orbiting station or at a Moon or Mars base. In addition to the technical aspects, the reader is urged to analyze the ten dimensions of this paradigm and use the imagination to expand upon the descriptions in the categories, or even to add new classifications. Then apply these insights to just one aspect of space development, such as industrialization or commercial activities aloft:

(1) *Physical*—the medical or physiological concerns for survival and development of spacefarers, including *space ecology* or the biological aspects of closed ecosystems; as well as *space ergonomics* or the design of equipment and facilities to ensure both safety and quality of life (refer to Chapter 5). It encompasses all forms of protection from radiation and other hazards aloft, from habitats to space suits. This dimension of study and research is concerned not only with weightlessness and ensuring the continuation of human life in a low or zero gravity environment aloft, but its enhancement through preventative health, fitness and wellness programs before, during, and after spaceflight. All aspects of **life sciences** impact this area of concern.

(2) *Psychological*—the adaptation and behavior of spacefarers, especially emotionally and in small groups, within an alien or hostile orbital environment. Initially, the focus should be upon human performance, productivity, crew team morale and management for long-term space living, including *stress reduction* (refer to Chapter 5). Special programs, some computerized, will be developed to counteract the negative effects of an isolated, confined environment and life style. Eventually, it will extend to the role of other animals who are introduced into space habitats and settlements. In time, all branches of psychology may contribute to improving space living and working, such as experimental, physiological, comparative, developmental, social, abnormal, clinical, counseling, environmental, industrial/organizational and educational/learning psychology.

(3) *Sociological*—the study and development of space human systems, whether aloft within stations, settlements or societies. The evolution of such space groups and institutions may require social innovations back on Earth, such as synergistic efforts by national space agencies or inter-institutional cooperation. Under this category, new sociological methods may be devised ranging from social engineering and computer simulation to adoption of futures research approaches of technological forecasting and environmental scanning. A key concern in this area of study is the relationship and interactions between **earthkind** and **spacekind.**

(4) *Cultural*—the creation of coping skills, customs, communications, and life styles appropriate to the space environment (refer to Chapter 4). Initially, this may involve the adaptation or convergence of Earth's many macro and micro cultures, as is already happening with international participation on both American and Russian spaceflights. Eventually, it will lead to formulation of unique free-fall cultures, suitable to different orbiting facilities, planets, or colonizers. All of the other nine categories described here, plus others, will contribute to this process, perhaps receiving some designation like "astro" or "space" (e.g., astrolaw or astroeducation, space management or space finances).

(5) *Financial*—the economic dimensions of funding and supporting outer space projects, both in the private and in the public sectors, as well as the money-like exchanges

and compensations to be provided between and among spacefarers. Although exploration began with public monies, there is growing evidence of privatization through investment and profit-making in space enterprises. Space financing through the private sector may range from public bonds and stocks to lotteries and leasing of real estate aloft. Space industrial parks and colonies will require not only new venture capital and mechanisms for joint ventures or limited partnerships, but innovative cooperative funding action by consortia of government/university/industry, as well as between and among industries, institutions, and nations. Global space trading corporations may be one way of funding macro space undertakings (see Chapters 9 and 10).

These may require new provisions for national or international charters that help to underwrite the investment costs of space consortia and cooperatives. When a space government is eventually established, it may also mean creation of **metamoney** and **meta-banks**.

(6)   *Political*—the building of national/international consensus and policies to obtain commitment of resources for exploratory, scientific, commercial, and defensive undertakings in space, as well as for the creation eventually of appropriate political systems and procedures of governance in space communities (refer to Chapters 7 and 9). International or national administrations and commissions may set space goals, but implementation depends upon facilitating the political will and spending decisions. It is in this arena that space agencies must develop new skills, and that the space activist organizations are beginning to learn about lobbying and political power; here particularly the contributions of statesmen, political scientists and historians are most needed, in addition to those elected or appointed to political office. In time, human ingenuity will be taxed to create social orders that not only ensure participation for space colonists, but also autonomy from earthly sponsors.[33]

(7)   *Legal*—the analysis and application of relevant national and international laws, regulations, and agreements relative to space matters, as well as the creation of interplanetary authorities and laws to regulate extraterrestrial conduct of human affairs (refer to Chapter 9). **Space law**, lawyers, and associations already exist; treaties, conventions, and agreements are in place relative to the exploration and use of outer space, including liability and damage from space objects, establishment of space agencies and contracting with them, commercialization of space activities and services. Space commercial law is especially needed to define rights, privileges, and responsibilities from regulation and decision-making to profit sharing for researchers, investors, and users. **Astrolaw** is the term used to designate the future practice of law in orbit. With law comes the need for order, so perhaps in this category should come eventually justice or security systems in orbit, as well as law enforcement and peacekeeping.[34]

(8)   Educational—in general, the preparation of people for the space age and environment; in particular, personnel with their families for successful deployment to and from outer space (refer to Chapter 6). The educational programs and technologies to be developed must be both ground-based (e.g., a space academy or university courses), as well as in orbit (e.g., in-flight training or self-learning provisions aloft or a space

university). Career development is necessary now for space roles and occupations that are recent or have yet to come into existence. Formal systems of education on Earth will have to completely revise their curricula and methodologies to deal with the emerging space realities and opportunities, while space educational systems and universities will have to be inaugurated for the high frontier. Satellite communications will then link both ground-based and orbiting or planetary educational programs, as is emerging at the International Space University and Challenger Centers (see Appendix C listings).

(9)   Management—the organization, administration, and leadership required for outer space systems, projects and installations, both on the ground and in orbit, manned and unmanned. This ranges from macro to micro management of space enterprises, including human, material and financial resources. This requires development of *new* management paradigms, technologies, and information systems, as well as space-relevant organizational models and leadership styles. Global management processes and systems will evolve to administer large-scale projects of great distance and cost in outer space (refer to Chapter 8).

(10) *Communication*—the multidimensional aspects of human interchange related to space exploration and enterprise, from the marketing of such programs to the operation of space projects on the ground and in space. The range extends from technology, such as communication satellites, to information systems and artificial intelligence, to man–machine interface, to interpersonal/intragroup/intergroup interactions. The whole area of cybernation, or communication and control in humans and machines receives new applications in space—telerobotics and expert systems being cases in point. Progress in the experiment of human living aloft requires communication systems which provide continuing, non-intrusive monitoring and feedback on experience (refer to Chapter 3). Innovative adaptations of both communication and communication technologies, as well as virtual reality, will enhance living and working aloft.[35]

The ten designations above are but convenient categorizations suggested for studying space developments beyond the scientific or technical dimensions. They are by no means exhaustive, indicating only possibilities for a new taxonomy. For example, should space commerce or industrialization be considered as a separate entity, or does astrobusiness cut across several other categories, such as sociological or financial or management? Suppose this were an eleventh classification, and one wished to include in this paradigm a twelfth category—the technical. Then one would have a more comprehensive model for systematic analysis of space planning, operations, and management.

2.5 REDIRECTING KNOWLEDGE AND WORKFORCES UPWARD

It should be clear by now that opening up the space frontier demands re-thinking as to who and what we humans are, how we learn and disseminate information, and how we prepare the future generations to live and work in orbit. Human development in space is not the sole province of aerospace industries, agencies, or career persons in the field. This challenge can be best perceived in terms of two emerging realities:

(1) *Refocusing academic disciplines and professions upwards.* Discoveries or developments in space stimulates not only cross- disciplinary endeavors, but require professionals to redefine the very scope of their activities and the learning provided to candidates for their discipline or profession. Space requires a more holistic, multidisciplinary approach to problem-solving and learning. That is already happening in many fields, such as in space medicine or psychology. For example, Dr Stewart Johnson, a professional engineer in Albuquerque, New Mexico, has begun such a re-examination with his wife, Mary Anis Johnson. The Johnsons have jointly prepared a paper, "The Civil Engineer and Space", which they have permitted us to quote. Their thesis—the effective use of Greater Earth resources (including the Moon and intervening space, particularly utilization of solar energy)—is a subject of increasing importance to the nation, the world, as well as the engineer. Among trends cited by this professional couple are the following.[36]

- The Civil Engineering Research Foundation will fund research related to space projects in robotic construction techniques, advanced material applications, computer-integrated manufacturing at construction sites. Space exploration and exploitation, especially a lunar base, is dependent upon such developments for habitats and facilities, resource and vehicle planning, and astronomical observatories.
- This convergence of civil engineering expertise and space exploration will benefit both terrestrial and offterrestrial construction through major technology transfer from the space program to commercial applications. But it requires cooperative research and education that will alter both engineering and the construction industry.
- The Mission to Planet Earth should be broadened to include electric power produced by solar energy transmitted from the Moon (refer to Appendix B). Similarly other space/lunar resources have implications for their profession as shown in the two volumes published by the American Society of Civil Engineers.
- The support of the civil engineering community for utilizing space resources can provide new energy sources, while mitigating environmental damage on Earth. But it also offers strong motivation for education of more youth in science, mathematics, and engineering, a people-serving profession.

(2) *The needs of the emerging space workforce.* To succeed with space development, the workforce must be made up of competent, high performing, and diverse personnel. The authoritative weekly, *Space News* (Feb. 4–10, Vol. 2:3, pp. 5–10), did a special report on the "Space Work Force in the 90's," some observations of which underscore themes made so far in this book:

- The aerospace industry faces two major workforce roadblocks—lack of qualified personnel, as well as lack of understanding between engineers and management. To correct these deficiencies and achieve an ambitious civil space program, an interdisciplinary curriculum has been designed which combine studies in engineering, policy, and management. Now if universities would only adopt such integrated programs in engineering, behavioral and life sciences!
- A NASA study shows an anticipated need for scientists, engineers, and technicians increasing 33% by the year 2000, while student interest in these fields is declining in the U.S.A. The agency's educational affairs division, therefore, is funding more cooper-

ative programs with universities in these fields—40% of NASA new employees come from such partnership.

- NASA's workforce of 28 900 in l990 (plus 66 000 contract workers) is diminishing. The agency seeks to obtain and retain a higher quality workforce through changes in Civil Service provisions, and improved compensation/benefits.
- According to the U.S. Bureau of Labor Statistics, l55 000 people were employed in the space industry in l988. During this decade of the '90s, labor analysts reported that work-force as a result of corporate cut-backs and mergers, as well as a downturn in market conditions. The Aerospace Industries Association expects space spending in the U.S.A. to remain at present levels for the next few years, with a consequent growth in space jobs from maintenance to technical to professional in the 21st century.
- The Planetary Society's executive director maintains that for professional space jobs an educational background in science, biology, chemistry or mathematics is basic. Al-though a bachelor degree is necessary to be hired, a Masters, Ph.D. or other specialized training will be necessary for advancement. The manager of the Center for Innovative Technology in Herndale, Virginia, predicts a change in the tools such workers use (e.g., supercomputers, mainframe and neural networks).

A transition is under way to a new work culture where **knowledge** and service workers will predominate, especially within space careers.[37] To prepare a properly skilled global workforce demands synergy not only between and among disciplines, industries, and government entities, but among nations. The trends are evident in news reports that the University of Akron's School of Law and Moscow's Institute of State and Law will cooperatively study legal issues relating to the exploration and use of space; or that Dr Roald Sagdeev, formerly U.S.S.R. Peoples Deputy and Academy of Sciences' member, has been appointed a distinguished visiting professor of physics at the University of Maryland, an indicator of a large exchange of Russian scientists in American universities!

Outer space is more than a place for travel. Space satellites have already proven the interconnectedness of our Earth for the maintenance of habitability on this planet. The high frontier presents the human organism with a profound challenge and need for a new homeostasis! Space habitation demands not only a systems approach, but the multinational and multidisciplinary contributions. If there are to be permanent settlements on the inexhaustible frontiers of space, then:

(a) the knowledge of biology must be joined not only to physics and chemistry, but with the social sciences;
(b) engineering must be linked not only to all of these, but information sciences;
(c) law and political science must have new applications not only to promote migration and commerce aloft, but to protect human rights in orbit.

2.6 SPACE IS A PLACE FOR SYNERGY

Obviously, the vastness, complexity, and costs of exploring the space frontier call for **synergy** as an integral norm of human behavior. Space technologies were spawned toward the end of the Industrial Age, a period of rugged individualism and aggressive competition, especially at the national level. Yet the success of the Apollo missions and lunar landings proved the effectiveness of collaboration between NASA personnel and aerospace contractors. Space achievements contributed to the emergence of a post-industrial, information, and technically oriented society. But to continue progressing, this new condition requires the practice of synergy, a concept of cooperative or combined action.[38]

Today we live and work in a global village and economy, an interdependent world brought closer together by advanced forms of communication and transportation. In that environment, **effective** human performance depends upon sharing perceptions, insights, and knowledge. We create cultural synergy by capitalizing on people's diversities and talents. Space synergy occurs when disparate groups work together on extraterrestrial macroprojects. The Space Age has demonstrated that more is to be gained by individual, organizational, or national cooperation than simply going it alone. The trend grows dramatically toward greater international cooperation in joint space technological and social efforts.[39]

If spacefaring nations and project sponsors were to become more synergistic, then their endeavors aloft would be characterized by:

- a win/win philosophy of cooperating to mutual advantage;
- promotion of individual growth through team or group development;
- support of programs that advance the common good, while developing human potential;
- values that espouse openness to change, mutual reciprocalness, non-aggressive behavior, and enhancement of human capability (such as through the use of robotics).

Collaboration among countries, institutions, and their representatives will advance the construction of a space infrastructure for the colonization of the Solar System. Here are three illustrations of a **synergistic space ethos** in practice:

First, *synergy among space scientists and other professionals*. This occurs in multilateral, cross-disciplinary, integrated endeavors that prompt scientists and academics, engineers and technicians, government workers and contractors, military personnel and civilians to move beyond traditional boundaries in the **macromanagement** of space projects. The ISY'92 World Congress manifested this when global professional groups gathered together under the joint sponsorship of the International Councils of Scientific Unions (ICSU) and its Committee on Space Research (COSPAR), the International Astronautical Federation (IAF), the U.S. National Academy of Sciences and its Space Science Board, plus numerous other associations ... . Two other organizations manifested this trait by studies and recommendations to promote **international cooperation in space**—namely, the 1993 reports on the subject by the Center for Research and Education on Strategy and Technology; and the American Institute of Aeronautics and Astronautics.[40] Both efforts brought together the best thinking on the subject from a variety of experts representing diversified affiliations ... . Synergy is also manifested within a space agency, such as NASA, when there is real collaboration *between* headquarters and field centers; natural and behavioral scientists in enhancing "manned" flight programs; between and among space agencies of other nations; between and among those agencies and other government departments, universities, and corporations ... .

Second, *synergy among space advocates*. The pro-space groups usually operate independently, but increasingly must engage in joint undertakings, such as in support of a Space Station or Lunar Base, with those responsible for funding such macroprojects, as well as with the public. The history of space development advocates is one of missed opportunities by not uniting for greater impact and improved quality of service. Leaders in the National Space Institute and the L-5 Society proved they were synergistic when they merged to form the National Space Society, which has many chapters, some beyond the borders of the U.S.A. Their affiliated entity, *Spacecause*, also reaches out to enlist the whole space community in their lobbying with the U.S. government. However, to raise the consciousness of any nation's citizenry about the importance of utilizing the space resources requires a coming together of many entities in the pro-space movement in more effective educational and media efforts. Perhaps this explains why among the 250 million Americans only 15 million or so pay any attention to the space programs, including the fans of *Star Trek*; or why the British Interplanetary Society has failed to gain the support of its government in vigorous support of space enterprise. Too many space associations talk only to themselves and serve their own constituencies—they do not reach out and synergistically relate or network with those outside their sphere of influence. If it is to formulate a **global space ethos,** then space proponents should actively **unite** in making a convincing public case for space investment which can strengthen economic, educational, and technological systems, while contributing to environmental wellbeing ... .

Third, *space synergy is necessary among nations*. The scope of human migrations into the universe necessitates transcendence of national and political differences. Photographs and remote sensing from outer space of planet Earth illustrate its unity and fraility, while enabling us to catalog its resources and catastrophes. Those who have been in orbit return with appreciation for the interdependence of our ecosystems, and urge us to think globally beyond local borders and concerns. The conceptualizer of communication satellites, Arthur Clarke, astutely observed that we are witnessing the rise of the global family or tribe—humans linked together electronically across the world who are challenged to transcend ancient frontiers in our loyalties and interests.[41] Space developments made this possible.

Most nations are perceiving that space cooperation is mutually beneficial, but they have not realized that it also means creating new social attitudes, relationships, and systems for sufficient accomplishment together. Since the Space Age began, we have had numerous examples of collaboration among nations on the high frontier—such as the **Apollo/Soyuz** linkup in 1975, **Intelsat** communications satellite, East/West space science exchanges such as the **Intercosmos** satellite programme, the formation of the European Space Agency, and multinational agreements to build an International Space Station, and the opening up of space facilities to foreign visitors. With the globalization of space undertakings in the decade of the '90s, a significant synergistic ethos became more evident, as the following demonstrates:

• Signing of cooperation agreements between the **European Space Agency** and the Russian Space Agency (1994) for joint activities in manned flights and associated space activities after ESA had successfully flown both experiments and astronauts on **EuroMir** missions; and this has led to establishment of an ESA Permanent Mission in Russia (1995) with recognition of ESA as a public-law governmental institution to conduct activities in-country with Russian institutions and industrial concerns. Within the European Union, ESA has been the focal point of multinational collaboration on that continent, as well as with the Americans in both the NASA shuttle and station programs, as Exhibit 16 illustrates.

Exhibit 16—Space is a Place for Synergy. *Source*: Photograph of STS-63 international crew onboard Shuttle Discovery: (front, l. to r.) astronauts Bernard Harris and Michael Foale (British); (back row, l. to r.) astronaut Janice Voss, cosmonaut Vladimir Titov (Russian); astronauts James Weatherbee and Eileen Collins, the pilot. Courtesy of NASA's Johnson Space Center.

● **United States** agreements with the Russian Federation (1994) for long-planned joint missions for exchange of astronauts/cosmonauts on each other's space vehicles, partnership in the **International Space Station**, collective research on environmental monitoring and and planetary exploration. A new level of synergy was reached in 1995 when NASA's Dr Norman Thagard flew on a Russian spacecraft to spend 115 days on the **Mir** orbiting station. Beginning in February 1995, the Shuttle **Discovery** (STS-63) flew around the Russian **Mir**, followed by **Atlantis** (STS-71) which actually docked in June 1995 with this orbiting station (see artist depiction of historic rendezvous in Exhibit 17).

Instead of the Cold War rivalry epitomized by the race into space,[42] Russian/American space synergy is evident now in both the private and the public sectors. We end this section with two examples of this positive trend.

(1)   **The Citizen Ambassador Program** (110 Ferrall St., Spokane, WA 99202, U.S.A.; fax: 509 534-5245)—sponsoring with InfoMOST, Inc., (fax: 703 448-5669) delegations of business leaders and investors who travel to Russia to gain better understanding of the direction and capabilities of its space programs and to meet top-level contacts. Guided by

Exhibit 17—Russian/American Space Cooperation: Station **Mir**. *Source*: Artwork of John Frassanito and Associates depicting the Shuttle docking to the **Kristall** module of the Russian space station in June 1995. The combined nearly 240-tonne spacecraft, Shuttle STS-71/**Mir 18**, comprised the largest, human space platform ever assembled, while the two crews exchanged visits. Courtesy NASA's Johnson Space Center.

Jeffrey Lenorvitz, former editor of *Aviation Week & Space Technology*, the visitors meet with counterpart professionals and companies to examine Russian space facilities, program funding, technology transfer, and how to cooperate for mutual benefit. Apart from visiting cultural sites, the teams confer with key executives at the Russian Institute for Space Devices, NPO Energia, Russian Space Forces Satellite Control Center, NPO Saturn/Lyukia, and NPO Lavochkin ... .

(2) **Cosmos International Center for Advanced Studies** (Volokolamskoe sh.4, Moscow 12571, Russia; fax: 095 229-3237; E-mail: cosmos@sovamsu. sovusa.com). This is a joint-stock company founded by enterprises specializing in aerospace technology. Founders include Moscow Aviation Institute, Cosmos Concern, Salyut Design Bureau, S. A. Lavochkin, Central Research Institute of Machine Building, INTERGRAPH Corp. (U.S.A.) and University of Alabama in Huntsville (U.S.A.), and others. Under its director general, Dr Oleg M. Alifanov, the company promotes:

- cooperative international programs in the field of personnel development for aerospace science, technology, and related disciplines;
- research and experiments in these fields to promote new market-oriented, economic development in Russia;
- promotion in commercial applications of findings from Russian aerospace enterprises and organization and abroad.

Their aim is to promote human resource development by aerospace education and research through consultation, training, data banks and improved marketing.

2.7 CONCLUSIONS—SYNERGIZING SPACE VISIONS

Space is a place not only of high risk, but of unparalleled opportunity *if humankind can adapt to cooperative actions there*. It is also a place for high leverage technology and research in space transportation and construction, energy and information processing, automation and artificial intelligence, as well as for habitats and social innovations that ensure survival and the quality of human life aloft. The high frontier represents an evolutionary departure in human culture that requires the merging of art and science, economics and technology, public and private sectors in the pursuit of free enterprise and human enrichment.[43]

But for actions of this nature to occur, individuals, institutions, and even nations must have a **vision** of what they wish to do and where they want to go offworld. In this chapter, we have reviewed some of these visions that will contribute to human development aloft. To build a spacefaring civilization, there must also be supporting **myths** that harmonize past and future. A myth is one attempt by people to explain unusual experiences in their lives, a way to make visions believable. Mythologist Joseph Campbell envisioned the space program as a kind of projection screen upon which new mythologies are created and displayed. Commenting on this insight, David Cummings, executive director for the Universities Space Research Association, wrote:[44]

> **Human exploration of space, for example, is an extension of the great exploration mythologies of the past, giving cultural guidance about the importance of courage and the spirit of adventure in our lives. The famous view of Earth from lunar orbit gave use another lesson about the importance of living harmoniously with the Earth's environment, as did the exploration of Mars and Venus.**

Cummings reminds us that the accomplishments of space agencies show us the power of combined efforts by thousands of gifted, dedicated men and women, scientists and engineers of diverse racial and ethnic origins. Space program achievements also teach lessons about the value of international respect and cooperation. This is the message we have tried to convey in the above section on **synergy**. Even apparent failures, as the Hollywood film **Apollo 13** recalls, can be transformed into heroic acts when people on the ground and in orbit work together as problem-solving teams to discover creative solutions! As Lt Gen. James A. Abrahmanson once reminded an audience, "That incredible spirit of reaching beyond all reasonable bounds is now an *accepted characteristic* and *attitudinal spin-off* of the space program."

Yet, the complexities in space colonization are beyond the capability of a single space agency or nation, or even one discipline or industry—space exploration and exploitation demand collaboration, synergistic relationships! Thus, **space settlement** represents a new direction and vision for humankind.[45] Pioneering there implies more than surviving aloft; it involves human procreation, distribution, civilization, and destiny. Space resources may not only provide the human family with a means for coping with species' overpopulation, but a way to channel our fecundity outward.

Before offterrestrial mass migration can occur, technical problems of space transportation and cost of access must be resolved so that *large numbers of people* can be transferred into orbit, a hundred miles or more up through Earth's gravity well and beyond. To that end research proceeds on advanced, reusable launch vehicles, aerospace planes, and even laser propulsion. Creative solutions may call for *in vitro* fertilization to populate other planets, or other innovative approaches such as described in subsequent chapters. However we transport ourselves to the high ground, the transition of human civilization into space marks an epochal transformation in our species' development!

Human Evolution has reached a critical juncture:
escalate earthly rivalries to the frontier of outer space, or acknowledge the unique characteristics and potential of the space environment, and consciously choose to build upon existing cooperative relationships based upon experiences of trust, instead of fear. Political, economic, and other forces have already begun to show that nations find it mutually beneficial to work together on Earth to research and develop the space frontier which is only 100 km above all our heads. There is a growing momentum towards international cooperation in space activities that has literally opened the skies -- and our minds to a new way of thinking.

Recall the beginning of the UN's 1967 Outer Space Treaty, "The exploration and use of outerspace ... shall be carried out for the benefit and in the interests of all countries ... and shall be the province of all mankind."
—Thomas E. Cremins and David E. Reibel
Spaceline, Institute for Security and Cooperation in Outer Space, Spring 1988.

3

Behavioral Science Space Contributions

By the end of this century, **Homo sapiens** might well be described as an *interstitial species*. We are not just between two time periods, but between two ways of life. As we transition into the 21st century, humankind is being transformed from a terrestrial to an extraterrestrial being. Dr George S. Robinson uses the Latin term **Homo alterios spatialis** to describe the future of human progeny in orbit.[1] In the process of this metamorphosis, our species will be changed both physically and psychologically, especially if genetics are engineered or synthetically induced characteristics are used to ensure survival aloft. Beginning now, but more so in the New Millennium, social practices, institutions, and knowledge organized for living on Earth are being profoundly altered to acclimate to offworld living conditions. The survival of the species in outer space demands significant adaptation to differing environmental realities. Technology and management have already discovered that both with spacecraft and spacefarers, innovative approaches must be created for dealing with the complexities of safely getting and keeping humans in orbit, as well as getting them back unharmed. Experience in the Space Age teaches us powerful lessons: (1) the necessity for **synergy** that goes beyond individual organizations, nations, and disciplines, as discussed in the previous chapter; (2) the **interdependence** of all life systems, especially between those of Earth and those considered nonterrestrial.

As a result of putting human technology and society into outer space, Jesco von Puttkamer suggests that we must transform not only our images of our species, but of our place in the universe. As he illustrates in Exhibit 18, our perceptions of the latter are further changed by adding space to land, sea, air, a new type of **super-ecology** is being created in our minds, which in turn will introduce many more interactive loops and replace the geocentric view of ourselves and our environment.

Harry Holloway, a NASA administrator for Life and Microgravity Sciences Applications, astutely observed in an *Aerospace America* editorial (April 1995):

> **We are in the process of becoming a global, spacefaring civilization. And space exploration is helping to blaze the way toward a new paradigm for life on Earth: life without boundaries. By exploring the planets in space, we can also begin to explore our own internal universe.**

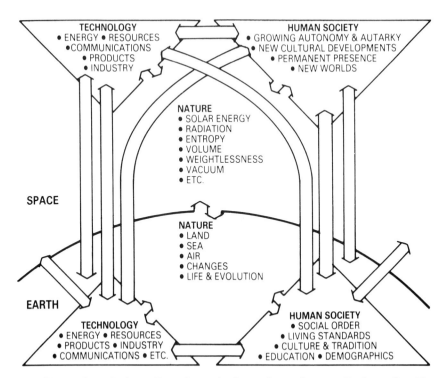

Exhibit 18—Super Ecology—Future Interactive Loops. Source Jesco von Puttkamer from his "Foreword" to the first edition, *Living and Working in Space*, Chichester: Ellis Horwood, 1992, p. 21.

This new reality and interdependence also applies to both the non-technical, and the technical aspects of manned spaceflight, especially as it involves missions of increasingly longer duration that lead to space settlements. In the search for integrative strategies for living and working in the orbital environment, the concept and processes of the **behavioral sciences** seem appropriate as a way to facilitate interdisciplinary research and development. Dividing knowledge and information into separate fields, academic professions and departments may have been suitable in past centuries. The latter will become of growing irrelevance in a **metaindustrial work culture**, particularly as we struggle to cope with extending human life off this planet.

The term "behavioral sciences" has been explained by *The Harper Dictionary of Modern Thought* in this manner:[2]

> Those sciences which study the behavior of men and animals (e.g., psychology and social sciences, including anthropology).

It adds that some practitioners believe that for such studies to be scientific, they must be observable and measurable. Behavioral science is relatively new terminology that is much misunderstood. The expression appears to have been originated at the University of Chicago in 1949 when James Grier Miller named a Committee on Behavioral Sciences, an interdisciplinary group of scholars concerned with developing a general theory of behavior. Later, Miller and his colleagues would found there in 1953 an Institute of Behavioral Sciences. When they relocated to the University of Michigan, some of these same scholars created in January 1956, *Behavioral Science*, an interdisciplinary quarterly journal which Miller, the founding editor, continues to edit over forty years later. In his inaugural editorial, Dr Miller wrote of different approaches to the study of behavior—mathematical biology, biochemistry, physiology, genetics, medicine, psychiatry, psychology, sociology, economics, politics, history, philosophy, and others.[3] Among the last-mentioned, I would include law, management/business sciences, communication/information sciences, as described in the previous chapter (Exhibit 15). Miller seeks a unitary behavioral science in which separate skills converge in the scientific study of human behavior. Originally, he hoped this new umbrella would cover both social and biological sciences, although many colleagues prefer to restrict the concept to the former. Eventually, his efforts were to climax in living systems theory and its application to space habitation, something to be discussed in the last section. For historical context, it should also be noted that the NTL Institute of Applied Behavioral Science and its journal of the same name has focused its action research on human relations, group dynamics, organization development and transformation.

3.1 SPACE HABITABILITY AND LIFE SCIENCES

After forty years of space exploration, there is a consensus among those who have been in orbit or studied microgravity that **outer space is indeed a unique place to live and work.**[4] NASA's Jesco von Puttkamer has summarized some of this uniqueness in Exhibit 19. Therefore, a critical concept throughout this chapter is **space habitability**. In their classic work on *Living Aloft*, Connors, Harrison, and Akins have clarified its meaning:[5]

> Habitability is a general term which connotes a level of environmental acceptability by potential users. The requirements for conditions to be 'habitable' change dramatically with circumstances. For brief periods, almost any arrangement that does not interfere with the health of individuals or the performance of their jobs would be acceptable. Over the long term, conditions must support not only individuals' physical, but also their psychological health.

Space habitability, then, implies a quality of life in orbit that ensures both survival and development of human potential. To achieve the latter, behavioral scientists will have to contribute more than they have so far to the space program.

Although the Russians have long used the behavioral sciences in their space studies, the National Aeronautics and Space Administration has been slow to adopt the term, the concept, or the expertise associated with this integrative strategy. Since the organizational cultures of both NASA and its aerospace contractors have been dominated by science, engineering, and technology, the emphasis has been on human factors from a narrow, industrial engineering and medical perspective. Until recently, even that was generally neglected by the agency, along with bioscience. Instead, NASA has preferred to adopt the designation "life sciences" with a limited concentration on applied medical and biological research relating to spaceflight. This approach has suffered from philosophical differences, as well as inconsistent direction and management.[6] Until the mid '90s, the NASA

Weightlessness (facilities, special manufacturing activities, construction of very delicate structures, and reliability of operations).

Easy gravity control.

Absence of atmosphere (unlimited high vacuum).

Comprehensive overview of Earth's surface and atmosphere, for communication, observation, power transmission, and other applications.

Isolation from Earth's biosphere, for hazardous processes: little or no environmental, ecological or 'localism' issues.

Readily available light, heat, power (10 times rate on Earth).

Infinite natural reservoir for disposal of waste products and safe storage of radioactive products.

Super-cold temperatures (infinite heat sink near absolute zero).

Large, three-dimensional volumes (storage, structures).

Variety of nondiffuse (directed) radiation (ultraviolet, X-rays, gamma rays, etc.).

Magnetic field.

Availability of extraterrestrial raw materials on Moon and possibly on asteroids.

Avoidance of many Earth hazards (storms, earthquakes, floods, volcanoes, lightning, unpredictable temperatures and humidity, intruders, accidents, corrosion, pollution, etc.).

Potentially enjoyable, healthful, stimulating or otherwise desirable for human well-being.

Exhibit 19—Useful Attributes of Space Living. *Source*: Jesco von Puttkamer from his "Foreword" to the first edition, *Living and Working in Space*. Chichester: Ellis Horwood, 1992, p. 10

management view of life sciences is illustrated in Exhibit 20. Note that psychology is submerged under "biomedical research," and no mention is made of anthropology, sociology, or other behavioral sciences.

In the mid '90s, **life sciences disciplines** are thought to include life support systems, medical sciences systems, biological sciences systems, human engineering, and extravehicular systems and techniques. The term "closed loop systems" is used with reference to long-duration human spaceflight requiring controlled ecological support systems. Both Russian and American space life scientists now share their experiences, and a five-volume series on space biology and medicine has been published by the American Institute of Aeronautics and Astronautics.[7]

Concern about the long-term requirements of space on biology and behavior goes back to 1960 when a NASA Biosciences Advisory Committee submitted a report urging the agency to undertake fundamental investigations in medical and behavioral sciences on human functions and performance aloft. The issue was not really addressed until 1987 when the National Research Council through its Space Science Board undertook to de-

Gravitational Biology
- Understanding the role of gravity in the development and evolution of life

Biomedical Research
- Characterizing and removing the primary physiological and psychological obstacles to extended human space flight

Environmental Factors
- Defining the space environment and habitat in which humans must function safely and productively, including air and water quality and the biological effects of radiation fields

Operational Medicine
- Developing medical and life support systems to enable human expansion beyond the Earth and into the solar system

Biospherics Research
- Developing methods to measure and predict changes on Earth on a global scale and the biological consequences of these changes

Physiochemical and Bioregenerative Life Support Systems
- Assembling the knowledge base needed to design, construct, and operate life support systems and extravehicular suits in space that are independent of major resupply

Exobiology
- Exploring the origin, evolution, and distribution of life in the universe

Flight Programs
- Conducting experiments in space, including the development of facilities and hardware for space flight, mission planning integration, and flight plan implementation

Exhibit 20—NASA's Perceptions of Space Life Sciences Programs. *Source*: Life Sciences Division, *Exploring the Living Universe—A Strategy for Space Life Sciences.* Washington, DC, NASA Headquarters, 1988.

velop a strategy for space biology and medical science.[8] In the following year's report, NRC was even more candid on the manned/unmanned mission debate:

> The National Research Council has examined prospects for intensive exploration of Mars. The Council concluded that while geologists can learn much from autonomous devices and sample returns, it is insufficient. Ultimately, to resolve important questions and to compare Mars in detail with the Earth will require **exploration capabilities on the surface of Mars possessed only by humans.**
> —*Space Sciences in the Twenty-First Century—Imperatives for the Decade 1995–2015.* Washington, DC: National Academy Press, 1988.

At the same time, a Congressional Budget Office (1988) analysis on NASA's future noted that the agency's **Pathfinder** technology initiative did contain more funds in the l989 budget for life sciences research related to manned exploration of the solar system. Yet, despite rhetoric about the importance of long-duration human missions, at this writing (l995), NASA continues to severely delimit its spending on life and **behavioral** sciences, devoting most research to the medical and biological studies of spaceflight.[9]

In a life sciences review, **"Learning from Living in Space"** (Aerospace America, March 1995, pp. 24–30), Theresa M. Foley offers these insights:

- Although research over the last forty years has not found a single, long-term, detrimental health effect when people return to Earth from orbit, there are concerns about extended missions aloft ... . Humans returning from spaceflights sometimes fall down or pass out; they walk more slowly as their bodies try to maintain balance in 1.g. They have fewer red blood cells, their bones and muscles are weak, and they are unable to keep their eyes naturally focused on objects as they move. In space, most people experience nausea and discomfort for a few days while suffering from **space adaptation syndrome**. Because they get taller in orbit as their spines stretch, most spacefarers suffer back pains. Since gravity shapes cells and tissues, little is known about the long-term effects of microgravity upon body chemistry and physiology—such as calcium loss and bone weakening, shutting down of the immune systems and its effects, regulation of body fluids, loss of muscle and protein, etc ... .
- NASA has finally recognized the importance of such factors to future goals by creating in 1993 an Office of Life and Microgravity Sciences and Applications (OLMSA). Yet former Apollo astronaut and U.S. Senator, Harrison Schmitt admits, "We've wasted at least 25 of the 30 years of manned spaceflight by neglecting to undertake serious, well-planned scientific investigations of the effects of spaceflight on the human body." He complains that the Agency failed to establish a protocol for life science experiments and to establish a data base on what has been learned. Schmitt also contends that NASA still does not have a good plan in place to take care of sick or injured crew members.
- In 1995, NASA will spend $183 million on life sciences as part of its new "strategic enterprise" called **Human Exploration and Development of Space**. Yet the issue is not just Lower Earth Orbit problems, such as on a Shuttle or Station, but longer missions to the Moon and Mars where high-energy particles can kill human cells.
- NASA Administrator Dan Goldin admits that this research must go beyond the experience of four decades which he called "pale, male, and stale," and be aimed more at women and minorities, as well as integrating findings with partners in Russia, Europe, Japan, and Canada. One area of study should be on how the reproductive system might work if a woman became pregnant in microgravity.

My critique of NASA's contemporary life science research is that it still neglects behavioral science studies on the psychological and sociological aspects of individual and group behavior in isolated, confined environments aloft.[10] This was admitted by Administrator Goldin publically after Norman Thagard, astronaut–physician–engineer, experienced adjustment problems during his three-and-a-half months on the Russian space station **Mir** in 1995. Goldin confessed:

> **We put all of our focus on the physical well-being of the astronauts and the success of the mission. We neglected their psychological well-beings,**

and Dr. Thagard made it clear to us ... . I think this was one of the major findings of the mission. If we expect to send people on missions of two or three years, we had better deal with the psychological aspects in addition to the physiological ones.

—Houston Chronicle, July 8, 1995.

Yet as this writer pointed out in correspondence to the Administrator, the isolation problems of Thagard could have been prevented if the medically dominated personnel in life sciences had utilized the findings of several agency-sponsored studies on the subject, including those of the Russian station experience.[11] On **Mir** and its predecessor station, for example, cosmonauts manifested interpersonal problems by fighting with one another and ignoring ground controllers' instructions when the stress of living and working in an isolated, confined spacecraft for extended periods got to them. Under such circumstances sociologist B. J. Bluth, now working for NASA, indicates that disturbing characteristics may arise, such as boredom, irritability, fatigue, depression, anxiety and mood fluctuations. In an agency response to the author Harry C. Holloway, M.D., OLMSA associate administrator, wrote (August 11, 1995):

> **NASA's Office of Life and Microgravity is actively engaged in efforts to address the psychological issues associated with extended human space flight ... . This means learning to address a wide range of organizational and transcultural issues. NASA is committed to developing a dynamic Space Human Factors Program that includes basic and applied research, tests, and evaluation to gain knowledge of human psychological and physical capabilities and limitations in space ... . As your letter suggests, these efforts have been intermittent and we recognize the need for a sustained, serious programmatic commitment.**

Holloway then invited proposals from behavioral scientists to participate in that agency's Space Human Factors research (OLMSA, Code UL, NASA, 300 E. St., Washington, DC 20024, U.S.A.; Tel: 202/488-2940).

In Europe, the German Aerospace Research Establishment (DLR) is also interested in Advanced Life Support investigations. Their director, Dr Marrianne Schuber, expressed interest in:

- joint ground-based research programs within a multinational framework ...
- support of ground-based research with already available inhouse infrastructure ...
- optimization of simulation methods and image processing ...
- support of spaceflight experiments ...
- development, design, and preparation of life science experiments ...
- development of Telescience studies.

(Contact: LLR Institute for Aerospace Medicine, Linder Hohe 45, D-51147 Köln, Germany; Tel: 49-2203-601-0; Fax: 49-2203-696.212.)

Since 1990, the European Space Agency has engaged in isolation research within the confined environment of hyperbaric chambers. This has been part of their Isolation Study for

Manned Space Infrastructure (ISEMSI) to prepare people for space station life and work. The word **EMSInaut** was coined for the initial six young European males, future astronauts, who were isolated for four weeks at the Norwegian Underwater Technology Centre (NUTEC), and analyzed as to their behavior, group performance, and interaction under these stressful circumstances. This was followed by EXEMI '92 when a crew of four volunteers, one woman and three men, spent 60 days in isolation in the **TITAN** hyperbaric complex at the DLR aeronautical medicine institute near Cologne, Germany (Exhibit 56, p. 159). In 1992, ESA astronauts were prepared for missions on both American and Russian spacecraft, and a European Astronaut Corps and Centre was established to prepare Europeans for long-duration crewed missions in the 21st Century. In 1993, ESA's Long Term Programme Office under the Directorate of Space Station and Microgravity Studies began **Human Behaviour Space Simulation Studies (HBSSS)**. In addition to ESA isolation studies, this endeavor seeks to learn from similar studies in polar stations, underwater habitats, and submarines.

Exhibit 21—Shuttle Docking at International Space Station: Year 2000—Providing Behavior Science Insights for Space Living. Source: Computer-generated scenario by John Frassanito and Associates of the Space Shuttle Atlantis docking at the newly constructed station to deploy a 6-person permanent crew made up of Americans, Europeans, Canadians, Japanese, and Russians. *Source*: NASA's Johnson Space Center.

Three powerful factors are driving space agencies globally to give more attention to habitability issues:

• One is the development of a *permanently orbiting* **International Space Station** by 1999. Early on, a British observer, M. H. Harrison of the Royal Air Force Institute of Aviation Medicine, succinctly forecast the issue:[12]

> When America's Space Station becomes fully operational, it will serve primarily as a laboratory, workshop, and servicing facility. As such, it will provide unique opportunities for scientific studies in the material and life sciences ... . The Soviets have also been first to take seriously the psychological aspects of space flight. During their missions of several months duration, problems of boredom, restlessness, and depression have been encountered ... . Traditionally, NASA has been dominated by engineers, and this has been reflected in low priority being given not just to space psychology, but to space medicine generally. With a new era dawning of sustained human operations in space, this attitude will have to change, as the Soviets have discovered.

• Two is that planning for **Space Exploration** underscores the need for more studies related to long-duration space living and its effects on individuals and groups aloft. In the words of President George Bush, going "back to stay" on the Moon and developing a lunar base there, as well as taking a "journey into tomorrow through a manned mission to the Martian surface," requires more information on space habitability than we now possess.

• Third, there is an increasing public interest worldwide in space colonization, but a woeful neglect by both space agencies and scholars of identifying what is required for real space settlement[13] As indicated in our opening chapter, an attempt was made to deal with this issue through a **Space Settlement Act** introduced in the U.S. Congress' House of Representatives in 1988 as legislation H. R. 4218. This measure, if enacted, would have incorporated the idea of settlements in space as official government space policy. It proposed to amend the National Aeronautics and Space Act of 1958, so as to set the establishment of space settlements as a long-term mission objective for that agency. Moreover, the bill would have required NASA to be the lead governmental agency in conducting a steady, low-level effort to explore all the scientific, technical, and **sociological** issues relating to the achievement of settling space with humans. When its proponent, Rep. George E. Brown, Jr, (D-CA), was unable to get the Act approved *per se* by his colleagues as originally written, the wily Congressman did manage to incorporate some of its language and provisions as an attachment to a NASA appropriations bill which 100th Congress passed in its second session. The wording is reproduced in Exhibit 22 as a model of what governments might do to further exploration and colonization on the high frontier. For the first time on the record, this 1990 funding legislation officially recognizes the inevitability of such offworld activity!

AN ACT

To authorize appropriations to the National Aeronautics and Space Administration for research and development, space flight, control and data communication, construction of facilities, and research and program management, and for other purposes.

SEC. 217. (a) The Congress declares that the extension of human life beyond Earth's atmosphere, leading ultimately to the establishment of space settlements, will fulfill the purpose of advancing science exploration, and development, and will enhance the general welfare.

(b) In pursuit of the establishment of an International Space Year in 1992 pursuant to Public Law 99–170, the United States shall exercise leadership and mobilize the international community in furtherance of increasing mankind's knowledge and exploration of the Solar System.

(c) Once every two years after date of the enactment of this Act, the National Aeronautics and Space Administration shall submit a report to the President and to the Congress which:

(1) provides a review of all activities undertaken under this section including an analysis of the focused research and development activities on the Space Station, Moon, and other outposts that are necessary to accomplish a manned mission to Mars;

(2) analyzes ways in which current science and technology can be applied in the establishment of space settlements.

(3) identifies scientific and technological capacity for establishing space settlement, including a description of what steps must be taken to develop such capacity;

(4) examines alternative space settlement locations and architecture;

(5) examines the status of technologies necessary for extraterrestrial resource development and use and energy production;

(6) reviews the ways in which the existence of space settlements would enhance science, exploration, and development;

(7) reviews mechanism and institutional options which could foster a broad-based plan for international cooperation in establishing space settlements;

(8) analyzes the economics of financing space settlements, especially with respect to private sector and international participation;

(9) discusses sociological factors involved in space settlement, such as, psychology, political science, and legal issues; and

(10) addresses such other topics as the National Aeronautics and Space Administration considers appropriate.

Exhibit 22—Legislation for Space Settlement. *Source*: from S. 2209 (Section 217), 100th U.S. Congress, 2nd Session, 1988. (For developments on this legislation, refer to Hopkins, M. "The Space Settlement Act," *Ad Astra*, January/February 1994, p. 16; Office of Spaceflight Policy and Plans, *Strategic Plan for Human Exploration and Development of Space*. Washington, DC: NASA Headquarters, 1995).

This legislation provides a framework for behavioral science research on this issue. NASA has funded some studies and conferences that resulted in publications that are beginning to address broader human issues in space. *Space Resources and Space Settlements* was a classic under the leadership of Dr Gerard K. O'Neill which NASA (SP-428) issued in 1979. One produced through Georgetown University resulted in *Social Science and Space Exploration—New Directions for University Instruction* (Cheston, Chafer &

Chafer, 1984). Two others in conjunction with the University of California—Davis produced ground-breaking volumes—*Living Aloft—Human Requirements for Extended Spaceflight* (Connors, Harrison, and Akins, 1985); *From Antarctica to Outer Space* (Harrison, Clearwater, and McKay, 1991), which was also underwritten partially by the National Science Foundation. Finally, another NASA special publication (#509) was released under the title of *Space Resources* (McKay, McKay, and Duke, 1992); the fourth of this six-volume series, *Social Concerns*, is of special interest to behavioral scientists. The next three chapters will discuss these and other NASA-financed studies on human performance in orbit.

One of the most definitive reports on habitability themes was produced by the NASA Life Science Strategic Planning Committee (1988). Entitled *Exploring the Living Universe*, it provided unusual findings, recommendations, and strategies, along with an extensive bibliography on human spaceflight. Among the former were proposals for increasing research, scientists, facilities, and funding in the space life sciences both on the ground and in orbit. Two of its strategies are of particular significance here:

- **Synergize the presently independent research activities of national and international organizations through the development of cooperative programs in the life sciences at laboratories of both global space agencies and universities.**
- **Complete and consolidate global data base consisting of basic life science information and the results of biomedical studies of spacefarers conducted on a longitudinal basis. This data base should be expanded to incorporate information obtained from all spacefaring nations and be available to all participating partners.**

One could envision the value of such a data bank if it became multinational in scope, containing, for example, the spaceflight experience of the Russian cosmonauts and the European Space Agency isolation experiments. In fact, the findings from all terrestrial analogs to space living would be a welcome addition to this stored information, whether collected in submarines, on offshore platforms, at Arctic scientific outposts, from foreign deployment experiences, or experiments like Biosphere 2 in Arizona [14] (Exhibit 57, p. 161).

The potential contributions that well-funded behavioral science research could make to advancing global space programs are manifold. A sample of habitability issues worthy of well- designed, scientific investigations are:

- the sociobiological implications of isolated, long-duration missions and space biospheres;
- the integrative requirements for personnel deployment and habitation for masses of people aloft;
- the training, communication, and perceptions of ground-based support teams of a multi-disciplinary composition;
- the educational strategies, programs, technologies, and institutions that will be required for large numbers of spacefarers both on ground and aloft;
- the reinterpretation of terrestrial knowledge and science for adaptation to extraterrestrial environments;
- the broadened study of space technology and development by social scientists as to their long-term impacts on human culture and society.

From the perspective of an astrophysicist, Dr David Criswell presented four space habitat dimensions, including research needs, to challenge scholars. In Exhibit 23, the Director of the University of Houston's Institute of Space Systems Operations summarizes the type of biospheric issues to be considered with humanity's expansion beyond our terrestrial home.

Criswell documents why continued growth of our industrial way of life can only occur if humans move off the Earth and permanently settle space. He makes a convincing case that this transition will create new wealth from solar energy and the common resources of our Solar System (see Appendix B).

A. *Uses*	B. *Long-term questions*
— Diverse colonies of people	— Maths based human language
— Solar system travel/interstellar flight	— *Homo sapiens* divergence
— Directed research centers for exploration, theory observation, experiments	— Si & III V alloys in intelligent systems
— Entertainment/tourism	— Sentient/inert mass ratios for Solar System, Galaxy, Universe
— Macromachines	
— Macrobrains (swarm)	

C. *Immediate research needs*	D. *Habitat research needs*
— NTM's & Growing Systems (mass multiplication	— Scaling, life support systems, closure limits, fire hazards, time constraints, control and passive mass uses
— Parallelism: how small a first system growth limits criticality	— Long-term residency: radiation shielding, toxicity infection
— Human extenders: teleoperators/robots	— Sp. Medical Procedures with low cost E-Leo Transportation
— Improved spacesuits	
— Food (via Chem. Sys.)	
— 1/6 (Lunar) Gravity Ops.	
— Independence from Earth	
— Training tech. personnel as physicians	

Exhibit 23—Space Habitat Dimensions. *Source*: Dr David R. Criswell, "Human Roles in Future Space Operations", *Acta Astronautica*, 1981, Vol. 8:9/10, pp. 1161–1171, Pergamon Press.

3.2 BEHAVIORAL ANALYSIS OF LIFE ALOFT

A pioneering psychologist in NASA's original experiments with animal behavior in space, Joseph V. Brady, has already set forth a research agenda for studying human behavior in space environments.[15] Further, this Johns Hopkins University medical school professor has also theorized on applied behavior analysis of life aloft. He calls for studies of the motivational factors essential for the maintenance of quality performance in long-duration operational missions. Brady and other researchers advocating the construction of programmed environments are concerned that such investigations on human group maintenance of satisfactory behavioral ecosystems be analytic, experimental in nature, utilizing a scientific method that is observable, manipulable, and measurable. Using a behavior management or modification strategy in space studies, Brady advocates B. F. Skinner's *contingencies of reinforcement* method for empirical analysis of both antecedent and consequent environmental events that influence behavior in confined, isolated situations akin to space living. Brady concludes that human space behavior and environmental research requires multidisciplinary inputs from wide-ranging fields such as molecular biology, environmental physiology, behavioral biology, architecture, political science, sociology, and others.

Two other factors propelling behavioral science space research are early 21st century plans for a lunar base and manned mission to Mars.[16] Anthropologist Ben Finney believes that going back to the Moon permanently requires that social research be part of the planning process, and that the space station should be used for prototype studies. He also proposes that Earth analogs and space simulations should be used now to design space communities that will not be inhabited for decades. Relative to the Mars' exploration by spacefarers, psychologists Connors and Harrison have produced insightful papers on the behavioral challenges in such missions.[17] Such an undertaking might involve a mixed crew of men and women, up to twelve people. If it is done with multinational sponsorship, the space travelers are likely to be multicultural in composition. Apart from technical considerations, the duration in orbit would be between two and three years. Connors and Harrison envision this not only as a challenge in human adaptability, but requiring serious pre-flight analysis of the psychological and social dimensions of Marsflight. To keep the mission within tolerable human safety and stress limits is not enough. These two behavioral science researchers seek to provide Marsfarers with an environment in which they can be fully productive and happy, where they can prosper and grow, particularly if we expect to establish eventually a base on the surface of Mars or its moons. Therefore, they contend, behavioral scientists need to examine a whole range of issues—human qualities and behavior to ensure a successful performance aloft. These and other psychological considerations will be identified in the next section.

Whether the emerging human missions are to the Moon or Mars, greater attention by space planners will have to be directed in the nearterm to:

- architectural and environmental interventions which assure habitation that is both survivable and liveable ...
- a combination of effects likely to occur in an offworld outpost or base—psychophysiological, psychological, and social;
- issues of governance, including command structure and roles.

The impact of living aloft on the evolution of the human species can be perceived somewhat by reviewing the model developed by Gerard, Kluckholm, and Rapport (see Exhibit 24). Analyzing its multidimensional content takes on added meaning in the context of space culture (refer to Chapter 4).

It is impossible in this chapter to review the wide-ranging prospects for behavioral science contributions in this regard. To illustrate the possibilities, the author will next attempt to synthesize studies and views of colleagues from the perspective of anthropology, psychology, and sociology, ending with the living systems paradigm which shows much promise for space habitation research. Then the next three chapters will expand upon these themes, while the contributions of other social sciences will be examined in Chapters 7 and 8.

CORRESPONDENCES BETWEEN BIOLOGICAL AND CULTURAL EVOLUTION	
Biological evolution	Cultural evolution
Distinct species and varieties	Distinct cultures and subcultures
Morphology, structural organization	Directly observable artifacts and customs distinctive of cultures
Physiology, functional attributes	Functional properties attributable to directly observable cultural characteristics
Genetic complex determining structures and functions	'Implicit culture' — i.e. the inferred cultural structure or 'cultural genotype'
Preservation of species but replacement of individuals	Preservation of cultures but replacement of individuals and artifacts
Hereditary transmission of genetic complex, generating particular species	'Hereditary' transmission of idea-custom-artifact complexes, generating particular cultures
Modification of genetic complex by mutations, selection, migration, and 'genetic drift'	Culture change through invention and discovery; adaptation; diffusion and other forms of culture contact; 'cultural drift'
Natural selection of genetic complexes generally leading to adaptation to environment	Adaptive and 'accidental' (i.e. historically determined) selection of ideas, customs, and artifacts
Extinction of the maladapted and maladjusted species	Extinction of maladapted and maladjusted cultures

CORRESPONDENCE BETWEEN CULTURES OR SUBCULTURES AND SUBSPECIES OF A SINGLE SPECIES	
Partial isolation of subspecies — 'cellulation'	Partial isolation of cultures — 'cellulation'
Cross-breeding through migration and limited interbreeding	Cross-breeding' through diffusion of ideas, customs, artifacts
Hybrid vigor	Hybrid vigor?

Exhibit 24—Correspondence between Biological and Cultural Evolution. *Source*: Gerard, R. W., Kluckholm, C., and Rapoport, A. "Biological and Cultural Evolution—Some Analogies and Explorations," *Behavioral Science*, January 1956, Vol. 1:1, pp. 6–34.

3.3 ANTHROPOLOGY AND SPACE HABITATION

Anthropology's greatest contribution to space development may be in its unique evolutionary macroperspective on human behavior, especially on our nature as wanderers, explorers and colonizers. While our technology gives our species the capacity to migrate off this planet, it is our explorer's bent, embedded deep in our biocultural nature, that is leading us to the stars.[18] As the leading space anthropologist, Ben Finney of the University of Hawaii reminds us, the field is a tool for humanizing space.

Again we turn to the *Harper's Dictionary* for insight into this discipline:[19]

> **While anthropology in the past centered its studies around evolution, embracing the biological, prehistoric, linguistic, technological, social and**

cultural origins about the development of humankind, modern anthropology is more focused on the concepts of biological endowment, environment and culture.

Typically, as an academic subject, the field is divided into sub-specialities, such as physical, social, and cultural anthropology. The latter two areas may yet prove to be the most
vital to space studies, as applied anthropologists become less past oriented and turn their
skills toward futures research, especially relative to space communities. Harrison, a psychologist, maintains that modern anthropologists have to become less Earthnocentric![20]

The survival of the wisest in space will depend upon the accumulated knowledge and
insight of interrelated sciences and multidisciplinary research as was intimated in Exhibit
15 and its explanation. To adapt successfully to offterrestrial living, humankind will require synergistic information, paradigms, and methods from all the sciences, particularly
the biological and behavioral. Unfortunately, the comparative studies of our species from
an anthropological viewpoint is too often ignored within the space community. A NASA
published report on *Social Sciences and Space Exploration* failed to even include anthropology.[21] On the other hand, anthropologists have generally neglected the implications of
their field and methodology for space. When a distinguished group of such scholars compiled a volume on *Anthropology Today* (CRM Books, 1971), a picture of an astronaut
working in space was used to illustrate the opening chapter on the "science of man," but
the contributors failed to discuss the emerging space culture in the five hundred pages
which followed. One can only speculate how many textbooks of the '90s on introduction
to anthropology mention the opportunities to apply this discipline offworld.

Interestingly, when a group of anthropologists and other social scientists did publish
Cultures Beyond Earth, their concerns in extraterrestrial anthropology were for contact
with ET species through interstellar travel.[22] That is still the space focus of many social
scientists.[23] Similarly, anthropologists lead in the development of the **Contact** movement
(see Appendix C), which holds annual public forums to prepare for "contact" with alien
cultures. At these simulations, participants break up into two groups as a Human Team
and an Alien Team to invent possible **contact** scenarios and experiences. Publication of
proceedings on their *Cultures of the Imagination* are available, and are also a rich source
of papers on space settlement.[24] As far back as 1966 the American Association of Physical
Anthropology organized a symposium and published proceedings in Berkeley, California,
on **Man in Extra Terrestrial Environments: The Role of Physical Anthropology**. By
the 1990s, Roland A. Foulkes was teaching a course at the University of Florida—Gainesville on **Astroanthropology and Futuristics** in which he used anthropological perspectives and futures research to span pre-human, human, and post-human evolution.[25]

Culture is a central concept for space exploration and settlement, as we will discuss in
the next chapter. Therefore, it is puzzling that cultural anthropologists do not focus more
research on the concept. Consider its implications for this purpose from this useful insight
of a prominent anthropologist, Edward Hall:

> **In physics today, so far as we know, the galaxies that one studies are all
> controlled by the same laws. This is not true of the worlds created by
> mankind. Each cultural world operates according to its own internal dy-**

namic, its own principles, and its own laws—written and unwritten. Even the dimensions of time and space are unique to each culture. There are, however, some common threads that run through all cultures. Any culture is primarily a system for creating, sending, storing, and processing information. Communication underlies everything ... .

[Extract from *Hidden Differences* by E. T. Hall and M. R. Hall. Garden City, NY: Doubleday/Anchor Press, 1987]

What other roles can anthropology play in space enterprise? In addition to the above prospects, Dr Finney suggests anthropology can contribute to space planning by:[26]

- providing of a vision of where we are going extraterrestrially, as well as where we have been terrestrially;
- joining multidisciplinary research teams examining interpersonal and group behavior in the space environment, especially when crews and societies in orbit are heterogeneous and international, requiring skills in cross-cultural relations;
- demonstrating how cultural resources and differences can be best utilized and cultural synergy created among space travelers;
- doing field work on small space communities, particularly examining issues of mating and reproduction, child rearing and education;
- designing cultures and principles for space settlements;
- anticipating contact and communication with other ET species in time through interstellar travel or technology, or with descendents of Earth cultures who have developed aloft as independent and quite disparate cultural entities.

Another anthropologist, Namika Raby, reminds us that anthropology can be used for cultural engineering in space colonization.[27] Our terrestrial human experience has used culture for developing rules of behavior, sets of values for judging others, and to provide meaning to group actions. This World Bank consultant would apply anthropology not only to study humans in orbit, but space agencies on the ground, especially with reference to the impact of bureaucracy on the exercise of power, authority, and decision-making; the emergence and resolution of cultural dissonance and conflict; the bonding goals in human exploration and settlement; the political, scientific, military and commercial goals of space exploitation. Raby, now teaching at the California State University—Long Beach, would further apply anthropological expertise to better understand sub-cultures of elite groups, such as astronauts, aerospace engineers and scientists; as well as differences in the organizational cultures of NASA and its contractors. Dr Raby believes that her field could contribute much to mission success by welding together and creating effective "space groups." She even thinks that anthropology can help in understanding space failures like the **Challenger** orbiter explosion. During a NASA Summer Study at the California Space Institute, Raby proposed additional anthropological research into:

(1) a charter that embodies the central cultural values for promoting cultural synergy among space groups;
(2) enhancing the symbolic valence for bonding space crews at a space station or lunar base, as well as for socially sanctioned outlets for the resolution of conflict aloft;

(3) studying space living issues of status differentiation, proxemics, structured interactions, rituals and ideology, and other such anthropological concerns.

In an address to the Society of Applied Anthropologists (March 29, 1986), Albert Harrison, psychologist, of the University of California—Davis proposed that anthropologists might engage in high frontier studies[20], such as:

- analysis of the social and technical systems developed to meet the challenges of the nascent space era, as well as creation of new systems to facilitate the transition from visiting space to living there for longer durations;
- minimization of problems and increasing human performance and production in space;
- strategies for improving the quality of life or habitability aloft for spacefarers, beginning with the space station and lunar base;
- application of the anthropological method of participant observation. Dr Harrison provided pragmatic advice to anthropologists so they could adapt their research proposals to NASA culture and spaceflight settings, urging them to enter their theoretical papers and reports into mainstream scientific literature. He forecasts that anthropologists may gain their greatest achievements by participating on interdisciplinary teams doing research on human habitability in exotic environments as terrestrial analogs for space living.

Exhibit 25—A Pioneering Mission for the 21st Century. **In this lunar settlement, *closed-loop* life support systems recycle and reuse air, water, food, and wastes in an integrated fashion.** *Source*: Art by Robert McCall in National Commission on Space, *Pioneering the Space Frontier*. New York, NY: Bantam Books, 1986, p. 2.

3.4 PSYCHOLOGY AND SPACE HABITATION

Psychologists have been involved in the space program since its inception over thirty years ago, both in the U.S.A. and in the former U.S.S.R. Working in conjunction with medical teams, they have helped to determine if both astronauts and cosmonauts have "the right stuff" and what the impact of spaceflight is upon their mental and emotional well being.[28] In collaboration with industrial engineers, psychologists have expanded human factors research about spaceflight.[29] In fact, a whole new specialization of **space psychology** has begun to emerge, among which this author includes himself. Furthermore, university departments of psychology, such as at the University of California—Davis and Georgetown University, are offering courses related to the psychological aspects of spacefaring. The Master of Space Studies at the International Space University has a multidisciplinary program in the Humanities featuring the behavioral science.

Before proceeding, it might again be useful to go again to *The Harper's Dictionary*, previously cited, to obtain a basis for understanding this field:

> **Psychology—the study of mind, behavior, or man interacting with social and physical environment. The ultimate aim is systematic description and explanation of man in the fullness of his powers, as a thinking, striving, talking, enculturated animal.**

Dictionary editors Bullock and Stallybrass remind us that psychology has yet to devise central concepts such as the laws of physical science, and can be characterized in terms of either its choice of processes to be studied, or its method of analysis. For the former practitioners concerned about input/output metaphors, their investigations center around perception and habitation, and the ways organisms transform energy; for those concerned about response processes, the focus is upon expressive movement, motives, drive states, language, and various forms of social behavior. With respect to the methods of analysis, psychology uses a variety of experimental approaches (e.g., clinical, observational) and tools (field studies, mathematical, and computer modeling). As a university subject, psychology has proliferated in its applications—industrial, educational, cognitive, existential, genetic, humanistic, mathematical, informational, and so forth. Perhaps psychology will only complete itself as a field of knowledge when it converges into the behavioral sciences, possibly within the context of human behavior studies and research in outer space? Harrison and Summit argue that recent reviews of psychology and spaceflight tend to focus on performance decrements, dwindling motivation, emotional stability, social conflict, and other adverse consequences of prolonged missions aloft. They argue for a "third force" of humanistic psychology that describes the psychological benefits of manned spacefaring, such as:

— enhanced competence and mastery;
— peak experiences or the **overview effect**;
— increased ability to cope with stress;
— high social cohesion and teamwork;
— role model and inspiration for others.

These psychologists contend that such an approach would emphasize human talent and resourcefulness aloft which has implications for space mission design.[30]

What are some of the unique contributions to be made by psychology in space developments and habitation? Connors, Harrison and Akins maintain that the answer to that question is in fostering a high level of spacefarer wellbeing and maximum productivity. But to do that, they think that the profession must gain greater understanding of the psychological issues of adaptation to outer-space life.[31] Cheston postulates that as a discipline, psychology needs to thoroughly address how crew members can be taught self-generated reinforcement strategies which enhance their sense of personal and professional accomplishment aloft.[32] Apparently in agreement with Joseph V. Brady's position stated above, T. Stephen Cheston reminds us that psychology has already made a remarkable start toward ensuring more effective human performance by manipulating schedules of reinforcement and punishment, but this research should be applied to orbital human factors.

Christensen and Talbot did a comprehensive review on the psychological aspects of spaceflight.[33] They have examined both the former Soviet and the American experience aloft with regard to perception, cognition, psychological stability, performance, small group dynamics, stress, as well as psychological methods and models. Their remarkable Exhibit 26 summarizes the key factors which influence orbital behavior and performance in terms of environment, space systems, and support measures. Furthermore, in this study for NASA's Life Science Division, Christensen and Talbot have offered far-ranging suggestions to behavioral scientists for near- and long-term research and development, especially in terms of the shuttle/space station operations and observed aberrations.

What are the major psychological issues that spacefarers face, past, present and future? Kanas addressed that question relative to both former Soviet and American space studies, particularly in light of longer manned space missions and more heterogeneous space crews.[34] He identified nine psychological and seven interpersonal issues requiring further research:

- *psychological issues*—sleep problems, time sense disturbances, demographic effects, career motivations, reaction stages to isolation, transcendent experience, postflight personality changes, psychosomatic symptoms, and anxious/depressive/psychotic reactions;

- *interpersonal issues*—interpersonal tensions, problems resulting from crew heterogeneity, anger displacement on outside personnel, need for dominance, decreasing group cohesiveness over time, task neutral interactions, and types of leadership.

Analysis of other space studies has provided me with this summary of additional behavioral issues worthy of further research:

- recruitment, selection, training and orientation of spacefarers, especially those who are not members of the astronaut corps, such as contractor personnel;
- individual and group adjustment to living in a zero or low gravity, unearthly environment; human performance in exotic environments which are isolated, confined and high-risk settings as an analog to space living;

A. Psychological Psychosocial, and Psychophysiological	B. Environmental	C. Space System	D. Support Measures
limits of performance (perceptual, motor)	spacecraft habitability	mission duration and complexity	inflight psychosocial support
cognitive abilities	confinement	organization for command and control	recreation
decision making motivation	physical isolation social isolation	division of work, man/machine	exercise selection criteria
adaptability	weightlessness	crew performance requirements	work-rest/avoiding excess workloads
leadership productivity	lack of privacy artificial life support	information load	job rotation
emotions/moods attitudes	noise	task load/speed crew composition	job enrichment preflight environmental adaptation training
fatigue (physical and mental)	work-rest cycles shift changes	spacecrew autonomy physical comfort/ quality of life	social sensitivity training
crew composition crew compatibility	desynchronization simultaneous and/or sequential multiple stresses	communications (intacrew and space-ground)	training for team effort inflight maintenance of proficiency
psychological stability		competency requirements	
personality variables	hazards		cross training
social skills	boredom	time compression	recognition, awards, benefits
human reliability (error rate) space adaptation syndrome			ground contacts self-control training
spatial illusions			
time compression			

Exhibit 26—Space Factors and Human Behavior. *Source*: Christenson, J. M. and Talbot, J. M. "Psychological Aspects of Space Flight," *Aviation and Environmental Medicine*, March 1986, pp. 203–212 (NASA Contract#3924).

- human factors in design of space stations and habitats, especially as these become more spacious and commodious;
- human performance effectiveness and productivity aloft in terms of workload, assignments and scheduling;
- orbital sleep, rest cycles, and recreation;
- socialization of new arrivals, inflight support, and preparation for re-entry to Earth life and culture;
- coping with problems of extended spaceflight, personality conflicts, and deviant behavior;

- communication aloft among mixed crews of government employees, military, and private contractors with sexual/cultural/national and professional differences, as well as between such diverse crew and ground-based personnel;
- crew morale aloft, and communications with ground-based families and friends;
- inflight biomedical and psychological monitoring of spacefarers' health and wellbeing;
- inflight exercise of authority and decision-making, leadership and management of space missions and emergencies;
- post-flight debriefing and psychological evaluation of crew adaptation and performance. The latter would require scholarly examination of the whole phenomenon of the "transitional experience" or **overview effect**—already well documented—and its implications for those who undergo offterrestrial living.[35]

What are some additional contributions that psychologists may undertake toward the furtherance of space development and habitation? Obviously, orderly investigations of the above issues by psychologists, in conjunction with other behavioral and biological scientists, would advance human presence in space. As Connors and Harrison so rightly conclude, psychologists should provide leadership in helping astronauts and other space travelers maintain a healthy psychological state, examining particularly how to promote an atmosphere of cooperation and mutual support within a group aloft. Furthermore, they urge that such research findings be published in mainstream scientific and space literature. Recall that Oleg Gazenko, head of Space Medicine in the former Soviet space program, expressed the belief that the limitations of living in space are not medical, but psychological.[36]

The case for interdisciplinary studies can best be appreciated by examining just one of the issues cited above—sleep. Dr Martin Moore-Ede, professor of physiology at Harvard Medical School, thinks that inadequate investigations have been addressing crew shifts and sleep patterns during long-term space travel (*Omni*, June 1987, p. 2). Both astronauts and cosmonauts report that they do not sleep deeply in space, and have difficulty in the early stages of missions. Despite the need to develop new sleeping patterns and arrangements more suitable to gravity-free living, Frank Sulzman, when NASA chief of biomedical research, admitted the agency then has no organized programs examining crew schedules, the physiology and psychology of sleep, and human circadian rhythms in space.

With a little imagination, psychologists should be seeking to engage in studies with other social scientists on the design and conduct of (1) new university courses in *space* psychology, anthropology, and sociology; (2) multidisciplinary education and training programs for spacefarers. Social and environmental psychologists, in particular, have only begun to apply their expertise to improving the spaceflight experience in the design of space habitats and the organization of space communities.

One facet of such research was undertaken for NASA on privacy and space habitat design by Harrison, Sommer, Struthers, and Hoyt.[37] These psychological investigators from the University of California—Davis examined the dimensions of privacy and social contact and their implications for life aloft, especially on a space station. After a review of the psychological literature on the matter, they made fifty specific recommendations to the NASA Ames Habitability Group. The latter, under the direction of environmental psychologist, Yvonne Clearwater, are engaged in research for the Space Human Factors Office at Moffett Field, California, to make a space-based workforce both comfortable and productive during long-duration flights—that is thirty days or more aloft. Other innovations in behavioral science space research will be discussed in the last section on living systems.

Astronomer Carl Sagan forecasts that **cultural diversity** will be a strength and key to survival in future space societies, each of which would take aloft aspects of their terrestrial worlds, its planetary engineering, its social conventions, its hereditary dispositions.[38] Psychological selection, therefore, should evaluate for spacefarers who would be more tolerant of differences in lifestyle, race, or religion. Eventually, psychological research should be directed toward the differences in **earthkind** and **spacekind**. In time, the adaptations of the latter will bring about biological, physical, psychological, and social changes that sharply distinguish them from their terrestrial counterparts. For now, such behavioral and biological science research starts on the Shuttle and Space Station as indicated in Exhibit 27.

Exhibit 27—Research on Living and Working in Microgravity. Artist's rendering of research on "free fall culture" and its effects. *Source*: NASA/TAD Corps.

3.5 SOCIOLOGY AND SPACE HABITATION

One recommendation of the National Commission on Space's report in 1986 advocated building institutions and systems that make accessible vast space resources and support human settlements beyond Earth orbit, from the highlands of the Moon to the plains of

Mars. But Professor Ben Finney believes that this demands planning for social organizations in space. Gerald Carr, commander of the 84-day **Skylab** mission, has gone on record that he expects that "the sociological problems will prove to be more difficult to solve than the technological ones."[39] Current plans under way for a multinational space station, as well as Moon/Mars bases, challenge sociology and its practitioners to address offterrestrial social issues and civilization.

To comprehend better the meaning of the concept, let us return to the explanation in *Harper's Dictionary* :

> **Sociology—the study of societies by observation and description within a coherent conceptual scheme.**

The dictionary's editors, Bullock and Stallybrass, inform us that sociology is less fully a discipline than its counterparts because of several competing schemes and failure to develop coherent sets of concepts which can be applied and codified as theories. The three principal perspectives for viewing society are:

(1) **mechanistic**, analogous to social physics which explains social phenomenon and variations by reference to climate, soil, population or combination of such physical attributes;
(2) **social evolution**—progress due to evolving consciousness or man's material powers;
(3) **systematic empirical enquiry** about the facts of social life. Contemporary sociology is heavily influenced by the teachings of Karl Marx on class structure and ideology; of Herbert Spencer on social morphology of societies in terms of structures and functions; of Emile Durkheim on social solidarity and segmentation; of Max Weber on social action and comparative method, especially in the study of authority, power, bureaucracy.

Sociology has developed analytical tools that may prove useful in space research, such as conflict/evolutionary/ecological/mathematical *models*, case study methods and scenario development. Sociologists may join forces with biologists in social biology, the application of biology to social problems, such as pollution, ecology, and overpopulation. Or they may participate in social engineering with other behavioral scientists concerned with planning social change. Or they may collaborate with social psychologists in studying human social behavior, by combining biology and the social science.

Extending human civilization to outer space is a challenge that may enable sociology to achieve its fullness. B. J. Bluth, a forerunner in the application of sociological methods to space development, stated her case in this way:[40]

> **Sociology organizes knowledge to identify and analyze more of the hidden potential in human behavioral systems ... . Systems of sociological concepts can be broken into three basic categories: social systems, or systems of ways of doings things; cultural systems, or systems of meaning (e.g., language, values, beliefs, ideas); and personality systems, or systems of need disposition.**

Bluth maintains that sociologists can contribute to space living because of their expertise in identifying patterns of relationships between events and systems, and their ability to analyze the consequences of behavior. As one trained in organizing knowledge of human behavioral systems, this sociologist is interested in diverse space activities, such as:

- astronaut survival and safety;
- crew size, compatibility, and constraints;
- quality of space life;
- social processes aloft;
- human experience in orbit, especially with gender-mixed crews;
- the impact of space technologies on Earth's social and cultural systems.

However, Bluth has long-term concerns as a sociologist relative to the design of lunar and space communities, and the intelligent applications of behavior systems approaches. When she was a sociology professor at the California State University, Northridge, Bluth taught courses in astronautical sociology. Now at NASA headquarters doing research for the space station, Bluth analyzes the Russian analogs that may assist in U.S.A. planning for long-duration missions either on **Mir** or on the future **International Space Station (ISS)**.[41] Currently she has a book under way on designing for human performance in space.

Another space sociological leader is William E. MacDaniel.[42] While at Niagara University, he taught courses in both future studies and "Living in Extraterrestrial Space." He co-directed there the Space Settlement Studies Project which published a journal and continues a newsletter. As a result of research at the Johnson Space Center in 1982, this former Air Force jet pilot undertook an innovative Delphi scenario study on *Extraterrestrial Space Humanization* in 1984. Now a professor emeritus in San Antonio, MacDaniel still believes that sociology can make a unique contribution to the humanization of extraterrestrial space, as well as in social inventions for the space frontier. His studies convince him that ET social organization and culture will differ significantly from terrestrial experience. The unique combination of factors in outer space living—from zero or low gravity to unlimited solar power—will force human adaptations and alternative "free fall cultural development." MacDaniel maintains that technologically oriented professionals have influenced space developments, while social scientists have abdicated their responsibilities to participate with research leadership on space settlement and societies.

In light of such observations, an interesting footnote is that a 1983 Smithsonian symposium and publication on human adaptation barely mentions our greatest challenge in that regard—space, and then only in terms of speculation on space colonies.[43] Although it views the species' adaptability as a biocultural odyssey, these proceedings are notable for emphasis on the biological bases for social behavior, as well as by the absence of the three social sciences under discussion here. In contrast, Rudoff has offered three frameworks for sociologists to contribute new insight on extraterrestrial human adaptation:[44]

(1) space sociology from a terrestrial perspective;
(2) the study of small groups in confined space and its implication for space living;
(3) space sociology from an emerging extraterrestrial perspective.

In "Reflections on the Sociology of Interstellar Travel," NASA's John Mauldin suggests that trips on **starships** with their **microsocieties** in the centuries ahead may involve the lifespan of generations.[45] He envisions both biotechnology and sociobiology impacting the sociology of such a high technology environment which might involve 30 generations for 1000-year missions! Mauldin predicts that such spacefaring will push human potential and species' knowledge to the ultimate limit!

If the unusual problems of survival in the offworld environment can be partially solved through societal mechanisms and pre-planning or social engineering, then sociologists should be involved. By using their competencies in designing space communities, sociologists will help not only to improve civilization aloft, but may prevent the transfer there of cultural, social, and biological dysfunctions of this planet.

Exhibit 28—Science in Space. Artist's rendering of scientific experiments being conducted at an orbiting space station; *Source*: NASA/TAD Corps.

3.6 LIVING SYSTEMS AND SPACE HABITATION

To delimit the problems of orbiting groups and societies, to promote synergistic relationships and cultures in orbit, there seems to be a need first for integrative studies among terrestrial behavioral scientists themselves. Since space experience alters human perception, only the convergence of behavioral science methodology and insight will permit the holistic interpretation as to the meaning of life experienced aloft. One way to achieve this might be if such professionals devoted their research to a summative theme, such as "human sexuality and reproduction in space," a subject generally avoided by space agencies. Another opportunity would be for behavioral and biological scientists to team up in utilizing a common paradigm for studying space habitation issues.

That is what Connors, Harrison, and Akins did in their monumental work on *Living Aloft,* when they adopted the recommendation of the Space Sciences Board to use a systems perspective to analyze space flight. Missions were then viewed as comprising of highly interdependent components (e.g., technical, biological, social), such that variations in one element have repercussions on the others. These researchers attempted to expand this conceptualization by incorporating features of open-systems theory as devised by James Grier Miller and his colleagues.[46] In "Lunar Bases—Learning to Live in Space," Finney also confirmed Miller's approach, emphasizing that in space studies we are dealing with living systems involving the biological, technological, and social.[47] As an anthropologist, Dr Finney called for social scientists to work closely with biologists, human factors specialists, architects and, ultimately, the engineers and managers who conceive, design, and operate the whole space system. Thus, he himself has done such research at the NASA Ames Research Center.

During the 1984 NASA Summer Study at the California Space Institute, Dr James Grier Miller, a psychiatrist/psychologist, directed a lifetime of research on living systems toward issues of space habitation.[48] At the same time, another Faculty Fellow in that endeavor, your author, was then examining matters of space management, culture, and people deployment.[49] When NASA officials there requested proposals from both researchers on their mutual endeavors, it resulted in an interdisciplinary, combined project entitled, *Living Systems Applications to Human Space Habitation.*[50] Originating through the School of Medicine, University of California—San Diego, Miller served as principal investigator and Harris as senior collaborator, along with a team of fifteen behavioral, information, and natural scientists. Unfortunately, the then NASA's Life Sciences Division found the approach too holistic and declined to fund this innovative study. The project team still envisions LST applications to isolated, confined environments anywhere, whether in Arctic outposts, submarines, or outer space. To collect and analyze system data, Dr Miller advocates the use of identification badges with infrared transponder in the form of microchips, questionnaires, and observation; computers for information storage and artificial intelligence expert systems for analysis.[51]

In an article on "The Nature of Living Systems" with his wife, a psychiatric social worker, the Millers maintain that planning for extraterrestrial living requires a primary focus upon the human beings who are to inhabit the projected settlements.[52] They suggest that Living Systems Theory (LST) can facilitate space planning and management because it is an integrated conceptual approach to the study of biological and social living, the technologies associated with them, and the ecological systems of which these are all parts. They describe living systems as open systems that maintain a thermodynamically improbable energic state. This occurs through a continual interaction with the environment in which these substances of lower entropy and higher information content input than they output. The LST school of thought views biosocial evolution moving in an overall direction of increased complexity.

James and Jessie Miller have developed visual encapsulations of 20 critical subsystems and processes, with some applications to space activities. The paradigm illustrates on its vertical axis eight levels at which living systems have produced cells, organisms, groups, organization, communities, societies, and supranational systems (such as the United Nations). On the horizontal axis, the essential subsystems are depicted as reproducer, boundary, ingestor, distributor, converter, producer, matter–energy storage, extruder, motor, supporter, input transducer, internal transducer, channel/net, timer, decoder, associator, memory, decider, encoder, and output transducer.

The LST strategy, according to the Millers, facilitates empirical cross-level research. Each system can be identified in terms of a set of variables describing its basic processes. At the level of groups or below, these represent aspects of flows of matter or materials, energy, and information or communication. At the level of organization and above, it is useful to measure two additional flows—personnel (individual and group), and money or its financial equivalent (e.g., costs). The Millers can document a wide variety of LST research applications by behavioral scientists over the past thirty years in a variety of fields from the military to health care.

Now James Grier Miller is proposing its application to space habitation, first in terms of the **International Space Station** and later for a lunar base. Dennis Davidson, astronomical artist, has created, under Miller's direction, a series of illustrations using LST symbols to explain through color usage how this complex theory would be utilized. To provide the reader with some insight into such applications, we have reproduced here three summary diagrams, but these are in black and white, and do not reflect the distinguishing and separate color flows. Exhibit 29 explains the Living Systems Symbols. Exhibit 30 visualizes Living Systems Theory applied to the **International Space Station**, with its five major flows for that open system which may become fully operational with eight spacefarers by the year 2002. Exhibit 31 does the same in the context of a lunar outpost which may be functional by 2010. The complex patterns for a Moon base show a command center, habitation unit, generating station, as well as provisions for storage, solar power, and nuclear power plant.

With renewed interest in space human factors research by the present NASA Office of Life and Microgravity Science, the **Living Systems** research team still hopes to employ this integrative methodology to collect both subjective and objective data on spacefarers at both locations, employing computers and sensors, as well as a centralized knowledge base for analysis by an artificial intelligence expert system. If the first two prototype applications of LST aloft prove valid and worth while, then the studies might someday be extended to a Mars base and other space colonies which emerge in the next century.

Living systems theory has been successfully applied for over three decades in terrestrial settings. Now it would seem appropriate to continue this line of multidisciplinary research to improve human performance and habitation exterrestrially. By using the LST template and five critical flow measurements in space communities, it would seem that more comprehensive planning information, and management, may result. In terms of this chapter, the living systems conceptualization provides a computer-based framework for analysis of biosocial behavior both on and off this planet. It represents an opportunity to bring together in meaningful investigations the combined talent of behavioral, biological, and informational scientists.

Note to readers: On the subject matter of this chapter, a new International Journal of Earth–Space entitled *Life Support & Biosphere Sciences* is available from Cognizant Communication Corporation (3 Hartsdale Road, Elmsford, New York 10523, U.S.A.; Tel: 914/592-7720; Fax: 914/592-8981).

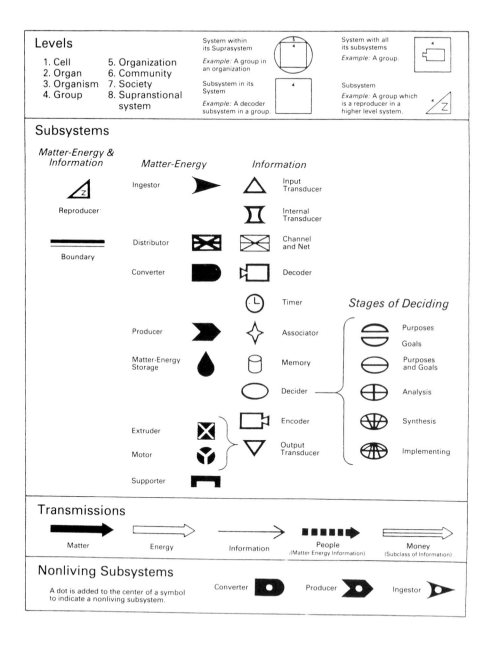

Exhibit 29—Living Systems Symbols*. *Source*: Miller, J. G. "Applications of Living Systems Theory to life in Space." in *Space Resources* (NASA-SP-509, Vol. 4, pp. 231–259). Illustration by Dennis M. Davidson.

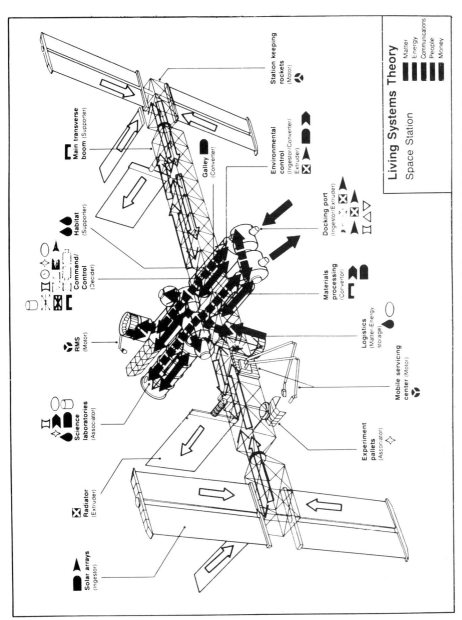

Exhibit 30—Space Station: Five Living Systems Flows*. *Source*: Miller, J. G. "Applications of Living Systems Theory to Life in Space," **Social Concerns** (Vol. 4, pp. 231–259), *Space Resources* (NASASP-509). (Washington, DC: U.S. Government Printing Office, 1992). Illustrations by Dennis M. Davidson.

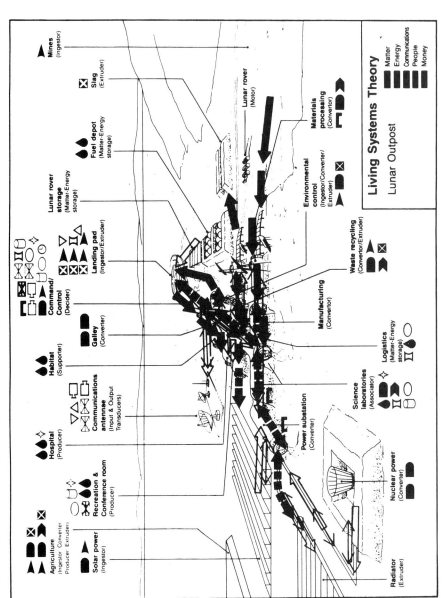

Exhibit 31—Lunar Base: Five Living Systems Flows* *Source:* Miller, J. G. "Applications of Living systems Theory to Life in Space," in *Space Resources* (NASA SP-509), Vol. 4, pp. 231–259). Illustration by Dennis M. Davidson.

3.7 CONCLUSIONS ON FUTURE RESEARCH DIRECTIONS

Writing on the subject of "Space and Society" in a previously cited work, Cheston made a case for space social sciences, or **spaceology,** as a new arena of study for integrating the methodologies and insights of mainline disciplines. Further, he defines this comprehensive term in this unique way:

> **Spaceology is that branch of knowledge that treats the origin, development, and varieties of interaction between human culture and the extraterrestrial environment. Spaceology would draw upon the humanities, social sciences, and natural sciences with equal facility.**
>
> [*Source*: *Beyond Spaceship Earth*, E. C. Hardgrove, ed. San Francisco: Sierra Club Books, 1986, p. 23.]

Cheston also recalls that a consortium of 52 universities already exists in the Universities Space Research Association (Appendix C). It now promotes synergistic projects with government agencies and industry in the applications of automation and robotics to space. UNRA's Lunar & Planetary Institute in Houston, adjacent to the NASA Johnson Space Center, also has a division of biomedicine. Perhaps that is the place for implementing the research agenda proposed by conferees at Johns Hopkins University's School of Medicine:

> To some considerable extent at least, the general lack of behavioral science impact on space research initiatives to date can be attributed to the characteristically narrow, oversimplified actuarial approach of the social and psychological disciplines with their traditional emphasis on statistical significance often at the expense of biological relevance.
>
> [*Source*: *Human Behavior in a Space Environment: A Research Agenda* by J. V. Brady. Baltimore, MD: Johns Hopkins University School of Medicine, 1982.]

Among the insights provided in that unique report was the recommendation for establishment of a free-standing Institute for Human Behavior in the Space Environment in close proximity to spaceflight operations at the Johnson Space Center! There is an idea whose time has more than come if behavioral science and technology data are to make significant contribution to extended space occupancy by human systems. Perhaps, then, outer space studies and research may foster more holistic scholarly efforts in contrast to the present fragmentation so evident throughout the academic and research worlds, as well as through global space agencies!

In another NASA publication cited above, Cheston and Chafers (1984) insisted that by its very nature, the social science study of space is an interdisciplinary endeavor. They not only called for a coherent course of university space studies in what might preferably be called the **behavioral sciences**, but proposed methods and materials. One recommendation would incorporate futures studies and research methodology in such courses. Space education and investigation might well benefit from the approach used by futurists and strategic planners relative to environmental scanning, identification of trends and issues, and forecasting, along with their tools for surveying, scenario development, and simulation.

(1)   To integrate through computerized data banks, worldwide findings and insights of diverse literature on human experience in exotic, isolated, confined environments, whether terrestrial or offworld; then to analyze this environment in preparation for long-duration spaceflight and space settlement.

(2)   To further a convergence of worldwide biological and behavioral research within the field of space life sciences on such matters as biospheres and closed systems, habitability and communications, human factors and genetic engineering, telemedicine and orbital health services, radiation and toxicology, immunologic and circadian changes, cardiovascular and regulatory physiology, group dynamics and team building, etc.

(3)   To apply such knowledge that may result from the above activities to the development of artificial intelligence and expert systems, diagnostic instruments and simulations, computerized and automated systems which could be used in the assessment and pre-departures orientation of spacefarers, as well as to facilitate their effective performance in orbit and upon re-entry to Earth.

(4)   To utilize and promote such research as will enhance the *quality of life aloft for spacefarers*, such as by creation of space deployment systems, including comprehensive, in-orbit support services; by designing communication technologies for data collection on human performance in orbit, and recycling of such information in the preparation of future *envoys of humankind* in space.

(5)   To foster studies for the creation of a spacefaring civilization, including *governance* of future space societies, human and property rights in space communities, legal/financial systems to facilitate scientific and commercial activities on celestial bodies, relationships between *earthkind* and *spacekind*.

Exhibit 32—Biological and Behavioral Space Research Challenges. *Source*: Harris, P. R. *Living and Working in Space*, first edition. Chichester: Ellis Horwood, 1992, p. 112.

Certainly, space is largely a study of the future, of what Thomas Paine, when chairman of the National Commission on Space, described as "the evolution of our species into the cosmos" (*Los Angeles Times*, May 25, 1987, p. 28). That is why George Robinson called for more structured studies of humankind in space or **Homo alterios spatialis** (*Space Governance*, June 1995, p. 17). Ideally, serious professionals in all fields from anthropology, sociology, and psychology to economics, political science, and law should be addressing how they can apply their expertise to expanding the offterrestrial human endeavor in the 21st century. It would be unfortunate for any discipline to miss out on contributing to what Krafft Ehricke so aptly described as the creation of a polyglobal civilization, observing on "The Anthropology of Astronautics":[53]

The concept of space travel carries with it enormous impact because it challenges man on practically all fronts of his physical and spiritual existence. The idea of traveling to other celestial bodies reflects to the highest degree the independence and agility of the human mind. It lends ultimate dignity to man's technical and scientific endeavors. Above all, it touches on the philosophy of his very existence. As a result, the concept of space travel disregards national borders, refuses to recognize differences of historical or ethnological origin, and penetrates the fiber of one sociological or political creed as fast as that of the next.

Hopefully, the spirit of International Space Year in 1992 may prove to be the catalyst for collaborative contributions by both the biological and the behavioral sciences to human habitation of the orbital microgravity environment. Exhibit 32 suggests the direction such research should take in the 21st century!

Spaceology defined:

The first contours of space social science as a separate entity are only now issuing from the jumble of other disciplines ... . Space social science, or *spaceology*, if we are to coin a word for economy of language, will become a recognized area of study that integrates the methodologies and insights of mainline disciplines ... . We might define *spaceology* as that branch of knowledge that treats of the origin, development and varieties of interactions between human culture and the extraterrestrial environment.—T. Stephen Cheston, 'Space and Society,' in *Beyond Spaceship Earth*, E. C. Hargrove (ed.) (San Francisco: Sierra Club Books, 1986), pp. 22–23.

Exhibit 33—Working in Space*. An original painting by Dennis M. Davidson, which was the cover art for this book's first edition, *Living and Working in Space*. Chichester: Ellis Horwood, 1992.

4

Cultural Implications of Space Habitation

During the past forty years of this 20th century, humankind has been extending its presence successfully into outer space either through automation or in person.[1] Landing a "Man on the Moon" through the Apollo 11 mission in 1969 broke our perceptual blinders—to repeat, we are no longer "earthbound" as our ancestors believed for centuries. Extra Vehicular Activity (EVA) in orbit dramatically reminds us of our new reality, as Exhibit 34 illustrates. Perhaps the real home and potential of the human species is on the high frontier. Just as application of fire and tools altered our forebears, so space technology and settlement forces modern men and women to change their image of the species. We can move beyond the "gravity well," free to explore and utilize the universe to improve the quality of human existence. The technological achievements of global space agencies and private enterprise aloft contribute mightily toward actualization of our potential. Space exploration and exploitation of offworld resources not only alters human culture on Earth, but contributes to the emergence in the 21st century of an entirely new **space culture**, a transformation into a being called **spacekind** or **Homo spatialis.**[2]

4.1 CULTURE—A COPING STRATEGY

Culture is a unique human capacity—a coping ability of *Homo sapiens* to the environment which facilitates daily living. Consciously or unconsciously, groups transmit cultural information and insights to future generations. By understanding the concept of culture, we gain better comprehension of human behavior whether on or off this planet in relation to one's physical environment.

Human beings create culture, or their social environment, in the form of customs, norms, practices, traditions and taboos for survival and development. The cultural conditioning and lifestyle of particular groups are passed on to descendants. Subsequent generations receive and adapt these "truths" of accepted behavior in society, creating their own situational standards, values, and ethics. Culture is communicable knowledge that is both learned and unlearned, overt and covert in practice, which influences all human systems, including technological. On "Spaceship Earth," culture has been remarkable for its diversity, and those who would be successful here or in orbit need to learn skills to deal with cultural differences and to create cultural synergy.[3] In the last chapter, we listed some of

Exhibit 34—EVA: We Are No Longer Earthbound!   Untethered astronaut with a mobile power
pack during extravehicular activity in orbit. *Source*: NASA's Johnson Space Center.

the environmental circumstances that distinguish outer space and will influence human
adaptation and culture aloft.

Space culture has been in formation for some four decades as humans traveled up
more than l00 km into orbit. Depending on one's perspective and academic discipline, one
might add to the listing provided previously in Exhibit 32. Certainly, those engaged in life
sciences research aloft would emphasize biological factors, while those in the behavioral
sciences would point to psychological and sociological influences related to living and
working on a space station or lunar base.[4] Although our unmanned spacecraft have probed
millions of miles around our Solar System, only a relative handful of humans have actually
been in orbit, and these mostly for short durations. For the most part, these pioneering
astronauts and cosmonauts came largely from three cultural entities—American, Russian
or former Soviet bloc countries, and European. Until the decade of the 80s, these first
spacefarers were principally white males from the subcultures of the military and science.[5]

The harbingers of future diversity among spacefarers can be seen in the composition of the late **Challenger** shuttle crew: two women and five men—five Caucasians, one Afro-American, and one Asian–American; six professional astronauts and one civilian, multi-disciplinary and multireligious in backgrounds (see Dedication photograph in front of this book). Whether sponsored by space agencies of the U.S.A, Russia, or the European Union, crews are becoming more multicultural and international in make-up (see Exhibit 35). Just as on Earth there are aspects of human experience that cut across cultures, it would appear that spacefaring is becoming a cultural **universal.** It is also evident that the next thousand people in orbit will have to possess greater cross-cultural competencies in dealing with differences.

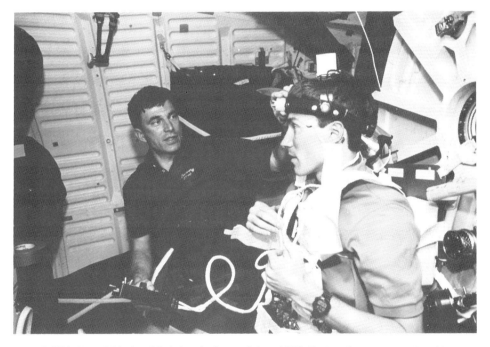

Exhibit 35—Multicultural Relations in Space. Onboard STS-60, American astronaut, Ronald Sega (left), and Russian cosmonaut, Sergei Krikalev, (right), work on a joint multinational metabolic experiment. Shown on the Shuttle **Discovery's** middeck, this 8-day mission occurred in February 1994. *Source*: NASA/Johnson Space Center.

In the next century, as we slowly expand our presence upward beyond one hundred miles, establish human communities in ever larger numbers, be it on space stations or Moon/Mars bases, the need for more cultural **synergy** is essential! Such synergy optimizes differences in people, fosters cooperation and teamwork, and directs energy toward goal accomplishment and problem solving in collaboration with others. The very complexity of transporting people into space stimulated the development of matrix or team management during the Apollo program.[6] While it takes but hours to place humans in lower Earth orbit and only 3 days to reach the Moon, opening up the trade and tourism routes to Mars means venturing many tens of millions of kilometers from the home planet on two-year round-trip journeys! That reality is the case for **space as a place for synergy**—only global cooperation will succeed on the space frontier, as our limited experience aloft already indicates.[7] To create successful and productive space habitats and colonies in zero or low gravity environment will require the practice of synergistic leadership and multicultural management.[8]

Current research in evolution indicates that harsh environments often result in innovation among species. The pattern of the past reveals that creatures are better at inventing and surviving when challenged by extreme environments. Fortunately, multidisciplinary studies are under way to apply terrestrial analogs to offworld living prospects that may accelerate human development.[9] The European Space Agency is using the Norwegian Underwater Technology Center, where scientists from the Russian Institute of Biomedical Problems supervise the ESA astronaut training for **EuroMir.** IBMP's director, Viktor Baranov, said the institute has prepared cosmonauts for more than 70 missions; now they are training ESA astronauts, such as the German Thomas Reiter, who was aloft for 135 days on Mir. Already the Japanese are not only planning for space tourism, but companies such as Shimizu Corporation are designing spaceports and orbiting hotels for 100 people that provide microgravity sporting events, while Ohbayashi Corporation is developing a lunar city for 10 000 inhabitants on the Moon.[10] Japan's Institute for Space and Aeronautical Sciences, like ESA and NASA, has studies under way on crew relationships and psychological effects of isolating small groups of people within closed ecological life support systems. Human movement from the home planet to the **high ground** will transform our culture and species! As editors Finney and Jones in *Interstellar Migration and the Human Experience*, previously cited, remind us:

> **Migration into space may be a revolutionary step for humanity, but it is one that represents a continuity with our past.**

But space scientists and engineers are only beginning to appreciate the implications of culture for space developments with reference to:

- human perception and behavior on this planet, particularly in terms of preparation of spacecraft and missions, as well as issues of safety;
- alteration of **earthkind** into **spacekind** as a result of living and working in the orbital environment;
- organizational cultures of space agencies and aerospace industries which influence success or failure in the deployment of spacecraft and people into orbit;
- technological systems designed for aerospace flight.

4.2 EMERGENCE OF A NEW SPACE CULTURE

In our last chapter, I suggested that space planners would benefit immensely by utilizing more effectively the insights and resources of behavioral scientists. Such strategic planning, especially for long-duration missions, would also be enriched through a more comprehensive use of **systems analysis.** A special issue on **human factors** in *Earth Space*

Review (January–March, 1995) confirmed these observations.[11] In writing there on "Safety Cultures and the Importance of Human Factors," Dr John K. Lauber highlighted **ergonomics** as that aspect of human factors concerned about our adaptations of machines to the abilities and capabilities of people. He described current efforts at **crew resource management** with its focus upon team-oriented approaches in aerospace systems design ... . In the next article on "Cultural Issues and Safety," Capt. Daniel Maurino of the International Civil Aviation Organization admitted that only recently has ICAO examined cultural issues related to accidents and safety management. He wisely observed that aviation technology is designed—culturally speaking—within a narrow segment of international industry, while it is used worldwide without any consideration of cultural differences. Maurino called for research that can help the ICAO develop strategic solutions to cross-cultural safety issues related to high technology aerospace endeavors ... . In the next piece of that *ESR* issue, Paul Sherman and Earl Weiner made a case for studying the interplay between culture and person–machine interaction because aerospace ventures are increasingly multinational in scope ... . Still another article in the same pages on "Implementing Human Factors Training for Space Crews" by a team of psychologists from the German Aerospace Research Establishment (DLR), decried that human factor problems of manned space flight until recently focused on biomedical issues, especially related to weightlessness. Manzey, Horman, Fassbender, and Schiewe maintain that the demands of today and tomorrow's space missions necessitate expanding human factor consideration to include behavioral and psychological issues, particularly to address greater crew heterogeneity and the rising risk of psychosocial problems arising from differences in crew roles, tasks, ethnicity, as well as personality attitudes, traits, and values.

All this input underscores why I proposed twelve years ago in a NASA Summer Study at the California Space Institute that a new paradigm is necessary for human factors in space developments.[12] Exhibit 15 previously described this broader approach that goes beyond technical systems, so dear to the engineering subculture. This model in the form of a lunar habitat proposes ten other dimensions to consider and defines each—physical, psychological, sociological, cultural, financial, political, legal, education, management, and communication. Each component would be important to the development of a spacefaring civilization. For example, under **legal** (#7), a body of space and astro law has been developing with the leadership of the International Institute of Space Law. Now, United Societies in Space, Inc., as explained in Chapters 7 and 10, proposes that a space **Metanation** be founded under the UN Trusteeship Council for New Territories which could act on behalf of the "common heritage of mankind" and administer a series of space authorities. The first would be a Lunar Economic Development Authority with the power to issue bonds that might finance lunar industrialization and settlement (category #5, financial), so as to facilitate private investment.[13]

The dozens of inhabitants on orbital platforms, such as **Skylab, Spacelab, Salyut 7/Mir** are precursors of a future human type called **spacekind.** Whether as astronauts or cosmonauts, these **envoys of mankind**, a United Nations designation, learn how to cope with **free fall culture**. These short-term missions—usually up to a year or less—have been mostly enjoyable though high-risk experiences. As far as we know only one space mission to date has been aborted by illness when a Russian developed the first psychological case of "**space culture shock**" (Soyuz T13, 1995). When Vladimir Vasyutin was reported listless, fatigued, uninterested in his work and spent long hours gazing out the window, ground control ordered the mission be terminated and evacuated the station; once back home, the cosmonaut recovered quickly. There will be many more who will suffer such "culture shock" aloft, a disorientation within an alien space environment. While 20th century spacefarers can be counted in the hundreds, the New Millennium will see that number expand to thousands and thousands, as bases are built on planets and asteroids or on orbiting colonies.[14]

With permanent extension of human presence beyond Earth, a true **space culture** will emerge, quite distinct from its terrestrial analogs. The high frontier can become a living laboratory to promote peaceful, synergistic societies. Imagine space communities which promote cultural norms that support cooperation instead of competition, group development over excessive individualism, mutual help in place of aggressive behavior. Such a space culture has a better chance for survival and development aloft, in contrast to the

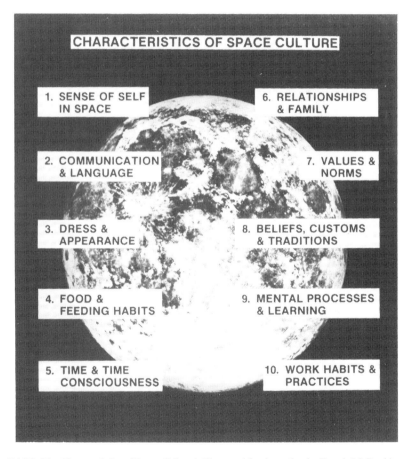

CHARACTERISTICS OF SPACE CULTURE

1. SENSE OF SELF IN SPACE
2. COMMUNICATION & LANGUAGE
3. DRESS & APPEARANCE
4. FOOD & FEEDING HABITS
5. TIME & TIME CONSCIOUSNESS
6. RELATIONSHIPS & FAMILY
7. VALUES & NORMS
8. BELIEFS, CUSTOMS & TRADITIONS
9. MENTAL PROCESSES & LEARNING
10. WORK HABITS & PRACTICES

Exhibit 36—Characteristics of Space Culture*. Illustrated for the author by Dennis M. Davidson, with a photograph of the Moon as a background. *Source*: Harris, P. R. and Moran, R. T. *Managing Cultural Differences*. Houston, TX: Gulf Publishing, 1996, 4th Edition

1584 "lost colony" of our English forebears at Fort Raleigh, Virginia! The hostile orbital environment requires a collaborative, regulated society to ensure the safety of the commonwealth—individuals will be regulated for safety's sake and that of their community ... . Trends all point to the first real space settlement which is likely to take place on the Moon within the next 25 years. The lunar environment will cause multiple adaptations of our Earth-bound culture. Remember, it lacks atmosphere and "weather" familiar to **earthkind;** various kinds of cosmic and solar radiation require protective cover—the Moon exposes future **Selenians** or lunar dwellers to 25–50 rems a year and other risks. Despite the recent **Clementine** unmanned probe which mapped the Moon in great detail, there is still much about it we do not yet know—such as its south pole and crater formation, the extent of available water and other natural resources. The rationale for returning to stay on the Moon ranges from scientific to industrialization.[15] (Refer to Chapter 10 and Appendix B.)

Exhibit 36 illustrates ten characteristics of lunar culture—yes, that is the Moon in the background. Let us briefly review these classifications which I have used to analyze **macrocultures**, such as a nation, or **microcultures**, such as a national space agency:

(1) **Sense of self and space** to be experienced. Culture contributes to one's sense of self-identity and acceptance, providing a sense of life space or comfort within one's group. Lunar culture offers an environmental reality of no atmosphere and one-sixth gravity, with a large sense of space and vast view of the home planet and the universe. This has been described as the "overview effect" with a perspective of a single ecosystem.[16]

We can only speculate as to whether a closed or open society will emerge on the Moon, but personally I recommend the latter—an open, friendly, informal, and supportive community devoted to the common welfare. We can also expect **re-entry** problems if long-term settlers on the lunar frontier wish to return to the home planet.

(2) **Communication and language** to be utilized. Culture groups develop verbal and non-verbal communication systems. Lunar settlers will utilize computers, satellites, and other information technologies extensively in their communications with ground "control" and families, as well as on the lunar surface. In the beginning, multiple languages may be used, principally English and Russian, but in time terminology will change and expand because of the environmental circumstances (e.g., construction, lunar "laws", new acronyms, and concepts). And when these spacefarers go to Mars, they will innovate with communication at or beyond line-of-sight distances, modulation frequencies and time message delays, transmitter power, data compression, and network development.[17]

(3) **Dress and appearance** to survive and function. Culture has always been expressed in outward garments, adornments and decorations. Astronauts and cosmonauts use spacesuits with life support systems for EVAs and wear mission patches, but their design today emphasizes using materials that offer more flexibility and lower costs. Lunar dwellers will live in habitats and wear clothing with helmets when they go outside which protects from radiation and provides life support—all this on a sustained basis will alter human culture (See Exhibit 37.)

Exhibit 37—Dress and Appearance of Humans and Robots Aloft. Offworld planetary exploration and exploitation requires that spacefarers wear protective suiting with life support systems, both on their backs or in the robotic vehicles which assist them. *Source*: Artistic rendering, NASA, Johnson Space Center.

(4) **Food and feeding habits** to be introduced. Food preparation, diet and eating procedures of groups of people set them apart from one another. Space agencies have pioneered in food technologies and compositions, altering earthkind's eating intake as a result. Transportation costs into orbit necessitate advances in hydroponic farming, organic gardens and seed development which will permit lunar dwellers to sustain themselves on lunar-grown foodstuffs within closed biological systems. Since microgravity affects taste buds, anticipates new foods, packaging, and preservation, as well as changes in diet, food preparation, and the manner of dining.

(5) **Time and time consciousness** to be adopted. Culture influences both one's sense and one's method of tracking time. Since lunar pioneers will be in touch with "ground control," in the beginning it is probable that a 24-hour time system will be adopted, but will it be set to Greenwich mean or Moscow time? Since the Moon has 14 days of light followed by 14 days of darkness, that may influence future concepts of "day, night, and seasons." The lunar environment will affect circadian rhythm, as well as work schedules and sleep patterns. Time schedules affect performance and productivity, so **Selenians** may learn to slow their biological clocks and adapt their biorhythms, creating in the process a new time sense—one likely to be less exact and more relative. Leisure time in lunar conditions offers unprecedented opportunities to develop human potential.

(6) **Relations and families** to be maintained and established. Cultures fix human and organizational relationships by age, sex, status, degree of kindred, as well as by wealth, power and wisdom. The first lunar dwellers are likely to be knowledge work-

ers or **technauts** with expertise in a variety of technologies and disciplines, so professional relationships will be important. Predeparture training is likely to emphasize team relations and group dynamics, so as to encourage teamwork and camaraderie aloft. Initially, husband–wife couples may be encouraged, but long-term lunar dwellers are likely to alter their understanding of marriage, partnerships, and families. At the start, support services from Earth will likely facilitate linkages and communication exchanges with families and friends on the home planet. But as sexual relations and lunar families develop in time, new generations of **spacekind** will differ significantly from **earthkind**, and interdependent relationships hopefully will prosper.

(7) **Values and norms** to be cultivated. Based on their needs and philosophy of life, cultural groups have always set unique priorities and standards. On the Moon, the need system will be focused on survival and development until a lunar infrastructure is in place. In such a high risk and high performing environment where on the Moon life support is paramount, competence may become the new norm, regardless of gender, race, color, creed, sexual preferences, or national origins. Thus, spacefarers there may value their fellows and their expertise more than sponsors located remotely on the home planet. Lunar living may dictate a commitment to shared values. These may include minimizing violence; physical/psychological wellbeing and safety of settlers; guaranteed level of social and political justice; maintenance and improvement of ecological quality and environment. New consensus will emerge on synergy, harmony, and nurturing within an isolated community, plus we may expect redefinitions of ethics, astrolaw, and governance principles.

(8) **Beliefs, customs, traditions** to be fostered. Culture manifests people's attitudes and outlook on life, motivating behavior influenced by spiritual themes, philosophies, and convictions. With a lunar population that is likely to be international in composition, a multicultural society will develop with diverse belief systems and a new sense of "cosmic consciousness." The view from the Moon will have a transforming effect that questions earthly customs and traditions, substituting mindsets and practices more appropriate to the new order and reality.

(9) **Mental processes and learning** to be cultivated. The way people think and learn varies by culture because of different emphasis on brain development and education. Space culture may cause humanity to focus on **whole**, not split, brain development.[18] With a well educated population at the start, lunar colonists will utilize a wide range of communication technologies to further information sharing and knowledge development. A lunar university and educational system may draw upon terrestrial data transmission and encourage self and small group learning through media, such as computer-assisted learning, teleconferencing and electronic networking. Cross-training in multiple disciplines may be common, with heavy emphasis upon use of simulations, virtual reality, and artificial intelligence. (See exhibit 38.)

(10) **Work habits and processes** to be established. One way of analyzing a culture is to examine how a society produces its goods and services, and conducts its economic affairs. On the space frontier, the work culture will be **metaindustrial**, featuring ex-

Exhibit 38—Education on the Moon. Whimsical artistic scenario of the first educational facility on the Moon attended by **Selenians**, lunar dwellers. Realistically, the building next century will probably be underground, covered by lunar regolith. The back-packs hold life support systems, not books! *Source*: NASA, Johnson Space Center.

tensive use of high/information technology, as well as robotics and automation.[19] Lunar vocational activities are likely to include habitat construction, human/material transportation, regolith mining, materials processing, helium-3 production, power beaming for solar energy (see Appendix B), and whole new industries, in addition to scientific research and astronomical observations. New career and work processes will emerge: operating space bubble machines and hemispherical concentrators; manufacturing from local resources thin metal mirrors for dish antennas and solar collectors. The principal ally to these multiskilled "ET" workers will be the "tin collar laborers" or robots. Of course, we can expect the tourism industry and its allied requirements for services to promote innovative occupations. (See Exhibit 39.)

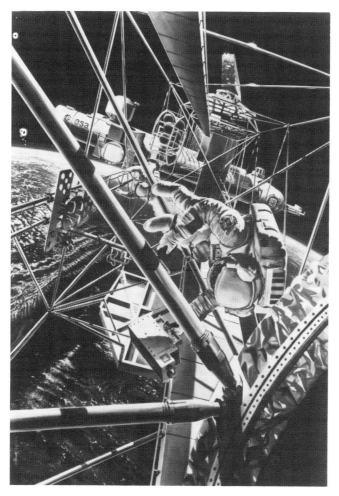

Exhibit 39—Orbital Work Habits and Practices. Artistic depiction of multinational crews building the **International Space Station** with the help of their robot assistants. *Source*: NASA/TAD
Corps.

Because culture is multifaceted and pervasive, other categories could be added to these
ten classifications. Nor can one form of **earthkind's** culture be imposed upon **spacekind,**
who will create their own adaptations. Krafft Ehricke, the late visionary rocket scientist,
left us a blueprint for a lunar city which he called **Selenopolis** after the Greek moon god-
dess of mythology. He viewed lunar settlement and industrialization as a process of so-
ciopsychological development and anthropological divergence which would cause humanity to transcend global confines to create a unique biosphere featuring new life styles
and social structures. This, Ehricke, concluded would lead to a polyglobal civilization
with no limits to growth![20]

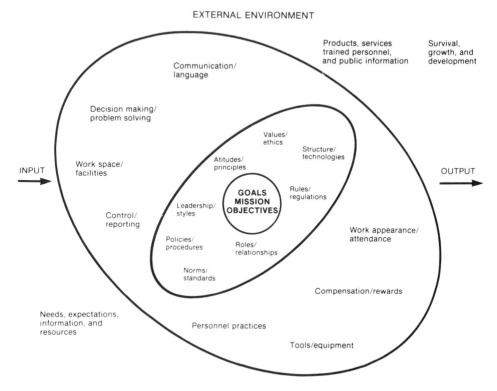

Exhibit 40—Aerospace Organizational Culture*. Whether in space or on the ground, aerospace institutions and teams develop a unique culture. Some of its dimensions have been illustrated by Dennis M. Davidson for the author. *Source*: Harris, P. R. *Living and Working in Space*. Chichester: Ellis Horwood, 1992, p. 143.

4.3 CULTURAL INFLUENCES ON AEROSPACE ORGANIZATIONS

Culture has already impacted our future in space. One dimension is the organizational cultures within the principal proponents of space technology and travel. These systems or institutions influence what goals to set, what missions to pursue, what launch systems and spacecraft to use, and what people to send, as well as how they should be trained and supported. Too often, their space scientists and engineers have ignored cultural factors while designing space vehicles and systems. The schemata outlined in Exhibit 40 may be applied to the analysis of organizational cultures like those of national space agencies, such as NASA, CSA, ESA, RKA, and NASDA. The same dimensions illustrated above also apply to aerospace contractors and private space enterprises, such as Lockheed Martin in the U.S.A., Innovation Enterprise **Lunokhod** in Russia, CNES/Arianespace in France, Alenis Spazio of Rome, or Spar Aerospace of Toronto. Such institutional cultures contribute to setting objectives, missions, roles, expectations, obligations, and boundaries which affect personnel morale and performance. These pervasive **microcultures** influence decisions, planning, operations, research and development, as well as all activities going into space transportation systems and missions. The space agencies, like their corporate partners, have their unique sets of values, myths, heroes, rites, rituals, and communication systems. The National Aeronautics and Space Administration, as a case in point, has its own space legends, beliefs, symbols, customs and practices which are embedded deeply in the system. In a number of contemporary management books, the research reported supports the conclusion that excellent organizations have strong functional cultures, constantly subject to renewal. Thus, it would be wise to heed the counsel of management consultants, Peters and Waterman, when they said:

> **In the very institutions in which culture is so dominant, the highest levels of true autonomy occur. The culture regulates vigorously the few variables that do count, and provide meaning. But within those qualitative values (and in almost all other dimensions), people are encouraged to stick out, to innovate.**[21]

In the early decades of the Space Age, many public agencies and private corporations in the aerospace field were quite innovative, but over time have become overly bureaucratic. In the public sector, the latter have found some of their budgets cut and functions stripped away, sometimes given to other government agencies; in the private sector, the industry has experienced downsizing and restructuring, with more of the innovation coming from start-up entrepreneurial companies. Today most space enterprises, both public and private, are involved in the process of globalization, which means more international cooperation in joint ventures and agreements. So organizational cultures and systems are becoming more transnational, each influenced by their counterparts in other countries.

As a demonstration case, consider the increase in multinational agreements among global space administrations, particularly with reference to the building of the **International Space Station**, assuming it actually gets into orbit by the end of this century. Such an undertaking means long-term missions, something the Russians have much experience in because of the **Mir** station now aloft for more than a decade. Thus, the Russian Space Agency has new agreements with NASA and ESA to exchange personnel and conduct joint projects. But again, the organizational cultures of the partners influence the outcome of such endeavors. Such multicultural interactions among the sponsors of this macroproject are bound to influence and change the organizational culture of each participant.

Take just one issue—the use of behavioral scientists in planning and participating in these undertakings. As indicated in the previous chapter, NASA's organizational culture previously downplayed the role of psychiatrists, psychologists, sociologists, and anthropologists, and focused on physical issues with the help of medical personnel and biologists. Now that "hard science/engineering" subculture's influence is diminishing as that U.S. agency places more emphasis on life science and biospheric research. NASA's culture is further impacted by the outlook and interactions with its new Russian partners. Vadim I. Gushin, a research psychologist at Moscow's Institute for Biomedical Problems, observes:

> **The Russian system of spaceflight psychology was created to give cosmonauts psychological support based on observed data of crew members'**

psychological state and work capability ... . Information about the cosmonauts' interests, tastes, family, and friends forms the basis for psychological support measures in space ... . There are also psychological experiments using questionnaires and special devices for operators' work capability test on board.[22]

But Gushin does admit that cosmonauts, like their astronaut counterparts in the West, try to avoid inflight psychological investigations and that the data collected are largely anecdotal; he advocates making crew members allies of behavioral researchers by training them through pre-flight simulations in astute observations of symptoms among their colleagues which might undermine safety, health, performance, or mission success. (see Exhibit 41.)

My point is that NASA's culture cannot help but be influenced by such approaches when the Shuttle flies to **Mir,** or the two agencies join with other international partners, in joint space ventures. As space organizations go global, they become less provincial and myopic, more cosmopolitan and ready to learn from their "foreign" allies.[23]

Exhibit 41—Learning from Russian Spaceflight Experience. With a space station in orbit for a decade, we have much to learn from Russian long-duration missions. Here, on **Mir**-18 in July 1995, astronaut researcher Norman Thagard watches an IMAX camera float to cosmonaut Gennadiy Strekalov. *Source*: NASA/Johnson Space Center.

4.4 CONCLUSIONS ON EMERGING SPACE CULTURE

Culture is a powerful concept and influence upon the future of space programs and development of the high frontier. Those engaged in human enterprises aloft would benefit from social science knowledge and research on the subject.[24] Specifically, behavioral scientists, especially cultural anthropologists and organization development specialists, can make contributions to the global space program by:

(1) Planning for the preliminary cultures of space settlements, beginning with a lunar base and industrial park ... .
(2) Consultation with aerospace agencies and corporations on renewing their organizational cultures so as to facilitate space industrialization and colonization through inter-institutional cooperation ... .
(3) Assisting aerospace designers and engineers to understand how (a) cultural differences between designers and users, particularly when the latter are "foreigners," may undermine safe operations of air/spacecraft, (b) complex automated systems reflect cultural assumptions that may lead to conflict, misunderstandings, and mismatch between the systems' and users' cultures which may increase risk and even lead to tragic failures.[25]

To this writer, it seems unlikely that global space agencies will provide the initiative and innovation to promote space settlement and industrialization. The author agrees with Professor Haym Benaroya of Rutgers University, previously cited, who concludes that space and planetary science concerned about the accumulation of knowledge and understanding must be decoupled from space engineering and industrialization which have more specific, immediate goals and realistic schedules; they are two different, yet parallel cultures. Furthermore, catalysts for lunar activity will likely come from world private enterprise, which appreciates the return on investment that can be realized by utilization of space resources, such as space-based energy.

Finally, perhaps the sheer challenge and complexity of space migration may turn the human race and cultures away from inward, self-destructive tendencies toward outward, self- actualization. **Human emergence** may truly occur when we leave our cradle Earth, to settle our Solar System![26] The "creeping" began a few hundred kilometers upward into lower Earth orbit (LEO); the "walking" begins when our species regularly, safely, and economically extends its personal presence 36 000 kilometers above into geosynchronous orbit (GEO). By practicing **cultural synergy**, we earthlings may mature through the formation of space communities that permit us to step into the universe and a new state of being!

The overall goal for the study of human behavior in space is the development of empirically based scientific principles that can identify the environmental, individual, group, and organizational requirements for long-term occupancy of space by humans.

(Committee on Space Biology and Medicine. A Strategy for Space Biology and Medical Science Washington, DC: National Academy Press, 1987. p. 169.)

5

High Performance among Spacefarers

Human performance in space, it is worth repeating, continues to be accomplished in two ways. The first is **unmanned** by extending ourselves out in the universe through automated spacecraft.[1] The premier achievement to date is undoubtedly the grand planetary tour of the two **Voyager** spacecraft. Now leaving the Solar System at over 50 000 kilometers per hour, these space vehicles have travelled more than 8 billion kilometers away from our earthly home where no member of our species has yet gone. The odyssey began almost thirty years ago, in 1977, when NASA's Jet Propulsion scientists launched the most sophisticated robot spacecraft ever built—each weighing nearly one tonne at launch with 65 000 parts. Of the two spacecraft in the $865 million joint mission, **Voyager 1** flew past Jupiter—and Saturn, making a close pass of Saturn's largest moon, Titan, and was allowed to end its planetary sojourn by zipping out to the stars. **Voyager 2** continued outward using the gravitational field of Saturn to fling it on to an encounter with Uranus. As a result of thousands of images sent back by **Voyagers 1/2** from throughout the Solar System, 20 new moons of the planets were discovered. The project scientists under the leadership of chief scientist, Edward Stone, vice president of Caltech, have been virtuosos of their complex computers with circuitry equivalent to 2000 color television sets. Through this unique man–machine interface, their explorations expanded eventually to Neptune and Triton, in 1989, from which radio transmissions travelling at 300 000 kilometers per second reached Pasadena four hours and six minutes later!

Such marvelous man–mechanical performance feats are being continually improved upon. As Exhibit 42 illustrates, the journey begins in Earth orbit.

Another milestone was the European Space Agency's **Ulysses** launched in 1990 from the Shuttle **Discovery** and monitored by NASA ground-tracking facilities in Madrid, Spain, and Goldstone, California. Following an encounter with Jupiter, this unmanned spacecraft was catapulted into an orbit from which it could survey the Sun's high-latitude regions and poles. It completed its northern polar pass in the summer of 1995, having made its first, southern polar pass a year earlier. Now it gradually descends in solar latitude, traveling out to Jupiter's orbit by 1998; then it will head back toward the Sun on a high-latitude trajectory, returning to the south polar regions in 2000 and another flight over the north pole in 2001. What a tribute to the human mind which conceived and executes this flawless mission! ... . Beginning in November 1995, the Euro-

Exhibit 42—**Eureca** Deployed from Space Shuttle. The **European Retrievable Carrier** is an automated, unmanned platform designed to carry instruments and experiments for life sciences, crystal growth, and astronomy from 6 to 18 months in orbit. Retrieved after its first mission in June 1993, it is reloaded at its control centre in Darmstadt, Germany, and then redeployed from the Shuttle Orbiter. Data collected are shared by scientists all over Europe. *Source:* European Space Agency.

pean Space Agency will launch an armada of unmanned scientific spacecraft to study the Sun further, our local space environment, and the far reaches of the universe. Both European and American scientists will use two observatories called ISO and SOHO to examine the inner workings of stars and galaxies and of our own Sun ... . Next a study team at the Massachusetts Institute of Technology proposes a new strategy that could represent a profound change in the space program's culture, according to Steve Bailey, manager of the Johnson Space Center's team studying Mars robotic missions. At a 1991 meeting in La-Jolla, California, America's leading space scientists agreed to *forgo* plans for one of two large Martian probes, controlled from Earth, to sample the planet's environment. Instead, they concurred on the approach suggested by Rodney Brooks and colleagues from the MIT mobile robotics laboratory. That is, 20 small Martian probes launched from Earth in batches of five, beginning in 1999. Some would contain small-scale surface rovers—mobile robots able to operate independently of direct control, not bound by NASA traditional criteria of 99%-plus reliability. The multiple simpler systems will operate at significantly lower cost, yet the robots will be capable of increasingly sophisticated behavior. By adding one capacity at a time, the scientists hope to evolve the activities of these roving robots as they learn from coping with the rigors of spaceflight and hostile environments. Later, Exhibit 51 on p. 138 illustrates how prototype unmanned robots might survey the lunar surface.

But our concern in this chapter is for that *other* form of human performance in orbit—"manned" missions, past, present, and future. As James E. Davidson of Microsat Launch Systems reminds us, a frontier is an imaginary boundary surrounding human activity. Davidson in a commentary on "More Than Missions" suggest that space exploration offers an opportunity to push beyond that boundary 600 km up on the space frontier by:

- **providing people on the ground and aloft with education and incentives to invest and create wealth through innovation and labor in building space infrastructure and homesteading;**
- **establishing outposts and permanent settlements on the Moon and Mars, so that their pioneering inhabitants are capable of innovative research and industrialization.**

("More Than Missions," Final Frontier, Jan/Feb. 1991, p. 6)

5.1 EXTENDING HUMAN PRESENCE ALOFT

The first human being to safely leave our planet provided a legacy of high performance in space on April 12, 1961. Yuri Gagarin was a 27-year-old Russian, a major in the Soviet Army, who flew a five-tonne space capsule, **Vostok 1**, for one hour and forty-eight minutes aloft. After liftoff, these were the memorable words of this top-performing cosmonaut:

Everything is working perfectly ... . I feel fine. I am in a cheerful mood.
C. Walker, *Final Frontier*, Mar/Apr. 1991, p. 8.

This technological feat of launch and return from orbit was not only significant for humanity, but it galvanized the Americans to meet the challenge with a series of manned missions called Mercury, Gemini and Apollo which culminated in the Moon landings of a dozen humans! (See Exhibit 43.)

Because of humankind's accomplishments in space for the past forty years, our species is posed now for dramatic growth in the 21st century. A gigantic leap in the development of human potential is likely to occur among spacefarers the next Millennium! For those who live and work in space, the zero or low gravity environment only offers both problems and perhaps opportunities for peak performance comparable to the first lunar landings. The challenges inherent in long-term space travel and the construction of habitats on the high ground will necessitate removal of earth-based physical and psychological blinders and binders. Despite numerous hazards of operating on that frontier, the work may extend individual capabilities and ingenuity. For planners of space transportation and living systems, the demand will be to move beyond technical and economic considerations so as to ensure more than survivability. To take full advantage of our humanness in such circumstances means that organizational sponsors, be they space agencies or private enterprise, must create a situation that ensures crew safety, productivity and quality of life through exceptional team performance, and human/machine interface. True space settlement means confronting a variety of psychosocial issues in space that range from preparation and training, to scheduling and "re-creation," to space ergonomics and ecology. The next hundred or thousand people in space will continue to be highly selected and probably top

Exhibit 43—Peak Performance: Humans on the Moon! **Apollo 17** EVA on the lunar surface by first scientist there, Astronaut Harrison "Jack" Schmitt. Dr Schmitt, a civilian geologist with a Ph.D. from Harvard University is pictured collecting rock samples at the Taurus-Littrow landing site. The photograph taken by mission commander, Eugene Cernan, shows the Lunar Module **Challenger** and the lunar rover. The last of twelve humans to visit there left a plaque with this inscription: **Here man completed his first exploration of the Moon, 1972 AD. May the spirit of peace in which we came be reflected in the lives of all mankind.** When will we return permanently? *Source*: NASA Headquarters.

performers in many ways. The prospects for extraterrestrial achievement and accomplishment are enormous.

Deaths in orbit of the three crew members of **Soyuz 11**, as well as the **Challenger** Seven tragedy not only convinced us that the space program is a high-risk activity, but that it demands high-quality performance by all involved in building the space infrastructure and making it suitable for human occupants. Further confirmation of this has been evident since the return to flight of the Shuttle **Discovery** in 1988 and the subsequent performance of the shuttle fleet with renewed emphasis upon safety. Indeed the international record of 3569 successful launches into orbit from 1957 to 1993 is a tribute to extraordinary human and technical performance. (See Exhibit 44.)

WORLDWIDE RECORD OF SUCCESSFUL SPACE LAUNCHES

Year	Russia/ CIS	U.S.A.	China	Japan	ESA	India	Israel	France*	Australia*	U.K.*	Total
1957	2										2
1958	1	5									6
1959	3	10									13
1960	3	16									19
1961	6	29									35
1962	20	52									72
1963	17	38									55
1964	30	57									87
1965	48	63						1			112
1966	44	73						1			118
1967	66	58						2	1		127
1968	74	45									119
1969	70	40									110
1970	81	29	1	1				2			114
1971	83	32	1	2				1		1	120
1972	74	31		1							106
1973	86	23									109
1974	81	24		1							106
1975	89	28	3	2				3			125
1976	99	26	2	1							128
1977	98	24		2							124
1978	88	32	1	3							124
1979	87	16		2	1						106
1980	89	13		2		1					105
1981	98	18	1	3	2	1					123
1982	101	18	1	1							121
1983	98	22	1	3	2	1					127
1984	97	22	3	3	4						129
1985	97	17	1	2	3						120
1986	91	6	2	2	2						103
1987	95	8	2	3	2						110
1988	90	12	4	2	7		1				116
1989	74	18		2	7						101
1990	75	27	5	3	5		1				116
1991	59	18	1	2	8						88
1992	54	28	3	1	7	1					94
1993	47	23	1	1	7						79
Total	**2415**	**1001**	**33**	**45**	**57**	**4**	**2**	**10**	**1**	**1**	**3569**

*   *Note that France, Australia, and the U.K. do not launch their own satellites. France and the U.K. are members of ESA, which built the Ariane, whose first launch was in 1979.*

Italy conducted eight launches from 1967 to 1975, and one in 1988, using U.S. Scout rockets from the San Marco platform off the coast of Kenya. Since the launch vehicles were American, NASA includes them in U.S. launch statistics and they are treated the same way here.

Exhibit 44—Worldwide Record of Successful Space Launches. Despite some failures, high performance is manifested in the 3569 successful launches from 1957 to 1993. *Source*: U.S. Congressional Research Service.

During the next 50 years, those persons who actually get to work in lower or geosynchronous orbit will be engaged principally in research and development, or in construction of facilities that will benefit those who will follow. Scientists, engineers, and the military are likely to dominate the space scene in the near term. Astronauts and cosmonauts who pilot spacecraft will be outnumbered by mission specialists who represent a wide variety of professional fields and disciplines, often under the sponsorship of nations, consortia, corporations, or even universities. The initial 21st century visitors from Earth will expand beyond professional astronauts and cosmonauts to include contract workers or "technauts" sent to build or service space stations, as well as eventually Moon and Martian bases.

Beyond the research and development phase, we can then expect occasional "space citizens" to encompass a wider range of the global community, whether politician, teacher, journalist, artist, or even tourist to LEO.

Expanded human presence in space will take place in two general ways. First, as indicated in our introduction, by those who remain physically on the home planet, but extend themselves by unmanned spacecraft deeply into the Solar System; or through those others on the ground who provide the technology and support services for the envoys of humankind aloft. Second, by those who are actually launched beyond the forces of the Earth's gravity for longer and longer duration flights. The complexity and scope of space activity requires in either circumstance people of high competency, both on Earth and in orbit. (Ironically at the very time that NASA is having difficulty recruiting the best engineering graduates from American universities, in the mid '90s, the Russians have serious difficulties in financing and employing their large workforce of scientific and engineering elite.) Furthermore, for progress to be made in recruiting skilled space professionals in the immediate decades ahead, considerable alterations must occur in space planning and management of human resources in the emerging work culture.[2]

More extended and permanent human occupation of space will bring role change from space operations of the past—the Space Station or Moonbase will function with less ground control and more autonomy for crewmembers. The extensive use of automation and robotics will enhance human capabilities, such as through teleoperations. Advances in communication technology for and from space will produce new capabilities resulting in expanded information and insights, while innovative expert systems or applications of artificial intelligence will enhance performance. Space industrialization and defense systems will engender novel technologies and occupational fields. Gradually, the numbers of people living, working, and traveling in space will escalate from the hundreds, to thousands, to millions until literally a mass migration occurs. In the process, humans themselves will be changed, especially the way they behave and perform.

The growing multinationalism and multiculturalism of spacefarers is evident in the European Space Agency pictures reproduced in Exhibit 45. The ESA astronaut corps came into being with the 1978 **Spacelab** that flew on the NASA Shuttle.

ESA astronauts fly on both American and Russian spacecraft, engaging in a variety of activities, from scientific experiments and manufacturing high-tech materials to deploying the **Eureca** platform and satellites. The longest manned mission in ESA history blasted off from the Baikonour Cosmodrome to the **Mir** space station while this second edition was being written. Called **EuroMir '95**, it began when ESA's astronaut, Thomas Reiter, lifted off with cosmonauts Sergei Avdeev and Yuri Gidzenko; during the 135 days aloft, Reiter undertook a five-hour spacewalk, and discussed *living and working in space* via live video link-up with the fourth Information Forum for Young Europeans (EURISY) meeting in Noordwijk, Holland. The U.S.A. Shuttle **Atlantis** also docked (for the second time) at the Russian station carrying four Americans and one Canadian. The mission landing by a **Soyuz** spacecraft occurred in Kazakhstan on January 16, 1996.

Perhaps it would be useful now to define the term **high performance** for its application to space work. obviously top performance in any job, whether on this planet or in orbit, is marked by accomplishment which is consistently above average. That is, the worker's behavior is deemed to be in the upper levels of productive effort, characterized by:

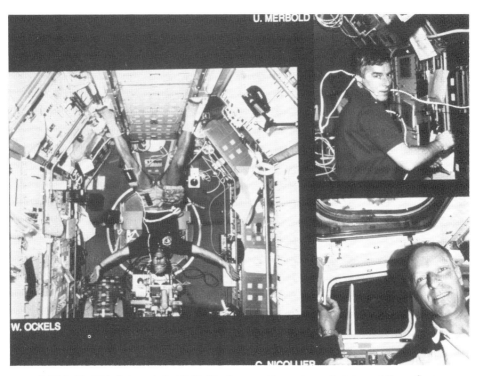

Exhibit 45—ESA Astronauts in Flight. The European Space Agency's Astronaut Corps fly on both American and Russian spacecraft, as pictured by the German, Ulf Merbold, with the most experience aloft. Dutchman Wubbo Ockels first flew on **Spacelab D-1** in 1985, while the Swiss Claude Nicollier helped to deploy from the Shuttle both the **Eureca** platform (1992) and the **Hubble Space Telescope** (1993). *Source*: ESA.

- high output in terms of quantity and quality;
- individual or team performance manifesting high levels of commitment, competence, and concern for completing tasks and achieving objectives.

Such behavior was very much in evidence among the thousands of workers involved in the Apollo mission series, as well as among their counterparts who have kept the **Mir** space station safely functioning for over a decade.

In a survey of the management literature on a high-performing work environment, the author identified what organizations and their leaders do to achieve such a state among their members.[3] Applying these findings, for example, to any space organization and its space missions, high performance would result *when*:

- **there is joint goal setting by both managers and workers, so that their targets and objectives are set beyond current levels to stretch human efforts without sacrificing safety requirements;**
- **standards of excellence are established which are based upon norms of competence, collaboration, and consensus rather than undue concern for minimum wages, hours and other benefits so prominent in disappearing bureaucracies of the industrial work culture;**

- an open communication system functions which espouses authentic interactions, feedback, and positive re-inforcement;
- management redirects work habits and activities from ineffective to effective means, helping employees to work smarter and learn from mistakes;
- the new work culture capitalizes upon human assets and potential by encouraging flexibility, responsibility, innovation, autonomy, and risk-taking among personnel, while maintaining accountability and being results-oriented.

This researcher found that many of the above qualities and practices were present in NASA and its aerospace contractors during the Apollo era of the 1960s, but diminished in the decades that followed (refer to Chapter 8). The Agency's current administrator, Dan Goldin, has yet another management reorganization under way to improve performance. Similarly, the new Russian Space Agency is attempting to replace the Soviet centralized planning and managed operations by becoming more market and people oriented.

Other strategies used by high-performing organizations can also be adapted by the space industry and agencies both on the ground and aloft:[4]

- recruiting, promoting, and rewarding to performers as policy, and then using them as a role model with ordinary employees;
- altering organizational structure and practice so that it is more de-centralized in terms of decision-making and responsibility, more mission-oriented and responsive, while reducing the levels of management;
- making work more meaningful and fun by cultivating informality, fellowship, and team effort within a context of productive achievement and joyful accomplishment;
- emphasizing human resource development, but trying a mix of benefits, rewards and incentives to talented performance.

Strategic planning, bold vision, future orientation, and high purpose capture the imagination of personnel, pushing them on to metaperformance.

Many of the above-proven approaches are directly applicable to ground-based operations that support space missions, and eventually can be adapted to space-based conditions. For example, in a NASA study on *The Human Role in Space*, Hall concluded that the human being represents a remarkable and adaptable system.[5] He recommended that systems designers develop improved workstations in space which enhance human productivity and value to the mission under way. Currently, human factor engineers are using biodynamic animated computer models to help design crew workstations for the future orbiting laboratory. One of my concerns as a psychologist is that industrial engineers in general, and NASA in particular, have been too narrow in their human factor perspective and focus relative to providing a stimulating work environment for increased performance.[6] Under the leadership of an expert in space psychology, Dr Albert A. Harrison, broader alternatives are being offered with reference to crew systems and support.[7] The immediate impetus for this comes from prospects of long-duration missions, as well as the design needs for an **International Space Station**, plus planned bases on the Moon and Mars. Professor Harrison, of the University of California—Davis, envisions several dimensions for enhancing human performance and settlement aloft:

(1) *protection* from acceleration, vibration, temperature extremes, near-vacuums, some poisonous atmospheres, radiation, and impact of meteorites and space debris ...
(2) *improving quality of life* in terms of support systems, equipment, supplies, habitats, amenities, communication and social relationships ...
(3) *accommodating diversity* among spacefarers in terms of differences in sponsor, role, gender, physiques, culture, national customs and language ...
(4) *inventing cultural adaptations* to the space environment to ensure survival and growth, as well as to meet emerging needs and tastes (such as creating new sociopolitical institutions and customs) ...
(5) *promoting self-sufficiency in orbit* because of the high transportation costs from the home planet; apart from better communication systems and technologies, this implies learning to use space resources and engaging in **terraforming** of asteroids and planets.

The matter of human performance has long been a centerpiece of behavioral science research—unfortunately, little of this has been centered on performance in space. Usually, psychologists have examined three interactive forces—the **person**, the **task**, and the **environment**. Positive changes in any of these factors normally result in greater productivity, while negative influences tend to decrease individual effort. The co-author of *Living Aloft*, Albert Harrison, has also proposed that this paradigm be employed with reference to performance in the Space Station.[8] This psychologist reminds us that the difficulty of tasks in orbit, the harshness of the environment, and the spacefarer's fitness impact performance. These three are dynamic, multifaceted, and multidimensional factors to be considered in all space work. Harrison rightly advocates that relative to space planning and management, consideration should be directed to these realities for improving human productivity in orbit:[9]

- **The space workers' characteristics which include** *competence* **(combination of ability, experience, and training);** *motivation* **(amount of energy and diligence that the person brings to a task);** *interest* **(degree of ego involvement and sense of participation experienced in the work, so that the task receives preference over other activities) ...**
- **The task' characteristics, whether on Earth on in space, include** *complexity* **(level of demands that the task places on the performer's perceptual, cognitive, and motor skills);** *duration* **(time required to complete the task); and** *repetition* **(relative frequency with which the task is performed over a period of time). Such variables combine to determine both satisfactory performance and motivation. Harrison observes that complex, challenging, and significant tasks that are consistent with personal values and interests are associated with high commitment and motivation.**
- **The environmental characteristics may be analyzed at several levels—primitive in terms of lines, forms, shapes, colors, textures, contours and patterns; higher or more synthetic levels which considers such factors as—***legibility* **(extent to which the environment provides an easily understood and coherent frame of reference);** *arousal potential* **(extent that environment increases or decreases drive or arousal**

of people's feelings and energies); preferability (extent to which environmental design elements are consistent or inconsistent with personal tastes and values).

Harrison suggests that the perceptual processing of the primitive levels of reaction leads to the emergent latter properties which, in turn, impact performance. This person/task/environment construct is useful for studying the performance of **Homo cosmonauticus** or spacefaring man, illustrated in Exhibit 46.

Exhibit 46—High Offworld Performance. In this depiction, a mission specialist conducts extra vehicular activity (EVA) for scientific experiments. *Source:* NASA/TAD Corps.

5.2 LEARNING FROM HUMAN OFFWORLD PERFORMANCE

During the past thirty-plus years of manned spaceflight, some insight has been gained about human performance in space.[10] Unfortunately, it has been largely about astronauts and cosmonauts on short-duration missions, and is of limited value for those who will live for much longer periods aloft. But we have learned that deprived of Earth's gravity for extended spaceflights, the body gradually acclimatizes to microgravity, and cardiovascular deconditioning sets in, human bones lose calcium, and a demineralization process begins, the "space adaptation syndrome" of nausea, disorientation or discomfort can be overcome by most spacefarers.[11]

The mission records for days aloft were first established by **Skylab** (3 astronauts in 1974 for 84 days and **Salyut 7** (3 cosmonauts in 1984 for 237 days). Since then the American experience in space has been limited to Shuttle flights, the longest being in 1995 when seven **Endeavor** astronauts spent 16½ stargazing days working two shifts with three ultraviolet telescopes to study some 600 galaxies, stars, planets, and moons. While aloft they chatted with their former crewmate, NASA's Dr Norman Thagard, the first American aboard Russia's **Mir**, where he spent 111 days on that orbiting platform ... . Currently, Russian cosmonauts hold the long-duration honors—of 100 manned space missions to 10 space stations culminating in **Mir**, the longest stay in orbit was Dr Valeri Polyakov's 14½ months as of June 29, 1995 when the Shuttle **Discovery** docked there. Such time spans offworld will constantly be extended to years and eventually decades.

Our extraterrestrial human experience to date proves that steps could be taken to improve crew efficiency, ranging from automation and environment control systems to exercising and scheduling alterations. In 1987, some of this accumulated information and insight benefited two Soviet cosmonauts on their **Soyuz TM2/3** missions—Alexander Alexandrov stayed in orbit for 160 days, and his companion lasted over 10 months! During his 326 days aloft, Yuri Romanenko grew one centimeter taller and lost 1.6 kg of body weight, but experienced fatigue, listlessness, and homesickness during his long flight.[12] In December 1988, cosmonauts Vladimir Titov and Musa Manarov spent over a year in space on the **Mir** station, returning on a **Soyuz TM-6** capsule with their four-week visitor, Frenchman Jean-Loup Chrétien. Radio Moscow reported that immediate medical checkups upon landing showed them feeling well, and subsequent physical evaluation of the two spacefarers at Star City revealed no serious difficulties after their prolonged weightlessness for 366 days. This was partially attributed to exercise twice a day by Manarov and Titov on both a treadmill and an exercise bicycle. They also had special spacesuits which forced blood to concentrate in the lower parts of their bodies, simulating the effect of gravity, so that they would not become accustomed to weightlessness. Despite these precautions against calcium loss in long spaceflights, they did shrink in size as their legs lost calcium, but managed to return to Earth in better shape by use of vitamin supplements. During their twelve months plus on the station, the then record-holders conducted extensive astrophysical studies, technological experiments, and medical tests, as well as studying sources of radiation and taking more than 12 000 photographs of the Earth's surface (*Los Angeles Times*, Dec. 22, 1988, p. 8). In January 1995, cosmonaut physician, Valery Polyakov, 51 years of age, blasted off for 438 days in space, circling the Earth over 7000 times in **Mir**.

Such accounts confirm **high performance in space** by both astronauts and cosmonauts, feats which will be continuously surpassed by spacefarers to follow!

An interview with NASA's director of life sciences, Arnauld Nicogossian, confirmed the paucity of knowledge concerning human physiology and performance in the weightless environment. Unless research is expanded on biomedical, psychological, and biospheric factors of spaceflight, long-duration manned missions may be jeopardized. Nicogossian made these pertinent observations:[13]

- **ground training in biogenic or autogenic feedback may precondition some people to spaceflight so they may master motion sickness aloft;**
- **after four days in space, the body adapts to weightlessness, yet we do not fully understand the influence of visual perception;**

- drugs have different effects aloft than on the ground, so use of that therapy for motion sickness or other ailments has to consider issues of metabolism in space;
- although the Russians have had missions as long as 237 days or more, NASA is presently confident operationally with scheduling for 90 days aloft in terms of providing adequate medical support;
- a health maintenance facility is proposed for inclusion on the International Space Station to deal with incapacitating illness or injury, as well as dental problems—all of which impact performance;
- ignorance exists on closed environments for long-term flights, especially relative to outgassed material and their contribution to headaches, irritability, drowsiness, and depression;
- meal times have been demonstrated as of psychological importance for on-orbit socialization;
- insufficient research has been done on matters of decor (color scheme, number of windows, etc.) and privacy influences on performance;
- inadequate data on human factors exist for a manned Mars mission.

This sobering assessment highlights what this writer has already reported in Chapters 2 and 3. Since that interview, NASA has begun to address some of these issues by expanding research in human factors, especially functional esthetics, as well as in life sciences, especially with reference to life support systems for the Space Station.[14] The prospect of building a Moonbase within twenty years has prompted some physicians such as Ron Schaefer to examine future medical care for lunar dwellers.[15] With proper selection and effective paramedic services, Dr Schaefer feels that explorers' illness and accidents can be coped with adequately for up to a six months' stay. Beyond that timespan, he is concerned with treatment of acute illness, severe injuries, and chronic conditions. Schaefer anticipates problems on the lunar surface related to bone demineralization, cardiovascular deconditioning, trauma, decompression sickness, and radiation poisoning. Obviously, the performance of a lunar worker would deteriorate with any such disabilities. Thus, ongoing orbital research on the physiological effects of microgravity, as illustrated in Exhibit 47, provides significant insights for future long-duration missions.

In preparation for living and working on the future space station and bases, it is wise to listen to those who have actually been in orbit. In the past, flight surgeon Douglas, for example, interviewed ten astronauts on their spaceflight performance.[16] Their comments confirm somewhat the validity of the above observations. Furthermore, they provided this McDonnell-Douglas researcher with these recommendations for operational changes:

(1) space flights have been constructed by NASA so that everything is dictated by checklists, which take away one's ability to think;
(2) a workday of fourteen or fifteen hours in space leads to inefficiency and mistakes;
(3) a space station design should take advantage of a real and unique environment, such as delta temperature and pressure;
(4) the suit used for extravehicular activity (EVA) should be better designed and have more provisions (e.g., from honey water to a toolkit);

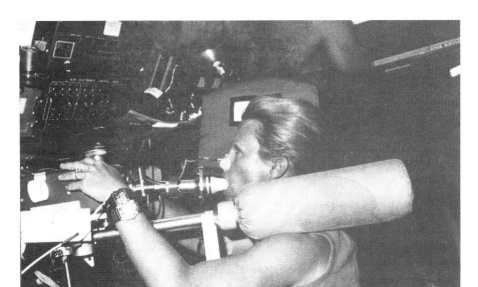

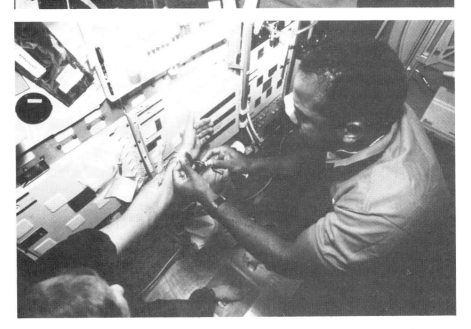

Exhibit 47—On-orbit Research of Microgravity Effects. Two actual on-board scenes are shown in these photographs of physiology experiments investigating microgravity conditions aboard the Shuttle Orbiter. For example, in one, astronaut-physician Dr Rhea Seddon, on the bicycle ergometer, breathes into the cardiovascular unit during an STS-40 spaceflight ... . In the latter part of 1993, STS-55 mission on **Columbia** during a ten-day **Spacelab** D-2 mission, Dr Bernard Harris, an African-American physician, draws blood from Hans Schlegel, a payload specialist from the German Aerospace Research Establishment (DLR). *Source*: NASA/Johnson Space Center.

(5) crew shifts should allow for group sleep, exercise, and play, as well as permit work occasionally on off-duty time;

(6) zero-*g* environment can be optimized by pre-departure training and conditioning, so that space sickness can be eliminated or limited.

This sampling of opinion from those who have "been there" included the spacefarer's conclusion that the qualities or competencies that make for high performance on Earth are also replicated in spaceflight:

> **If a person is exceptionally good as an observer or working out something analytically, or if he is good at doing any kind of detailed task, you can put him up there where is looking at the real thing. He can do just as good a job up there as down here. There is no reason for not using a person's intellect fully.**

Such a commentary underscores this writer's preference for more utilization of feedback from those who have worked in orbit in designing space deployment systems, as will be discussed in the next chapter. It also demonstrates the importance of the above model proposed by A. A. Harrison, which utilizes the dynamic interplaying forces of person, task, and environment as a means for maximizing performance whether on space platform or at a Moon/Mars base.

Hall is also concerned about the amount of stress which human operators in space can handle, and urges more research on "cognitive ergonomics" or mental workload assessment and its effects on man–machine systems when there is overload (see section 5.7).[17] Spacefarers will be extremely vulnerable to disruptive internal and external forces; stress may be reduced when crews can control and adapt technology to meet contingencies. Within the living systems of an orbiting station, for instance, the behavior of the individual can affect group performance, contributing to either mission success or mission failure. Since open or living systems methodology is more holistic, its advocates hypothesize that it can contribute to the control of stress in space habitats and enhance the human performance by diagnosing system pathologies which lead to inefficiencies and errors (refer to Chapter 3).

Although prior manned mission experience contributes to our understanding, there are also terrestrial analogs to draw upon, as will be discussed in Chapter 6. Other helpful information about human performance may be gathered from analyzing how workers fare at remote, alien, or foreign locations on this planet. Social scientists have sometimes described these as "exotic environments marked by severe climate, danger, limited facilities, isolation from family and friends, and enforced interaction with others." In a review of such research literature, Helmreich and associates discovered that a high level of performance under such circumstances may be dependent on the cohesiveness of the isolated and confined work group.[18] The investigations of Furnham and Bochner on psychological reactions to unfamiliar environments also provide performance data for those on extended sojourns and may help to prevent or lessen **space culture shock**.[19] Such developments explain, in part, why the National Science Foundation joined with NASA in sponsoring a 1987 conference on "The Human Experience in Antarctica; Applications to Life in Space."[20]

The best hope for the future of extraterrestrial performance would seem to be in multilateral research on the subject by various space organizations. For example, in the mid '90s, agreements have been worked out among the American, Canadian, Russian, European, and Japanese space agencies to share not only their facilities and equipment, but also their findings on humans aloft. Twenty years after the historic **Apollo 18** and **Soyuz 19** linkup, in June 1995 a linkup of the 37.2-meter-long **Atlantis** shuttle with the **Mir** station formed the largest spacecraft ever to orbit the Earth. The synergy lasted for five days as six American astronauts, including two women, physician Dr Ellen S. Baker and biomedical engineer Dr Bonnie Dunbar, exchanged with four Russian cosmonauts and conducted medical experiments together. As *The New York Times* (July 11, 1995) wrote:

> **Experts say the new East–West teamwork is an ideal remedy for deficiencies of each side, creating a synergy that will produce the first real understanding of man's destiny as a space traveler. It is also seen as a prerequisite for sending astronauts aloft for many months at a time aboard an International Space Station, which NASA and its Russian counterpart want to build from 1997 to 2002 and operate at least until 2012 ... .**
>
> **Dr Larry R. Young, a professor of astronautics at the Massachusetts Institute of Technology, who has also trained as a NASA payload specialist, said "With the combined program we hope to get the best of both worlds—Russian length with American strength ... We stand the chance of combining the successes of two national traditions. It's a turning point for the life sciences."**

Dr Norman Smith, a Van Nuys, Californian management and systems consultant, commented in a memorandum to this author (5/11/86) that the human must become the central focus of the space program. Spacefaring extends human capabilities, even when accomplished by our artifacts, such as remote sensing satellites. As the duration of space missions increases, surviving and living there become more complex. Because of the unknowns involved, Smith believes that manned offterrestrial systems have to be devised in such a way as to deal with changing conditions, emergencies, and opportunities as they are confronted aloft. He reminds us that humans are learning systems which operate simultaneously on the physical or biological, psychological and social levels. For a human to perform effectively in space over long periods, the total systems design must provide for crew members at all three levels. In examining with James Grier Miller and others how living systems theory may facilitate such space planning and management, Smith also cites the possible build-up of stress at any of these interacting levels during extended space missions. Performance impairment, he observes, may occur with stress caused by man–machine interface, group interactions, disparities between the social and technological systems.

Although human performance in preparation for and in space has generally been outstanding during the last four decades, we need to **learn from our failures and mistakes**. Some of this has been caused on the ground by inadequate or incompetent space workers, as well as fraudulent contractors. Some of the insufficiencies happened in orbit, as in the traumatic flights of **Apollo 13** and **Challenger**, though in the former case, the high-performing crew managed to return home safely. A recent media report about astronaut

Bonnie Dunbar for instance, indicated that she nearly died in space because of a sloppy medical experiment. This problem on shuttle **Discovery** was caused by deficient laboratory procedures and protocols that permitted Dunbar to inject herself with a drug that triggered a life-threatening allergic reaction. Misguided investigators got more interested in their experiment than in their human subject (*CBS News*, July 7, 1995). Similarly, one of the major findings from Dr Norman E. Thagard's 115-day orbital mission in 1995 was NASA's neglect of the psychological aspects of his preparation and support for isolation and confinement aloft ... . Sometimes it is a technological defect which is overcome by human performance. As ESA's Ulf Merbold commented, "It's amazing what humans can do when technology fails." He was recalling his sojourn on **Soyuz TM-20**, when the spacecraft's automatic docking mechanism would not work and there was only three minutes left to dock manually, which is exactly what woman cosmonaut Elena Kondakova did, driving in at high speed (refer to Exhibit 48a)!

Multidisciplinary studies of human performance in space are just beginning and open up vast research possibilities. Missions aloft will be more international both in scope and in composition of personnel. The next wave to create a space civilization will have more diverse backgrounds and expertize, so trainers, commanders, and ordinary spacefarers will require cross-cultural preparation. 21st century space settlement and industrialization will be followed by interplanetary exploration in the New Millennium. Author Alcestis Oberg envisions that these pioneering performers on the high frontier will become ancestors of more remarkable descendants to come, as different from us today as our generation is from the people of the Middle Ages!

5.3 CREW SYSTEM PRODUCTIVITY

With the globalization of space involvement, crew diversity will become routine while sponsors shift their research toward human performance aloft and enhancement of quality of life there. Dr Mary Connors, previously cited, has provided another helpful conceptualization in the term **crew system**, which she defines as the combination of human and technical subsystems. This NASA Ames behavioral scientist describes crew systems as wholes, entireties or totalities whose overall performance cannot be understood from looking at one subsystem alone. More recently, Harrison and Connors have been examining the phenomenological, structural and functional dimensions of crew systems, so as to obtain better fit among human, automation, and communications aloft.[21]

The next few hundred persons in space are likely to be crewmembers of a spacecraft or base under the immediate supervision of an American, Russian or European space agency. Although an expanded and mixed group of mission specialists, more varied in sex, race nationalities, and competencies than previously, their performance may set the mold for the thousands who will follow them into the Solar System by establishing the norms for long-duration space adaptability. Preliminary research indicates that a generalist is preferred, one who has multiple skills or is cross-trained in required fields of expertise.

Experience from the previous three decades of manned spaceflight has offered some clues as to what fosters productive behavior. In *The Human Role in Space* study (THURIS) by Stephen B. Hall cited above) the following insights are offered:

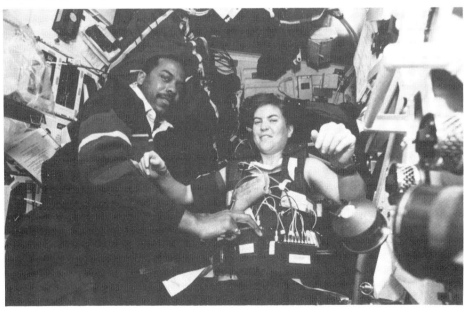

Exhibit 48—Crew Diversity and Ingenuity. Two more spaceflight scenes which emphasize the **globalization** of space missions. (Top) ESA astronaut Ulf Merbold is pictured between crew partners Talgat Musabayev and Elena Kondakova on-board **Mir** in November 1994 ... . (Bottom) In this **Discovery** photograph, payload commander and physician, Bernard Harris, with his test subject, astronaut Janice Voss, who wears a biomedical harness for the experiment to check her muscle responses to microgravity. These middeck **Spacelab** experiments in the Shuttle Orbiter's cargo bay also included James Weatherbee, commander; Eileen Collins, pilot; and Russian cosmonaut, Vladimir Titov. *Sources: Ad Astra*, June 1995 ... . NASA Johnson Space Center.

- hardware and systems improvements do enhance human performance;
- procedure and operational changes will allow for more effective use of the human element in man–machine systems;
- traffic flow through manned modules will be a consideration for improving human productivity;
- better designed and implemented storage and stowage abets productivity, permitting quicker retrieval and replacement;
- other architectural factors have been identified which can advance performance by changes in habitat module dimensions and surfaces, windows and partitions, configurations and spatial requirements, noise control and restraint systems, as well as improved anthropometrics (e.g., workstations, common areas, waste management, etc.).

These McDonnell-Douglas Corporation researchers under contract to NASA undertook analysis of past flights, discovering a number of ways to maximize crew productivity in future manned systems. Their findings ranged from improved personal hygiene equipment and provisions to better lighting, health maintenance, food and water systems, communications systems, and even housekeeping. The THURIS investigators made recommendations, based on such experiences as **Skylab** and **Soyuz**, for both pre- and in-flight training, in-flight maintenance, planning and scheduling activities, organizations and management. The ground support group, for instance, should be composed of mission control and payload operations personnel, as well as principal investigators of experiments on board the spacecraft, but they should function in such a manner as to monitor flight activities while allowing for greater crew autonomy. Ground communication and authority impact human productivity both on the ground and in space. *The Human Role in Space* analysis pointed up many systems improvements that enhance space performance—computer modeling, IVA (internal vehicular activity), IVA/EVA interfaces, remote systems management, among others.

Having examined both the U.S.A. and the Russian space programs to the mid '80s, the THURIS conclusions recommended more provisions for crews to deal with in-flight contingencies, in-orbit assembly, and *insitu* emergency management. This analysis confirms that the more crews are permitted to participate in on-board research, working together with principal investigators on the ground, the more they are energized and the success of mission objectives is enhanced. To make the most of the remarkable and adaptable human system in space, the THURSIS study (Volume II) offers criteria of performance, cost, and technological readiness, plus descriptions for thirty-seven generic activities identified so far for those who work in space.

As indicated above, in 1984 another McDonnell-Douglas researcher, William K. Douglas, interviewed about ten percent of those astronauts who have been in space, concerning performance issues which have implications for a manned space station.This expert sample offered valuable commentary, such as:

- *Space Station Management and Performance*—the commander there should have final authority and ground control should provide strategic planning and resources, such as data analysis—orbital veterans also counseled that communication from the ground should be informative not protective. They advised that while the station commander has the ultimate responsibility, that person will have to be a strong but flexible leader, able to deal with diversity, to listen for varied input, and to obtain consensus (a matrix-type organization was implied when they proposed that the commander delegate responsibilities to a chief scientist).

• *Improved Productivity Aloft*—suggestions from the astronauts were summarized in the last section, but ranged from equipment redesign to provisions for multipurpose toolkit, from work schedule routines to crew cohesion (assigning and rotating them as a unit). They warned that long-duration flights are a new experience and people will be unwilling to make the sacrifices and put up with the inconveniences present in past missions of short duration. These spacefarers observed that personality and human relations issues among the crew will have greater impact on group productivity; optimal stays in orbit have yet to be determined before fatigue, inefficiencies, and mistakes undermine performance.

Interestingly, at a 1991 NASA/NSF conference on isolation and confinement, Dr William Douglas raised additional issues of what could undermine performance aloft. He carefully examined reports and volumes on *Mutiny on the Bounty* (e.g., C. Nordoff and J. N. Hall, 1932), comparing them to interviews and writings of astronauts, especially in the books of H. S. F. Cooper, entitled *House in Space* (Bantam Books, 1978) and *The Flight that Failed* (Dial Press, 1973). In examining aberrant behavior caused by psychological and sociological aspects of spaceflight, he focuses on the 84-day mission of **Skylab** which led to a so-called near-revolt of its crew. Crewmembers are described by their own as lethargic, negative, irritable, complaining, bitching, grumbling, and given to bursts of anger. Some of this was caused by the burdensome schedule provided by ground control, climaxing in a request from mission control on the fourth Sunday that they had some additional tasks to perform. Having already worked the three previous Sundays, when the crew was meant to relax. Mission Commander Carr said that he had had enough. The first "mutiny" in space had occurred when Col. Carr responded "We better not work today. We better do our own thing and get some rest." Douglas concludes that NASA schedulers are not always aware of the physical and psychological changes that occur aloft, and the need to provide enough time for astronauts to gaze out the window, to enjoy the magnificence of the Earth and the heavens.

Performance might very well be enhanced on long-duration missions if there is less *control* from the ground. For crews of a Space Station or Moon/Mars base, it might be better to mutually set broad mission objectives and milestones before launch, and let the commander on site have the leadership responsibility for the accomplishment. The spacefarers would set their own schedule and task assignments.

In the several volumes produced by the Presidential Commission on the Space Shuttle **Challenger** are recommendations to improve personnel performance in the space program both on the ground and in orbit.[22] Two areas of the report underscore the above astronaut observations, namely management and safety. The Commission's proposed reforms adopted after endorsement by three generations of astronauts include:

— more access for the crew to flight decision-makers regarding their feedback, especially on safety matters;
— role elevation of the flight crew operations director;

— appointment of a NASA safety director who reports directly to the agency administrator;
— encouragement of senior astronauts to enter management posts.

Alcestis Oberg also interviewed space experts and astronauts, as well as reviewing the space literature. Among the most interesting performance insights obtained:[23]

- **Engineering concerns** (power, propulsion, protection, etc.)and mission objectives, not a happy environment for people, may initially drive the design of space structures and habitats ... .
- **Performance criteria** should begin with basic competencies—ability to contribute to group survival, to service mission objectives and crew needs, to operate coolly under life-threatening circumstances (thus a track record of being able to function with "grace under pressure" is an important selection factor) ...
- **Long-duration spaceflight** does present physical challenges to human performance that may cause disorientation, vestibular dysfunction, and bodily degradation. Protection is required from hazards like space sickness, toxic and noise contaminations, inadequacies of life-support systems, solar radiation, zero temperatures, and meteroids. Furthermore a continued supply of adequate food, water, and shelter must be provided in orbit. However, the real limitations as the Russians discovered may be psychological (sensory deprivation, spatial restrictions, isolation, lack of privacy, and dysfunction caused by prolonged confinement which, in turn, may lead to insomnia, depression, hostility, and carelessness). For long-duration flights, new interpersonal skills will be required (cosmonauts who have flown the longest valued "professionalism and kindness" ...
- **Group composition and interactions** impact space performance (such as, presence of women and their numbers, training and emphasis upon cooperation and synergy, group cohesiveness and camaraderie, style of leadership, ground-based communications ...
- **Zero-g work habits** involve developing new skills and instincts (putting on a spacesuit and operating effectively in it, piloting spacecraft, handling tools or practicing a scientific speciality in a new way).

All of the negative aspects of space living also have their positive sides or compensations—by careful planning, the downside may be countered. Here are some examples of what has been learned from spaceflight which could be used to increase productivity:

- **Space nausea**, for instance, may be relieved by medication, plus curbing excessive exercise before departure.
- **Vestibular system balance** in the inner ear is influenced by information from the eye; in zero-g, the vestibular organ fails to work at first because lack of gravity sends mixed signals as to what is up or down, but after two days in orbit the brain learns to ignore the input from the vestibular and accept the information from the eye as correct; upon return to Earth, astronauts are visually programmed, and it again takes two days to readapt to maintain an upright body position (blindfolds can accelerate the process).
- **Bone loss in microgravity** causes osteoporosis—data reveals that in sustained space travel upwards of 3% of total skeletal bone mass is lost within the first two months aloft. Mechanical forces, such as hydrodynamic flow and elongational strain, play a

role in the normal physiology and pathology of blood vessels, bone, and muscles. Dr John Frangos, professor of bioengineering at the University of California—San Diego, hypothesized that the hydrodynamic forces in bone induced by compression might stimulate bone formation, so his research team developed both *in vitro* models and *in vivo* studies which support that thesis. To increase blood volume before entering the Earth's atmosphere, antidotes are proposed, such as salt tablets and more fluid intake, as well as vigorous exercise before re-entry.[24]

- **Decompression sickness** prior to an EVA which requires a spacesuit, can be countered by pre-breathing pure oxygen from one to four hours to purge nitrogen from the body or by gradually lowering the pressure in a spacesuit or lunar habitat.

- **Computer monitoring** of medical patients in space by an on-board Health Maintenance Facility, measuring and administering intravenous fluids over extended periods, exercise devices for countering bodily changes. In addition NASA plans for the **International Space Station** provide for another Human Research Facility for data collection in the orbital laboratory which includes scientific instruments for use by crew members with paramedic training. Both Lockheed-designed facilities and information systems are to be linked not only to the large Space Station Data Management System, but to a Life Science Medical Operations Computer used by consulting flight surgeons on the ground. These systems will provide unique performance records improving future mission productivity and life quality.

Experience has demonstrated that humans generally find space work exhilarating for the most part, and over periods of many weeks can maintain high-work competence. Many astronauts and cosmonauts, like the first American woman aloft, Sally Ride, have written positively in books about the spacefaring experience.[25] Its uniqueness, the importance of the function, and the challenge of exploration are among the many factors driving spacefarers to peak performances. Yet in the immediate decades ahead, more studies need to be directed toward broader issues of crew performance. The analysis of traditional human factor specialists will have to be combined and integrated with broader research of behavioral scientists. Combined life science experiments of global space agencies are necessary for future joint missions regarding the safety, well-being, and productivity of space explorers. The fact that only 15 astronauts and cosmonauts to date have lost their lives in the performance of space duties is remarkable. Continuing research and precautions by space organizations, however, should ensure survival, but also consider how people can become more productive and comfortable aloft.

As far back as February 1966, the U.S. Space Science Board issued a report on advanced manned programs.[26] Because of the unknown stresses and unexpected responses of the space environment on crew health and performance as the duration of flights increased, the Board urged ground-based simulated research. Even then their findings underscored that psychological and behavioral investigations were largely being neglected in the manned space program. There is sufficient evidence that today this situation still is deficient in the American but not in the Russian and European space programs. This, the urgency for **more comprehensive international** human studies takes on new meaning in light of current planning for a space station, lunar outpost, and manned Mars mission. Perhaps the time has come for national space agencies to support a single Institute for

Human Behavior Research in the Space Environment, possibly in conjunction with the International Space University at Strasbourg.

Wubbo J. Ockels, a former ESA astronaut (refer to Exhibit 45), has made an impressive case as to why increased life science research should be conducted in space, as well as on the ground. Dr Ockels, ESTEC chair of aerospace engineering at the Technical University in Delft, observed that for two billion years, life and gravity have been together, and that relationship is a major scientific and philosophical question. Thus, he proposes more orbital studies on the matter because:[27]

- microgravity adds a new dimension for humans that is five to six magnitudes of change in its effects ...
- limited spaceflight findings justify further studies of the effects of weightlessness on the body processes and immune system behavior, especially with regard to our perceptions of up and down, the differences between horizontal and vertical (anisotropy) ...
- effects of gain are indicative of our flight with gravity, so how does life interact with gravity and is it a necessity for life? Ockels pursues such matters related to European manned spaceflight through the ESA's Space Research and Technology Center (ESTEC) associated with his university in the Netherlands.

There are many performance areas related to the orbital environment yet to be researched, such as human sexuality and family life. Georgetown University researchers Dr and Mrs Angel Colon have suggested five such areas as illustrated in Exhibit 49.[28]

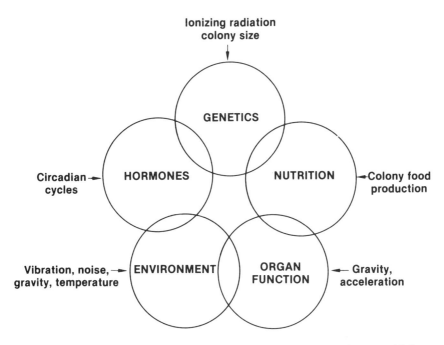

Exhibit 49—Human Factors Research for Space Family Living. *Source*: Dr and Mrs Angel Colon, The Children's Medical Center, Georgetown University Medical School, 3800 Reservoir Rd. NW, Washington, DC 20007, U.S.A

5.4 ORBITAL TEAM PERFORMANCE

The dangers and complexities of living and working on the space frontier make group cooperation and cohesion necessary for survival, no less for high performance. As woman astronaut Rhea Seddon so aptly put it: "We are no longer the solitary warriors of the early space program; we are more team players." In his interview with veteran astronauts to improve manned space station planning, Douglas (previously cited) also learned that they recommended that:

- crew selection include some peer review and input;
- that combined groups should work together for a long period of time before launching;
- consensus was that it is desirable for crews to have some group dynamics training, some noting that a lack of psychiatric or psychological support for both astronauts and their families in the past was a NASA shortcoming.

In his conclusions, Dr Douglas suggested that future station crews would benefit by survival training as a means of fostering social interaction. In a survival situation, even simulated, this physician rightly maintains that people have to learn to cooperate, as well as to make allowances for the emotional needs, personality idiosyncrasies of their colleagues. During such training, he thinks it is easier to identify insensitive, self-centered, and selfish persons who should be excluded from the crew. Since human performance is dependent on more than mechanical or environmental improvements, Douglas, a specialist in aviation medicine, proposed that some type of psychological orientation would be valuable for future station crews, so that the spacefarers could deal with individual and group needs at a human level.

A management scientist, Dr John Nicholas of Loyola University, Chicago, has published several interesting papers on space station crew performance and productivity.[29] He is particularly concerned about interpersonal and group behavior aloft, and how pre-departure training can facilitate team performance. Nicholas contends that earthbound analogs of group behavior studies in isolated environments are consistent with reports from Soviet **Salyut** experience—e.g., 30 days in to the station mission, and interpersonal hostilities grow between crew members, as well as between crew and ground control. Thus, he argues for more research into:

(a) crew group interactions and its impact on outputs, such as performance and morale;
(b) group communication with controllers;
(c) group norms in orbit, such as treatment of outsiders of visitors;
(d) group leadership in terms of designated commander and real leader/s who emerge on long missions.

After a careful literature review, Nicholas argues that current manned space program focus is upon individual astronaut and environmental inputs, which is insufficient. This professor wisely observes that attention to human factors, task design, and mission planning cannot alone ensure effective group interaction, and may set to the detriment of performance. Dr Nicholas concludes by advocating that space station crews be trained in interpersonal, emotion-support, and group interaction skills.

Dr Nick Kanus believes that physicians assigned to the space station need special skill training for dealing with psychological and social problems aloft.[30] He is convinced that when space settlements are established, space psychology and psychiatry will find there fertile fields for practice.

Apart from issues of mental health and therapy, there are forms of **group learning** sessions that can improve personal and vocational performance. As on long associated with the approaches of sensitivity, organization development and leadership training by the National Training Laboratory (NTL Institute of Applied Behavioral Science), it is obvious to this writer that educational programs have been successfully designed to bene-fit group interaction and performance among average executives, managers, and profes-sionals. In modern industry, there is another group technique called **quality circles** that has immediate implication among space agency and aerospace workers here on Earth, and possibly downrange for those in orbit. Quality circles were invented in the U.S.A., and then transported successfully to Japan before being re-exported to American organiza-tions. This team strategy could do much to counteract the type of inadequate workmanship that went into solid rocket booster o-ring seals (Presidential Commission on Space Shuttle **Challenger**, 1986). It could ensure high quality in the construction of future orbiting sta-tions and planetary bases. The International Association of Quality Circles (801-B West Eighth St., Cincinnati, Ohio 45203) has defined the term as follows:

> A quality circle is a small group of people who voluntarily meet on a regular basis, to learn and apply techniques for identifying, analyzing and solving work-related problems. This process is also known as employee involvement.

There is a whole body of social science research literature and technology in group dynamics and leadership development that is directly transferrable to preparing people for long-duration spaceflights. Among those centered on work performance, these studies and techniques center on issues—**team development** or team building, self-directed teams, collaboration in organizations, and improving work groups.[31] My own research and expe-rience as a behavioral science consultant confirm that *team* development and management are central to the new work culture, as well as emerging space culture.[32] If major corpora-tions can utilize **team building** with their personnel, surely it is safe for space organiza-tions to consider such training for both its ground-based and its space-based groups. In the practice of matrix management—a process pioneered by NASA and its aerospace partners in the 1960s for the **Apollo** missions—team development is standard practice with project teams in such contractors as Hughes Aircraft and TRW Systems. Viewing Exhibit 50 is a convincing case for team building among spacefarers.

The long-duration mission research confirms the importance of teamwork, coordina-tion, and conflict resolution to survival and performance of space crews operating in their high-technology orbital environment. Teamwork leverages human performance through the sharing of talent and by cooperative action. When functioning effectively as a team, people focus more brainpower and energy on task achievement. Today, high-output man-agement and the high-tech work environment both require the development of group inter-action skills, especially when dealing with cross-skill training. This is exactly the situation one is likely to meet in space, but with the added dimension that teamwork is necessary for survival.

Exhibit 50—Orbital Team Performance. Historic first salvage mission of a satellite from the payload of the Shuttle **Discovery**, spaceflight 51-A. During the 8-day mission beginning on November 8, 1984, two astronauts demonstrated superb teamwork by retrieving two communications satellites for repair. *Source*: NASA/Johnson Space Center.

High team performance, for example on a space station, means that the work unit is a cohesive group with common purpose; that members develop mutually helpful relationships in the accomplishment of agreed goals, objectives, and tasks. We have seen prototype behavior among astronauts on shuttle flights, as well as cosmonauts at the Russian station. For example, a week into a 1995 mission on **Mir**, ESA astronaut Ulf Merbold reported a potentially disastrous loss of electrical power. The problem was caused by having six spacefarers on board when the life-support system was only designed for five:[33]

> **We lost all systems. We could not use the radio. my colleagues calmly analyzed the situation and managed to reactivate one of the little** Soyuz TM-**19 capsules that are attached to Mir.** Soyuz **has its own batteries and we used the capsule's power to communicate with the ground.**

Then the cosmonauts maneuvered the station so that solar generators could convert sunlight into electricity, but the power loss cost lost time for experiments, some of which had to be canceled. The Russian space agency said the problem originated from excessive power consumption by video and photographic activities.

The challenge is to sustain such creative problem-solving, fellowship and coordinated efforts on long-duration missions. Assume that a station or lunar base were to have a crew of eight divided into two shifts or teams of four. Before launching, group dynamics or team training could ensure that these people were functioning and achieving together as a unit.

Two British management consultants, Dave Francis and Don Young, described these characteristics of a high-performing team which are pertinent to space crews:[34]

Output—combined results produced beyond any individual contributions;
Objectives—shared understanding of purpose and mission;
energy—motivated members who take strength from one another and promote synergy within the group;
structure—creative mechanisms evident for dealing with organizational issues (procedures, roles, control, leadership, etc.);
atmosphere—manifest spirit and culture of mutual interpersonal risk-taking and confidence-sharing.

Before going aloft, such characteristics can be cultivated in groups by skilled consultants to create a team culture. Leaders can learn to facilitate, foster collaboration, seek consensus, and even promote participative decision-making when appropriate. On the ground, crewmembers can be trained in human relations and shared leadership behaviors. The latter operates at two levels—task and maintenance. *Task leadership* refers to getting things done effectively on schedule, and includes such activities within the group as initiating and proposing tasks, defining and solving problems, seeking and obtaining relevant information, clarifying and elaborating, analyzing and interpreting, action planning and implementation. *Maintenance leadership* in teams contributes toward group cohesion and morale, while promoting supportive behaviors such as showing acceptance and friendliness, learning to share oneself and others, giving recognition and empathy, harmonizing or reconciling differences, lessening tensions and resolving conflicts, exploring differences and feelings, negotiating and obtaining compromises, communicating and receiving feedback sensitively. All crewmembers can learn to exercise their unique talents for one or both forms of leadership, and at times to exercise both sets of behaviors. There is little evidence that any space agency is providing such training to spacefarers.

In a sense, space missions encompass a sophisticated from of project team management. Top performance can result when the groups involved receive both technical and interpersonal training. The technical skill development centers around exercising task leadership in defining and analyzing the work to be performed in space; in planning for and wisely utilizing resources; in setting performance objectives, priorities, and standards. In time, such space team management could include budget development, funding and allocation; recruiting and developing team members; establishing controls and meaningful supervision; facilitating communication, reporting and evaluation systems. On the other hand, *interpersonal skills development* for group maintenance leadership involves team building, which helps members to be more authentic with on another, more experimental and flexible, more spontaneous and sensitive, more collaborative and mutually supportive. For extended stays in space, ground-based research should combine both sets of skills to strengthen self-confidence, improve group morale, and enhance problem-solving capabilities among crewmembers. This is but another example of the contribution that behavioral scientists could make to improving the quality of space life, as was suggested in Chapter 3.

There is ample evidence from behavioral science research that high team achievement and synergy occur when members:[35]

— take interest in both individual/group accomplishment;
— tolerate ambiguity and seeming lack of structure, especially during times of uncertainty;
— give and take feedback freely and non-defensively;
— contribute toward an informal, comfortable, and non-judgmental atmosphere in the group;
— are capable of establishing short intense work relationships, and then disengaging upon mission completion;
— encourage group participation, consensus, or decision-making;
— can manage change and promote innovation;
— value listening and authentic communication;
— can periodically clarify roles and relationships.

To avoid psychological and social disasters on future long-term space missions, it would behove organizations sending spacefarers into orbit to build upon such behavioral science research with groups and to enlarge its "human factors" concerns so that considerations such as described above be incorporated into:

(a) the recruitment and selection of future crewmembers;
(b) the preparation and training of spacefarers;
(c) the evaluation of space station management experience with a view to developing strategies for future space settlements.

Hooper has described in detail *The Soviet Cosmonaut Team* of the past, including the program of the Gagarin Cosmonaut Training Center; Bluth and Helppie identified many practices on Russian stations which contribute to more effective crew performance.[36] These range from policy and environment to technology and organization. Under the latter systems, these researchers discovered that the general two-year training program for cosmonauts includes psychological training upon which much emphasis is placed. When preparing for specific missions, intensive group dynamic preparations are given to station crews before departure. The latter include development of joint action skills, particularly in critical and stressful situations, group interactions and games to enable cosmonauts to work together as a team. Crews are selected for compatibility, and then trained to work together effectively. Not only is a psychological support team provided on the ground, but behavioral scientists may visit the station. The ESA and NASA astronauts who go aboard for long missions should benefit from similar preparation, and the approach should be adapted for those assigned to the forthcoming **International Space Station**.

In summary, the aim is to develop a space team worker who is more concerned with the crewmembers as a whole than his or her own ego needs. Such a spacefarer is capable of building upon individual strengths, while delimiting weaknesses; of facilitating team achievements, while minimizing frictions and antipathies. Management can install the mechanisms and means for fostering team skills and culture which optimize individual and group performance.

5.5 HUMAN/MACHINE INTERFACE ALOFT

Another way to improve human performance in space is through superbly crafted machines, tools, and habitats. In this category is the extensive utilization of automation and robotics (A&R), and the preparation of human operators or monitors so that information systems and "tin-collar" workers are used effectively. In this high-technology environment, whether on the ground or in space, the new technologies are transforming work, roles, and relationships, while helping people to be more productive. When due regard is given to work ergonomics, the new computers, robots, cybernated or expert systems can be designed to enhance the human performer. Robots in space, for instance, can take over undesirable, dangerous, tedious or automatic jobs—they can go where humans present cannot or dare not go. Exhibit 51 illustrates these possibilities on the Moon—there SAIC's artist, Pat Rawlings, envisions an exploration program employing robotic rovers to conduct a lunar site-survey. The automated workers help in selecting a suitable place for a piloted landing. Launched on expendable rockets, the **Artemis** lunar lander is the size of a compact car.

Unmanned spacecraft have already proven how they can probe the far corners of the universe at immense distances without jeopardy to human sponsor and supervisor. The automatic arm developed by the Canadians for use on the shuttle has demonstrated its potential in launching satellites or for space construction. Space has become a laboratory for pragmatic tests or *teleoperations* (manipulator controlled continuously and in detail by a human operator either remotely or with "hands on") and *telepresence* (when a remote systems operator feels as if he or she were actually present in the workplace through sensory inputs and feedback).

The Universities Space Research Association (Appendix C) under the leadership of Dr David Criswell conducted a NASA-funded workshop and report on *Automation and Robotics for the National Space Program*.[37] It was congressionally stimulated and mandated with a view to the construction of the Space Station in the next decade. The research provided A&R guidance based upon previous investigations of major aerospace contractors relative to operator–systems interface (Boeing Corporation); space manufacturing (General Electric); subsystems and mission ground support (Hughes Aircraft); autonomous systems and assembly (Martin Marietta); satellite servicing (TRW Systems); and advanced automation and technology designs (SRI International). That significant report had a foreword by the futurist John Platt, who viewed a space station as a new tool for augmenting human capabilities. Dr Platt described how the unique environment in space can be used for creating new structures, such as automated factories powered by

Exhibit 51—Human/Robotic Lunar Prospects. Artist Pat Rawlings envisions a near-future
exploration program when robotic rovers conduct a lunar site survey for piloted landings. Through
its telescopic eyes, millions of earthly eyes may soon witness humankind's return to the Moon. A
comparable mission is actually being planned by ISLEA and the Lunar Enterprise Corporation for
1997 (see Appendix C). *Source*: NASA/Johnson Space Center.

solar energy and operated continuously and unattended while economically processing
asteroidal or lunar materials. The noted professor and humanist wrote that the design and
maintenance of balanced ecological systems for space habitats provide learning regarding
the ecological systems of our own planet. If A&R in space is human-centered, Platt believes these systems can greatly amplify the powers of human beings and their technology.
In the findings section of this California Space Institute report, expert panelists concluded
that by building A&R into space station design, there will not only be many spin-off applications on Earth, but the knowledge gained will be used to accomplish, better and at lower
cost, an even wider range of space missions. For our purposes here, their most astute
observation is:

> Most important, humans can progress to increasingly higher-level tasks as au-
> tomated systems take over routine jobs. Such achievements require fundamen-
> tal knowledge of how to organize exceedingly complex, even somewhat "self-
> aware" systems. NASA Space Station programs can demonstrate complex ser-
> vice functions in "smart" computer systems or extend the electronic reach of
> humans.

Finally, the Criswell report also envisions that by 2010, robot teams will be competent aides to humans, carrying out work assignments with high-level crew supervision. These automated workers may even by human-like in appearance, capable of human-like dexterity and grace, who move mechanical arms quickly, smoothly, and precisely while performing required tasks deftly and unerringly. The robots are expected to perform payload deployment, inspection, maintenance, and repairs. Human communication with them will through efficient, user-friendly interface equipment—sometimes at a detailed, high level of exchange.

When the United States Congress passed Public Law 98-371 in 1984, it provided for the establishment of an Advanced Technology Advisory Committee (ATAC) to further A&R technologies on the space station systems, requiring NASA to make progress reports every six months on implementation of ATAC recommendations. In March 1986, the second report was issued (NASA Technical Memorandum 88785) detailing A&R accomplishments at the agency's various field centers. Among the advances was the beginning of a Technical and Management Information System (TMIS), which uses advanced computing technology for space station design and management, facilitating information exchanges between NASA centers and contractors. This shrewd statement was included:

> The basic philosophy adopted by NASA is to use advanced A&R technology for "people amplification"

Rather than replace people, NASA plans a human/machine system that integrates standards and expands human capabilities so that A&R operators move to supervisory and managerial positions.[38] The same report cited the previously mentioned THURIS study which NASA sponsored, and which described eight levels of technological readiness in the evolution of the Space Station.[5] In this way, increasingly sophisticated A&R applications will assist in the building of the station, furthering the technological base, while enhancing human involvement. For example, NASA's Office of Aeronautics and Space Technology has been working on telerobotic control by which the human operator wearing a special helmet will someday be able to operate a remote manipulator system or orbital maneuvering vehicle. Work under way at NASA field centers and among station contractors indicates a variety of uses for automation and robotics, ranging from electrical power systems, docking and payload inspection to station monitoring, navigation, laboratory management, and virtual reality systems. In addition, this second progress report to the Congress also predicted that crew hour costs in space will be considerable reduced by A&R applications. Finally, it described on-going A&R research, and workshops will not only advance the knowledge of American participators, but that international participation is being encouraged.

John Hodge, when director of the NASA Space Station Task Force, stated:

> Our experience in space tells us that some tasks are best performed by automated equipment and that other tasks require the unique responsibilities of Man. Tasks that are routine and which can be preprogrammed tend to be best performed by machines. Tasks that require initiative and judgment are more suitable for Man. The Space Station will have plenty of both kinds.
>
> (*Aviation Space,* Spring 1986, p. 23)

Steven Skaar of the University of Notre Dame and Carl Ruoff of Caltech's Jet Propulsion Laboratory recently edited a book for the American Institute of Aeronautics and Astronautics on *Teleoperations and Robotics in Space*.[39] In this significant volume, they discuss **space telerobotics** and teleoperators who take the place of astronauts on planetary and lunar science missions so as to reduce risk and cost. The editors note that space robots differ from the terrestrial counterparts because of weightlessness, vacuum, thermal environment, and the need to minimize mass. Thus, there is an economic tradeoff to consider when it comes to using manned versus telerobotic space operations, dynamics, and control.

Harrison and Connors[7] maintain that crew systems depend upon the success of human and technical combinations. In the previously cited study for NASA Ames Research Center, they explained that this means combining human and artificial intelligence; matching technical systems to behavioral propensities of their human users, and coordinating efforts by means of telecommunications. For these behavioral scientists, technical systems must meet the user's emotional, social, and cognitive requirements. In this fusion of "man and machine," they view a crew system as a *social cyborg*— a social system with expanded capabilities, an entity which can survive and perform where anything else would falter.

The Economist (February 1, 1986, p. 11) pragmatically editorializes that although many jobs can be done by robots, taxpayers will not support a space program that denies them the thrilling spectacle of human beings cavorting in weightlessness or drifting on slender tethers far above a distant Earth. While some space work may be symbolic, other jobs can only be performed by humans. Yet, continuing innovations with automation and robotics in space will not only have direct benefits for improved quality of life on Earth, but contribute to the actualization of human potential and performance.

5.6 HUMAN RESOURCE DEVELOPMENT OF SPACEFARERS

If history is an indicator, an experience in orbit itself advances one's personal and professional growth. As the numbers going to the high ground increase exponentially, more comprehensive and systematic human resource development (HRD) will be undertaken. In both the public and the private sectors today, HRD refers to organized activities that produce behavioral changes within a specific timeframe, such as happen through education and training. The American Society for Training and Development (ASTD) described the human resource areas as concerned with improved/increased quality of work life, productivity, job satisfaction, career development, and readiness for change.[40] According to ASTD, the HRD arena encompasses a wide scope of organizational activities—selection and staffing; human resource planning, organizational/job design; organization development; training and personnel development; union/labor relations; employee assistance, compensation, and benefits; personnel research and information systems. These activities on the human side of enterprise impact performance. Relative to present and future space programs, whether under government or industry sponsorship, HRD may be viewed in terms of space workers on the ground or in orbit.

With regard to the latter, this writer has proposed that within the emerging space culture, consideration be given to establishing *space personnel deployment systems*, which will be discussed in the next chapter. Essentially, this is an HRD strategy comparable to an Earth-based analog of foreign deployment systems used in the relocation of people to overseas locations. Space personnel deployment systems (SPDS) are planned and ordered means for the exchange of people to/from Earth and the space frontier. These systems are meant to both ensure the safety, satisfaction, and development of spacefarers, and further their space missions and settlements. This strategy centers around four compo-nents—assessment and orientation before departure into orbit, in-space services, and re-entry counseling. Chapter 6 will include further information on human performance in space within the context of the SPSD paradigm for relocation of people to and from the high frontier.

Obviously, national space agencies already have elements of such systems in place for their astronauts or cosmonauts. However, some of the approaches are either too narrow in scope, such as the de-briefing process, or unsuitable for larger groups of civilians who will eventually go into orbit. The contract workers who will build the **International Space Station,** and the varied visitors to it, make that station a perfect laboratory opportunity to study ways for improving living conditions not only for station inhabitants, but for the many who will migrate in the next century to other space bases and colonies. Under the Living Systems research strategy described in Chapter 3, this writer is among a group of behavioral scientists who also propose development of a data or knowledge bank with a pattern recognition and retrieval system relative to the human role, performance, potential, and challenges in space. In addition to storing findings from Earth-based analogs of signif-icance to space living, this information system could be a comprehensive storehouse on the human experience and habitation in isolated, confined environments, such as space. By using the Living Systems template of Dr James Grier Miller for cross-disciplinary analy-sis, such a framework and data-bank would enable planners and researchers to have easy access to knowledge necessary for human systems and management in space.

Space agencies have been circumspect in their analysis and release of information about the human experience in space during the past twenty-five years. Admittedly, NASA has been rather forthright about medical and biological knowledge gained from previous spaceflights, and does publish an internal history. Yet this same agency has been delimited studying or releasing information on the psychosocial experience of its personnel in space. NASA generally has limited access to the astronauts by social science researchers, even its own psychologists and psychiatrists; the agency has failed to exploit the data it has collected which could improve spaceflight and living for others to follow. There are indi-cations that transcripts of crew communications going back to the Mercury flights have not been analyzed from a behavioral science context so as to obtain clues to improve future missions and avoid tragedies. In the course of my own research, I have proposed that NASA sponsor an anonymous mail survey of past and present astronauts who have flown in space; using questions based on both space and ground-based deployment analogs, valuable input could be provided from this expert group to better human perfor-mance in space. If they cooperated with the Russian Space Agency in a comparable study of cosmonauts, then the combined findings would have global implication for HRD of spacefarers!

Although professionally a flight surgeon, Douglas[16] did attempt to gain such insights from astronauts, albeit with a very limited sample (10) and some inadequacies in terms of methodology (e.g., construction of questions and use with total group). But he understood that pilots and astronauts have often been described as individuals who distrust or even dislike psychologists and fear physicians. In a letter to the author (11/21/86), Douglas indicated that although he did not subscribe to that stereotype, he used a methodology that would best fit his understanding of the flyer's personality. Thus Dr Douglas ferreted out a sampling of astronaut opinions, beliefs, and experiences with a view to improving human performance on manned space station missions. The point is that persons who have been in orbit are a resource, both for information and as trainers, for those preparing To go aloft. We should capture their insights through instrumented surveys, audio and video cassettes, as well as computer systems for the purpose of further human resource development. From this, simulations, virtual reality, films, and other training aids could be developed for use with future spacefarers.

Exhibit 52—Lunar Ergonomics. Not only will a lunar infrastructure have to be built if people are to live and work on the Moon, but special protective suits and gear will have to be ergonomically designed. *Source*: Boeing Corporation in Philip R. Harris, "Why Not Use the Moon as a Space Station?" *Earth Space Review*, October–December 1995, Vol. 4:4, pp. 9–12.

5.7 SPACE ERGONOMICS AND ECOLOGY

In this chapter's introduction, attention was directed to the impact of environmental factors and man/machine interface upon performance. Dr Peter Hancock, when at the USC Institute of Safety and Systems Management, expressed concern that some designs for the space station are unacceptable from an ergonomics and living environment viewpoint; that there is the potential for design failure that would make the **Challenger** disaster look like a minor accident.[41] The development of the whole space infrastructure opens up a new field of applications for ergonomics and ecology.

The Harper Dictionary of Modern Thought describes **ergonomics** as the study of the physical relationships between man and machine relative to the reduction of human strain, discomfort, and fatigue, including the layout of machine tools, seat design, and even the positioning of dials on the dashboard.[42] Workplace ergonomics defuses employee and union opposition to using new technologies by designing equipment and workstations that are more compatible with human needs and safety. In orbit, ergonomics extends from spacecraft and space habitation to space suits and use of automation/robotics. Thus, space ergonomics can be defined in the context of human performance as it relates to the design, positioning, and functioning of equipment, furnishings, machinery technology, vehicles, or habitats which affect human survival, productivity, and comfort in space. In September 1995, for example, astronauts from the Shuttle **Endeavor** took spacewalks to test gear designed to protect against the intense cold aloft. In a situation comparable with that in Exhibit 50, Dr Michael Gernhardt and Lt. Col. James Voss hovered at the end of the robot arm raised 10 meters above the cargo bay. Working in temperatures ranging from –60°C to –90°C, the astronauts successfully tested new heated gloves, thermal socks, boot inserts, and redesigned long underwear for future use in constructing the space station. On an EVA the previous February, the orbiting workers fingers froze, so NASA produced these battery-heated gloves and $10.4 million spacesuit since hundreds of hours will be spent in spacewalks to build that facility. Undoubtedly, this R&D investment will have terrestrial spin-offs in equipment useful to workers in harsh environments on this planet.

In the long history of human performance, the ultimate test to date may be the building and operating of an **International Space Station** in which 16 nations are participating from 1997 to 2001 (see case study in Chapter 8). Since the U.S.A. is providing the leadership in this macroproject, its Congress, concerned for astronaut safety, has directed that as much as possible of the construction work on this facility be accomplished though automation and robotics. NASA announced that the facility would provide for the physical and psychological wellbeing of the visiting spacefarers with high-reliability, space-maintainable life-support systems. As Wendell Mendell of the Johnson Space Center noted (AAS-1984-160), the orbiting station will contain amenities for comfortable, productive working and living conditions ... . At MIT's Laboratory of Orbital Productivity, Dr David L. Akin and his group are presently studying design approaches that best fit the relationship between "man and machine". Their premise is that humans with a superb self-adaptive system are a valuable resource in space. They are examining those functions which can be done by robots alone, by machines under astronaut supervision, and finally only by humans. Numerous operations on the future station are being analyzed for specific man–machine mix.[43]

NASA seemingly employs the term "decor" rather than "ergonomics" with reference to the Space Station. A. A. Harrison, in a proposal for the agency on this subject, defines decor as the environmental design choices that remain viable given the constraints imposed by the station's anticipated weight, volume, and function. He inserts under this decor heading critical performance variables that affect the efficiency, speed, and quality with which work gets done; the accumulation of fatigue and the restorative effects of rest and leisure; and the continuing commitment to mission activities. To Harrison, as a psychologist, decor refers to elective and optional variations in form, shape, color, patterns and the like which affect overall environment and performance. It impacts both work and other living activities (such as rest, recreation, and self-maintenance). Therefore, if the environment and tasks in spaceflight are to be matched, Dr Harrison believes that there are advantages in making habitat interiors definable and redefinable—that is, a flexible environment which provides occupants with an array of alternatives to meet a variety of needs for solitude, interaction with a limited number of persons, or open interactions with all crewmembers. Space habitability specialists must be sensitive to everything from the color coding of displays and the rounding of equipment edges to providing attractive and functional meeting areas.

Dr Yvonne Clearwater[44] of the NASA—Ames Research Center joined with Dr Albert Harrison in examining crew support for a one-to-three-years' expedition to Mars. Given the prolonged isolation, confinement, hardships and dangers of such a round- trip for such spacefarers, they reviewed prospects in the context of habitability and food, physical and psychological health, interface with automation and robotics, as well as crew autonomy. In considering selection and training of astronauts for these journeys, they recommended extensive use of simulations, whether by technical means or in analogous environments. Clearwater and Harrison urge high-quality habitats and environmental design features that assuage stress and increase comfort aloft.

In a lecture at the California Space Institute in LaJolla (March l0, l988), Dr Mary Connors revealed that increased crew size, heterogeneity, and mission duration are among the forces moving NASA to make use of behavioral science resources. Research on exotic environments like outer space indicates that they provide both stressors and stimulations, downside risks and rewards. On the latter is total immersion in one's work, so persons can be judged primarily on their performance. In addition to more study of all aspects in space communications and selection/training, this human factors scientist urges investigations relating to environmental manipulation.

While **space ergonomics** impact productivity aloft, human performance in orbit will not be optimized without consideration of a twin concept—**space ecology.** Eugene Konecci, when director of Biotechnology and Human Research for NASA, described the latter as bionomics, the branch of biology that deals with the relations between organisms and their environments, which in this context is space or the zero-gravity environment.[45] Konecci, now a professor at the University of Texas in Austin, explained space ecological systems in the context of physical, biological, chemical, and situational environments required for orbital, lunar and interplanetary systems (refer to Exhibit 53). They are dependent upon the length of the manned mission or time in orbit, as well as the constraints of the spacecraft or living module.

Ecology comes from a Greek word meaning "home or living place", so here we would be using it with reference to a space vehicle, station, or habitat which are closed ecological systems. Ecologists are concerned with why a living organism can exist under one set of environmental conditions and not another.

Even the littering of our Solar System with space junk is within the purview of their ecological regard.[46] To prevent space from becoming an orbiting junkyard, builders of the

Space Environment:

Physical   — Temperature;   Pressure;   Humidity;   Radiation;   Gravity
                    (Acceleration); Dust; Air Ionization; Vibration; Acoustic Energy;
                    Illuminations ... .

Biological — Microorganisms; Metabolism; Waste Disposal ... .

Chemical   — Atmosphere; Water; Food; Drugs; Odors; Toxicology; Fuels; Fire ...

Situational— Missions; Logistics; Cabin Space; Tasks (physical and mental);
                    Personalities (on ground control and aloft); Emergencies ...

Exhibit 53—Factors in Space Ecological Systems. Four factors within the environment of outer space which impact human performance aloft. *Source*: Eugene B. Konneci, "Space Ecological Systems," in K. E. Schaefer's *Bioastronautics*. New York, NY: the Macmillan Company, 1964, p. 281.

future **International Space Station** are trying to develop technology to clean up these hazardous projectiles, such as a "sweeper" that would collect debris caused by explosion, collision, mechanical or human error. At the Johnson Space Center, the Advanced Program Office is also attempting to "harden" the material that goes into spacecraft and stations, so that they can better withstand external impacts, as well as to design shields, and to create warning systems to avoid debris collision.

In the Shuttle Orbiter, for example, miniaturization was a factor; all materials and supplies had to be carried abroad. With a Moon or Mars base, some of the latter may be transported in unmanned cargo carriers, and on-site materials may be utilized (such as lunar dirt for radiation protection of habitats). With the erection of a station platform or other large space structures, a whole array of ergonomic and ecological issues emerges.[47] Consider how waste management will become a critical factor there (one estimate is 4½ tonnes a week will be generated on the station). Rather then dump waste into space, ecologists are examining solutions such as, returning it to Earth by Shuttle or tether, recycling it for other uses, or storing it (the Shuttle's external tank if taken into LEO could be used for such purposes). The ecological logistics involved in a mission to Mars are staggering to conceive—for a three-year round-trip with a crew of three, the Russian estimate would require four and a half tonnes of food, ten tonnes of oxygen, and 17 tonnes of water.[48]

Waste management and toilets have been problem since we began to send people into orbit. *The New York Times* (September 18, 1995, p. A6) had a recent headline, **For Space Shuttle Astronauts, A Troubled Mission's Messy End.** This Associated Press story went across the world detailing how a clogged toilet caused Lt. Col. James Voss and Dr Michael Gernhardt to become "plumbers" on this **Endeavor** mission. When they were unable to bypass the clogged filter, the explorers had to drain urine and other waste water from a storage tank into smaller emergency containers. Obviously, as the number and diversity of spacefarers increases, more complex and sophisticated life-support and waste management systems will have to be designed and developed.

Research on ground-based groups in exotic environments has evolved an ecological model that seems to add another ecological dimension that fits within the situational category of Exhibit 53 under "personalities." Altman posits a fundamental interdependence under such circumstances among group members and between the group members and their environment.[49] Applied to space living, the ecological model, in this sense, would be concerned with the processes and products of social interaction, especially as to how they affect performance aloft. Interpersonal processes here include verbal and non-verbal communication, and manipulation of environmental props; "products" refers to levels of intimacy, degree of interpersonal accommodation, attraction, conflict, and the like. This theory of interpersonal dynamics based on investigations with isolated and confined groups bears further study for its applications to performance in space. It is an illustration of how behavioral science perspectives can be aligned with biological insights to create a synergistic "envelope of life" system that might enhance performance in the remote and alien environment on the space frontier.

Because the space culture involves an integrated environment for life and work, those concerned about human performance there cannot afford to ignore matters of space ergonomics and ecology, both fields of knowledge likely to grow in importance. Environmentalists and conservationists have only begun to direct their attention aloft as can be seen by the Sierra Club seminal conference and proceedings on *Beyond Spaceship Earth*.[50] Within that volume on environmental ethics in the Solar System, there are chapters on orbital pollution, space exploration and environmental issues, extraterrestrial ecosystems, military/nuclear use of space, and the social/physical environment of space stations and colonies!

One other dimension of space ecology may be the use of space technology to benefit the home planet, such as through earth-sensing and observation. This is the whole purpose of a NASA program called "Mission to the Planet Earth (MTPE)." Many are concerned about global environmental change, such as warming and pollution of the planet, and point to the growing role of space developments in understanding our own world.[51] The U.S. Global Change Research Program seeks to describe and understand the interactive physical, biological, and chemical processes that regulate the total Earth system, including the Moon, as influenced by human action. The MTPE centerpiece is the Earth Observing System, a series of satellites to be launched from 1998 to 2006, designed to gather data for observing global warming, ozone depletion, deforestation, and other planetary problems observed from outer space.[52]

5.8 SPACE PERFORMANCE RESEARCH ISSUES

This review on offterrestrial human performance led me to certain conclusions on areas for further interdisciplinary research:

- The need to broaden both perspective and data gathering beyond the traditional, narrow human factor and life sciences approach of industrial engineers. Global improvements might be forthcoming by better utilization of the biological/behavioral sciences. More attention should be directed to the space biosphere or envelope of life in terms of both physical and psychosocial considerations … .
- The complexity of long-term space living implies that more study should be directed toward social and group interaction among spacefarers, as well as the possibility of

their experiencing "space shock" (a form of culture shock manifested through depression, withdrawal, and other alienated behaviors ... .

- The value of creating knowledge bases or data banks on the global experiences of human living/working in space, or in similar ground-based exotic environments, so that cross-disciplinary insights will evolve and expert systems will develop ... .
- The necessity for designing a more complete space personnel deployment system to accommodate a larger and more diverse population aloft, including a feedback mechanism so that those who actually live and work on the high ground may contribute to the preparation of future spacefarers (e.g., broadening of the debriefing process and using astronauts as trainers/educators). This will be discussed further in the next chapter ...
- The expansion of cross-functional investigations into the human role and performance in space. Such research should address issues raised in this chapter, such as crew productivity, team performance, human/machine interface, human resource development of spacefarers, space ergonomics and ecology, as well as topics discussed elsewhere in this volume (e.g., space culture and macromanagement ...
- The planning for an **International Space Station**, as well as for future bases on the Moon or Mars, should include more comprehensive studies aimed at improving the quality of space life, work and performance. This would enhance the space explorer's environment and experience not only in these prototype communities, but also in space settlements and colonies to come.

In critiquing this chapter, sociologist William MacDaniel[53] raised two other matters for future research:

- The concept of productivity in space needs to be defined. From his perspective, productivity is a product of the total social and cultural environment in which people live, and is not to be limited to on-duty activity. Since productivity standards may or may not be internalized by individual spacefarers, overemphasis on team productivity could contribute to stress. MacDaniel points out that while on Earth we can get away from the stressful work situations, but that is more difficult aloft; therefore, the space culture should provide crewmembers with relief from job-related strain when not on duty. A Space Station culture and social structure, for instance, might contain values, attitudes, and norms that support productivity and competent performance, while also offering mechanisms for tension relief and relaxation from intensive duty activity. Such an environment should lessen prospects for the individual or the group to become dysfunctional. Perhaps the earthly dichotomy between work and play needs to be eliminated in space, so that productivity aloft encompasses leisure activity as well as duty hours?

- The role of conflict and competition among spacefarers requires more study, especially since they impact performance. MacDaniel, a professor emeritus from Niagara University and former military jet pilot, speculates about space workers who are trained in cooperation for survival aloft, who then must return to a home planet culture which emphasizes individualism and competition. He also inquires if a modicum of competition and conflict are necessary for innovation, invention, and original thinking in space communities? Would dependency on a norm of survival cooperation lead to a stagnant and uninspiring environment? However, to me, the latter would not be the same as the synergistic society aloft put forth in Chapter 2. I agree with MacDaniel that a cooperative community in space implies development of a wholly different set of cultural values, norms, and attitudes than is typically experienced now on this planet. Perhaps it is one of the lessons to be learned by **earthkind** from further **spacekind**. Furthermore, conflict to behavioral scientists is not necessarily a negative experience, but represents energy to be utilized and channeled. The 21st century challenge of offworld living is to innovate and to achieve levels of human potential currently not attainable on Earth. Exhibit 54 offers some additional recommendations.

- Synergize the presently independent research activities of national and international organizations through development of cooperative programs in life sciences at space agencies and university laboratories.
- Complete and consolidate the unique national data base consisting of basic life science information and results of biomedical studies of astronauts and cosmonauts on a longitudinal basis. This data base should be expanded to incorporate information obtained by all spacefaring nations and be available to all participating partners.
- Increase the frequency of life science data acquisition on the Space Shuttles, Space Stations, and international missions ... . The knowledge obtained by space life sciences will play a pivotal role as humankind reaches out to explore the Solar System.

Exhibit 54—Strategies for Life Sciences Research. *Source*: *Adapted from "Executive Summary," Exploring the Living Universe—A Strategy for Space Life Sciences.* Washington, DC: NASA Headquarters/Life Sciences Division, 1988.

5.9 CONCLUSIONS ON SPACEFARERS' PERFORMANCE

Superb human performance aloft has been manifest among both cosmonauts and astronauts from the first man in space, Yuri Gagarin in 1961, to the first loss in flight of both a spacecraft and its seven crewmembers with the **Challenger** tragedy in 1986; to the latest "manned" mission to be launched. With the emergence of space stations, the capacity to stay longer in orbit or to have permanent habitation on the high frontier is now feasible. Each pioneering effort contributes to the offworld migration of our species, as does each successful orbiting performance by forerunners of tomorrow's space colonists.

Charles Darwin once observed that the human mind evolves in response to our environment. The further exploration of space with consequent expansion of human presence there will advance our psychosocial evolution and may eventually even alter the biological evolution of the species. As Robert Powers reminded us—human dreams and ideas can span centuries before being translated into realities; the future Martians may be living among us now as our children and grandchildren! [54] Finally, it is worth recalling the words of the ancient sage, Lucretius:

> The mind wants to discover by reasoning what exists in the infinity of space
> that lies out there, beyond the ramparts of this world.

6

Personnel Deployment Systems for Space

Since ancient times, humans have left their homelands to explore and migrate to far places. Embarking on a **hero's journey** is the way that renowned mythology expert Joseph Campbell describes the relocation challenge of living or working in a foreign region:[1]

> **Furthermore, we have not even to risk the adventure alone, for the heroes of all time have gone before us. The labyrinth is thoroughly known. We have only to follow the thread of the hero path, and where we had thought to find an abomination, we shall find a god. And where we had thought to slay another, we shall slay ourselves. Where we had thought to travel outward, we will come to the center of our own existence. And where we had thought to be alone, we will be one with the world.**

For the expatriate does indeed dare to go outside the known boundaries, and may in the process experience a "shock" to his or her psychological construct, our perceptions of our "world" as we know it. But consider the demands upon the terrestrial who goes **offworld**, up into outer space beyond gravity! Surely that is the ultimate hero's voyage and adventure.

Within the human family today, only a select and small group of people have actually flown into orbit. Whether these are called astronauts or cosmonauts, our "manned" extraterrestrial experience is limited to thirty-five years with only about 200 humans, mostly Americans and Russians, who have stayed aloft for relatively short periods. Each year the record is extended. Presently, at the Russian's **Mir** station, Valery Polyakov holds the lead with 438 days aloft, surpassing 366 days in an orbital environment for cosmonauts Vladimir Titov and Musa Manarov. Before long, we will be measuring space expatriation in terms of years and decades. But if human extraterrestrial presence is to be significantly extended and expanded, then behavioral science research on personnel deployment and acculturation needs to be increased—in tandem with life science research on support systems and the orbital environment. This chapter will consider some of the relocation issues and strategies that will be necessary before a mass human migration into space can take place, possibly beginning midway in the 21st century. Policies, practices, and programs that have been suitable in the past for small elite groups in orbit will not be sufficient as missions increase in duration and complexity, crew size and heterogeneous composition, and diversity of required personnel competencies.

If we examine the near-term plans of spacefaring nations, the mission forecasts are for longer Shuttle flights, completion of a **International Space Station** by 2002 with crew rotations every 3–6 months, and outposts on both the Moon and Mars before 2050 AD, as illustrated in Exhibit 55. To build the required infrastructure for this, the growing population aloft will begin with a variety of contractors and specialists, expand with scientific and technical representatives from international partners in these macroprojects, and eventually end with space settlers and colonists. The near-term endeavors center upon orbiting laboratories for studying personnel deployment, life sciences and other human factors, so as to design improved human habitation systems for long stays beyond the Earth's atmosphere.

Exhibit 55—Lunar Deployment: Technauts on the Moon. Artistic rendering of 21st century lunar workers in protective spacesuits and habitats to provide life support, while assisted by robotic heavy equipment. *Source*: NASA/Johnson Space Center.

To plan for humankind's next "giant leap" into the Solar System, NASA embarked on "Project Pathfinder"—development of the enabling technologies for such missions, including engineering systems for effective performance and health of human inhabitants. Beyond such hardware and engineering concerns, research should also be expanded on "peopleware"—not only in terms of predeparture selection and training of space explorers, but their support on the high frontier—physical, psychological, social, and financial. Furthermore, what guidance and support should be provided to the family dependants left on the ground while the spacefarer is away for longer periods? Finally, what assistance should be rendered to space travelers and their families to help them cope with re-entry and consequent problems of terrestrial adjustment?

Wally Schirra, former astronaut, raised the fundamental question most simply: *how long can humans endure in space*? Speaking before the National Conference of the Aviation/Space Writers Association in San Diego (May 13, 1987), Schirra urged more research on human survival issues, such as what would be involved in a manned Mars' spaceflight of several years. Dr Bruce Cordell when at General Dynamics Advanced Space Systems also addressed this matter of 2–4-year missions to Mars and its moons. He noted that NASA planners are concerned about such spacefarers in terms of these five dimensions:[2]

(1) *adaptation and readaptation* to space and planetary environments, particularly with reference to physiological effects (e.g., from atrophy of bones and muscles, the need for gravity stimulations) ...

(2) *toxicological safeguards* against possible crew poisoning or exposure to toxic substances (e.g., from life-support systems, which need to be more bioregenerative) ...

(3) *radiation risk assessment and management* because of solar particle events or galactic cosmic radiation—the radiation environment of Mars, for instance, is one hundred plus greater than Apollo astronauts encountered on the Moon ...

(4) *health maintenance capability* for coping with medical emergencies and maintaining wellness—provisions must be made to cover everything from safety and diagnosis of illness, to integrating visitors and coping with death ...

(5) *psychological issues and human factors* related to extended spaceflight by mixed crews as to gender, nationality, and disciplines—the unprecedented experience of isolation and confinement requires programs in place to counteract impairment of mental/emotional performance.

The concerns of this book include all of the above and more if space habitation on a permanent basis is to be realized. First, we can learn much from many precedent experiences on the ground, so as to better prepare to send population masses aloft.

6.1 TERRESTRIAL ANALOGS

There is a growing body of literature and experience on the terrestrial relocation and adaptation of people to strange and alien places in the interest of adventure, science, defense, or trade.[3] Part of this information relates to exotic environments which are defined as remote, harsh, and potentially dangerous.[4] Some of the studies report various personnel problems of living in Antarctica.[5] But much of the insight centers on foreign deployment of executives, technicians, scholars, and volunteers.[6]

There is more to be learned from this human experience—whether undersea on a nuclear submarine, or in remote park service areas, or overseas on an international assignment, offshore rig, at a polar science or military base—which can help us in the management of people in orbit.[7]

To illustrate this point, consider the research on expatriates in the world of business conducted by the following.

● *Dr Joyce Sautters Osland* of the School of Business Organization, University of Portland.[8] As part of her doctoral dissertation, she developed an interview protocol which she utilized with 35 returned U.S. business executives from overseas assignments of at least 18 months. The process included an **Awareness of Paradox** instrument for data gathering. Osland found:

- a lack of adequate information on the expatriate experience;
- the majority of U.S. companies do not adequately prepare personnel for assignment abroad;
- human resource specialists lack expertise on how to prepare, support, and debrief expatriates.

As a result of this investigation, her book, *The Adventure of Working Abroad*, provides

- understanding of the **transformational** nature of this experience;
- words of other "heroes" about their international assignment and repatriation;
- a framework for helping current and return expatriates make sense of their cross-cultural adventure;
- ideas for improving the orientation of those going on such foreign assignments.

● *Drs Stewart Black and Hal Gregersen* of Dartmouth's Amos Tuck School of Business Administration, along with *Dr Mark Mendenhall* from the University of Tennessee, have researched and published for years on global assignments, premature returns, and turnover within multinational corporations. In their book together on the subject, they offer:[9]

- a guide to managing all phases of the global assignment process from employee selection to repatriation;
- qualifications necessary for foreign placement, and methods for training candidates to work effectively abroad;
- solutions for typical problems and conflicts faced by expatriates and their families as a result of such deployment;
- ways to ensure that knowledge gained in the cross-cultural experience is not lost to the sponsoring organization upon the return of the expatriate.

These scholars focus on **new mental road maps and behaviors** required for assignments to alien environments, and emphasize the need to reward expatriates while abroad, as well as to facilitate their readjustment and retention upon return to the home culture and organization.

● *Dr Rosalie L. Tung*, now a professor at Simon Fraser University in Canada, studied 80 U.S. multinational corporations, and discovered that more than half had failure rates on expatriate assignments of 10–20% (7% reported as high as 30% recall rate). The Americans who were unable to perform effectively abroad had adjustment problems centering in these human behavior areas:[10]

- inability of the manager and/or spouse to adjust to a different physical or cultural environment;

- family-related difficulties;
- personality or emotional immaturity of the overseas manager;
- inability to cope with managerial responsibilities abroad and lack of motivation to work there;
- lack of technical competence by the multinational manager.

Tung then compared her American results with comparable surveys which she conducted of 29 West European and 35 Japanese transnational firms. The former reported an average expatriate failure rate ranging from under 5% to 10%, while the latter companies reported largely under 5% (only 14% of the Japanese had failure rates as high as 19%). She discovered these reasons for lower failure rates among European and Japanese expatriate managers:

- long-term orientation regarding overall planning and performance assessment;
- more rigorous training programs to prepare candidates for overseas' assignments;
- provision of a comprehensive expatriate support system;
- overall better qualification of candidates for overseas work including foreign language capability;
- moral support both from the international orientation of the sponsoring organization and from the family (especially among Japanese nationals).

Tung's findings are consistent with similar relocation studies which attribute such maladjustment and high premature return rate on foreign postings to: (a) poor selection and training for the overseas assignment; (b) overemphasis on technical competence to the disregard of other qualities that ensure effective cross-cultural performance; (c) short duration of such assignments and overconcern with repatriation.[11]

The matter of premature return for terrestrial expatriates is a critical managerial issue that has serious implications both for assignments to polar regions and into the space frontier. Based on place and circumstances of foreign assignments, researchers estimate that it costs sponsoring organizations from $50 000 to $200 000 per employee or family to bring such persons back to a home base ahead of schedule. The Business Council for International Understanding, on the other hand, maintained that from their studies international personnel who go abroad without proper cross-cultural preparation have a failure rate on such assignments ranging from 33 to 66%, in contrast to less than 2% of whose who had such training. Such findings, plus the right costs of relocation and premature return, have caused a whole new industry of relocation specialists to emerge. The Employee Relocation Council is the clearinghouse for such professionals and their consulting firms, and their organ of communication is the magazine, *Mobility*.[12] Furthermore, researchers have developed instruments, from paper and pencil questionnaires to computerized expatriate profiles to use in assessment of those being sent out of their home culture.[13]

Unfortunately, there has been limited "technology transfer" of the above research insights to those agencies and corporations who currently send personnel into the Antarctic or into space. For example, whether in polar science outposts or in orbit, there do not appear to be any studies comparing U.S.A. and former U.S.S.R. expatriates in both arenas relative to their predeparture screening and training, on-site support, premature return rates and costs, and re-entry provisions? The U.S. Navy, for instance, seemingly has done the most comprehensive investigations of its Antarctic winter-over personnel, but the focus has largely been on follow-up studies related to mental and physical health upon return. Palinkas, previously cited, did review data collected from Antarctic research stations of different nationalities relative to sociocultural influences on psychosocial adjustments.[14] NASA has generally centered its investigations on limited human factors and medical reports, rarely comparing the results of its experience with the Soviet experience aloft as did Bluth and Helpple.[15] From the perspective of deploying people to polar or space sites, the sponsors have much to gain by examining the issues and findings in the foreign relocation or expatriate literature. The movement of a large number of people around this planet has spurred increasing interest and investment by international corporations and associations in the phenomenon of culture/re-entry shock. When people are rapidly transported from their home culture to strange and alien environments abroad, they may experience severe disorientation, confusion, and anxiety as discussed in the last chapter.[16] There is increasing evidence on the psychological reactions of spacefarers in unfamiliar microgravity environments and the shock that may impact their psyches.[17] Admittedly, the length of stays in Antarctica or space until now have been too short for studies to emerge of "polar or space" cultural shock, though both astronauts and cosmonauts aloft have given indications of the phenomenon. There is also a neglected area of terrestrial investigation that has implications for space living—returning home from a cross-cultural experience.[18] Ample reports exist from the Peace Corps, the military, and transnational organizations that some personnel experience "re-entry shock" when re-assigned from a host to a home culture. It is reasonable, then, to project that readaptation to Earth after long-duration spaceflights will necessitate innovative programs to cope with the physical and psychological effects of living aloft.

Basically, a major deployment issue then becomes how to reduce stress on a person when leaving or returning to a home base, whether on this planet or another.[19] For the traveler's sense of identity is threatened when they are removed from the comfortable and familiar and thrust into the unknown and uncertain. Such expatriates, especially when away for many months or years, usually go through a transitional experience that may include phases of growing awareness of differences, rage, introspection, and eventually integration or disintegration. As a result, many multinational businesses and foundations have developed relocation services to facilitate acculturation of their personnel to new cultural changes and challenges.

Although it is unlikely for the New Millennium that spacefarers will have to deal with extraterrestrial "foreigners" or aliens, they will have to increasingly cope and adapt to the unique cultural environment of space (refer to Chapter 4 and Exhibit 36, which illustrates some of the dimensions in the context of humans at a lunar base). What has been learned so far about both terrestrial and space relocation is transferable to long-term space habitation and interstellar migration. From this knowledge base, strategies, policies, and programs can thus be devised that facilitate the extraterrestrial deployment and delimit the psychological shock of isolation, loneliness, and strangeness of prolonged living on the high frontier.

When people enter dangerous and exotic environments, the primary concern is for sheer survival and then safety, as well as the factors which contribute to this for mission achievement. Next is upon productivity and satisfaction, conditions that further adjustment and quality of life.[20] Whether in remote terrestrial or offworld environments, planners should be studying ways to reduce social and health problems (both mental and physical), while optimizing performance in isolated and confined circumstances.[21]

The next four examples illustrate the possibilities of applying findings from ground-based human factor research to the preparation of spacefarers.

6.1.1 Polar Regions as Laboratories for Space

On this planet, the polar regions have long recognized by behavioral and biological scientists as natural laboratories for studying how humans cope and perform in remote, extreme environments.[22] We have yet to appreciate fully the learnings to be gained in this regard by investigations of native peoples in the Arctic, whether called Eskimos, Inuits, or Indians. We are more cognizant of the adaptive skills of Europeans and North Americans who have gone into these far northern regions as explorers, scientists, workers, or settlers. Even with modern comforts and new technologies, the situation is still harsh and hazardous for such hardy persons.

A recent issue of *Canada Today* contains two vivid anecdotes that highlight the challenge of isolated and dangerous living that eventually may have similar replications in space:[23]

- **Scientists in the Polar Continental Shelf Project—in 1986, this Canadian project supported 229 field groups, including ten from U.S. universities, at permanent bases in Resolute Bay and Tuktoyaktuk. The peaceful scientists live there from the beginning of March until mid-October in twelve, pre-fab buildings with a nearby runway for planes. The most ambitious site is on an ice island that is four miles long and 148 feet thick, weighing a billion tons. This natural base enables them to conduct long-term surveys of oil reserves and to study the Arctic Ocean and structure of the continental plate by monitoring ice conditions, water currents, and the shape of the ocean floor.**

- **Miners on Little Cornwallis Islands—since 1981 these Canadians work at the Polaris lead and zinc mines, some 875 miles below the North Pole—one of the northernmost and richest mines in the world. The $150 million investment is expensive because the 200 workers must be protected from loneliness and isolation, as well as the elements. Despite temperatures of 50 below, winds at 70 mph, mining work at 500 feet below surface, and severe outside travel, there are waiting lists of hundreds to apply for positions from cooks and miners to metallurgists and geologists. This is due, in part, to very high pay, two-week vacations flights to civilization every ten weeks on assignment, and luxurious living conditions and food in the residence building which includes basketball court, pool saunas, lounges, and jogging tracks. The annual turnover of personnel is 12%.**

Within decades we may be reading comparable stories about scientists at a lunar outpost or workers at an asteroid mine. The survival and adjustment of such space pioneers could be enhanced if contemporary scholars would devote more investigations on how people get deployed to polar regions, how they survive and perform when there, how they re-adjust after they return home from such assignments. Presently, Antarctica may be the best deployment laboratory for behavioral research analogous to the challenges inherent in space habitation. Scientists, technicians, military personnel and visitors go to these South Polar research stations representing different nationalities and sponsors. In the case of the U.S.A., for instance, the sponsors range from government agencies, such as the National Science Foundation; to defense services, such as the U.S. Navy; to private contractors, such as ITT. People who "winter-over" there experience unusual living conditions in small groups—psychological stress is exacerbated by prolonged isolation in a harsh environment, resulting for some in impaired health and performance.[24] Since the turn of this century, humans have been coping there with the windiest, coldest, and harshest environment on Earth.

In a lecture at the University of California—San Diego Medical School where he serves on the faculty (October 15, 1986), Dr Lawrence A. Palinkas stated that the Antarctic people experience may serve as a model for the Space Station, facilitating future commercialization and colonization of the space frontier. As a sociologist, Palinkas has analyzed data collected on thousands of naval Antarctic personnel, particularly the long-term effects of isolation and exposure to extreme environments, the relationship there of stress and illness, and the health behaviors of enlisted naval servicemen. Palinkas notes that similar types of stressor exist in Antarctica as in space, both in terms of *external* factors (inability to personally maintain contact with family and friends, pressures from real or imagined unpleasant events at home, feelings of rejection through delays in relief parties or shortages in supplies or interference in station routine by outside authorities) and in terms of *internal* factors (lack of privacy and cramped quarters, lack of stimulation and boredom, emotional and sexual deprivation, uncertainties over status and role). The "winter-over syndrome", a form of culture shock, may prove to be comparable to that to be experienced with long-term space living—depression, hostility, sleep disturbance, impaired cognition, physical and psychological distress. Coping with six months of light and darkness in Antarctica has its counterpart with light/darkness periods on the lunar surface. Despite the risks and that some took six months to readjust after their return, naval personnel generally were able to perform well, stay healthy, and experience no major adverse effects after the Antarctic assignment.

In 1987, The National Science Foundation joined with NASA to sponsor a conference on "The Human Experience in Antarctica: Applications to Life in Space." The conference organizers, Dr A. A. Harrison of the University of California—Davis, and NASA-Ames scientists, Drs Y. A. Clearwater and C. P. McKay, edited a significant volume of proceedings, previously cited, which expands our understanding of Antarctic research beyond survival to how humans may thrive under comparable conditions as are found there and elsewhere. Such studies may prove to be prototypes which contribute to maximizing human performance in outer space. There is much to be learned from the contemporary Antarctic selection programs for personnel, their on-site need and support, their re-entry problems and experiences. Furthermore, a body of research literature on this subject has been building up for three decades on the emotional and mental health of Antarctic personnel. Both past and on-going behavioral research of such human experience, combined with analysis of previous investigations of both foreign and space relocations, should provide immeasurable insight for the improvement of large-scale space deployment.

However, the scope of such human behavior studies in the South Polar regions not only needs to be expanded, but becomes more systematized and comparative, especially among international sponsors. For example, within the various institutions sending Americans to Antarctica, there does not seem to be synergistic planning and analysis of deployment issues among the sponsoring agencies. Sharing and cooperation on recruitment and selection, for instance, seems to be very limited between ITT, an NSF contractor, and the Navy, or even with other contractors? ITT Antarctic Services advertises in the press for those with engineering or mechanical backgrounds who can manage plant facilities in remote, self-sustaining research stations; who can exercise organizational and leadership skills consistent with the requirements of a small, closed community; who have the education and academic interest to interface successfully with diverse scientific investigators; who have the flexibility to react and manage effectively the fast-paced, changing, and unusual site operations on a seven-day per week basis. This advertisement for a year's Antarctic assignment is not unlike that which may appear someday in the media for technical personnel to go to a Space Station or base on the Moon or Mars.

One wonders what does a corporation like ITT do in its assessment process to ensure that persons with such qualifications are indeed hired? Furthermore, what does it do to assess the effectiveness of that new hire once on site, and how does it help that employee upon return to normal civilian life? Is there a system to this process? Is it applicable not only to space, but to other organizations who send personnel to the Antarctic? NASA has begun some joint human factor and life science research with NSF's Polar Biology and Medical Program, but is the information and insight shared with other public and private agencies sending personnel to the North or South Pole? How much ongoing behavioral data-gathering is occurring at polar research installations of other international partners? Are sponsoring entities even aware of the cross-cultural literature on relocation? To illustrate the applicability of this to living in closed communities of Antarctica and space, consider that the Canadian International Development Agency has a predeparture program for its volunteers which instills these seven skills:

(1) to communicate respect;
(2) to be non-judgmental;
(3) to personalize knowledge and perceptions;
(4) to convey empathy;
(5) to practice role flexibility;
(6) to demonstrate reciprocal concern;
(7) to tolerate ambiguity.

6.1.2 European Isolation Studies

There are several research endeavors under way in Europe which relate to isolated, confined environments (ICE), while providing insight for spacefaring. One was a simulation experiment which took place in 1989 under the leadership of COMEX, a French diving company in Marseille. With the participation of the European Space Agency (ESA), the study involved divers in a hyperbaric chamber. Another took place in England in conjunction with the JUNO mission to fly a British cosmonaut aboard the **Mir** space station.[25] In this 1990–91 undertaking, the focus was on selection and training from 13 000 hopeful British applicants to 150 chosen to take a succession of psychological and medical tests. The latter group took an executive physical, followed by tests of symbolic, mechanical, and spatial reasoning abilities. From these findings, 35 then underwent a day of further psychological tests covering personality, working style, and public presentation skills. By this means, hopefuls were reduced to 22, who received five more days of intensive medical tests, further reducing the number to l6. These were permitted to take the stress tests, centered around a centrifuge, at the Institute of Aviation Medicine in Farnborough. As successive individuals were eliminated, the assessment process climaxed for l0 with final examination by a team of Soviet doctors who came to London. From these, two were finally picked—a woman and a back-up man. Scheduled to actually fly on the eight-day, Anglo-Soviet mission, beginning May 12, 1991, was Helen Sharman, a 27-year-old research chemist, who was trained in Star City with other Russian crewmembers, Anatoli Artsebarski and Sergei Krikalev. JUNO project manager, Christopher Hayes, admitted that the private enterprise did not fully raise the $12 million required by the former U.S.S.R., but the Moscow Narodny Bank had committed $9.75 million toward underwriting the project (*Space News*, March 3, l991, p. 2).

The most significant ESA-sponsored project is the Isolation Study for European Manned Space Infrastructure. Called ISEMSI, it was partially funded by the contractor, Norwegian Space Centre. The first in a series of research studies began from September l7 to October 15, l990 with six young European males confined in the hyperbaric chabers at the Norwegian Underwater Technology Centre (NUTEC). Within a chamber comparable to a space station environment, this experiment studied the psychological and physiological impact of isolation and confinement and their implications for long-term space missions. During four weeks within the two cylinders, participants conducted 37 experiments, including an operation simulation, for some 23 different groups, coordinated by ESA's directorate of Space Station and Microgravity. Under the leadership of their Long Term Programme Office (LTPO), those in the facility had contact with the external world only through communication links (audio/video) while following a schedule such as envisioned for the station aloft. The group from ages 25–30 included five engineers/physicists, a physiologist, medical doctor, and professional pilot—the national composition was two Norwegians, two Italians, one Swede, one Dutchman, one Frenchman, and one German. They described themselves as **ESMInauts**, coined from the acronym ISEMSI (the European Programme of Isolation Study for Manned Space Infrastructure). Throughout, they were on duty for a 13-hour day and monitored as to words, movements, and behavior; they measured medical, physiological, and environmental parameters. They spontaneously organized themselves into very structured patterns and roles, allowing time in the morning for a group meeting on the day's work assignments and in the evenings for recreation. When the group emerged from their 28-day confinement in Bergen, Norway, they appeared healthy and reported:

- **they stepped into the chamber as individuals, but came out bonded as a team ...**

- **cultural differences, such as diversity in diet, did not interfere with crew cohesion, as they gradually became an entity working together and trusting each other ...**

- **minor conflict within the group centered around tidiness, and a mutiny almost arose when the crew objected to the demands of external principal investigators ...**

- **asked what they missed most in isolation, one replied, "The Dutch landscape."**

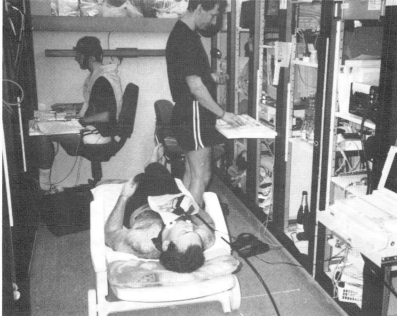

Exhibit 56—ESA Ground-based Isolation Studies. In 1990, the first of 37 experiments involving 23 groups was undertaken by the European Space Agency in an Underwater Technology Centre at Bergen, Norway. Called **ISEMSI-90** (Isolation Experiment for the European Manned Space Infrastructure), the **EMSInauts** conducted isolation studies for four weeks to prepare for future long-duration missions in space. In a later experiment, EXEMI '92, a crew of one woman and three men (pictured here) spent 60 days in isolation in the TITAN hyperbaric complex at the DLR aeronautical medicine institute near Cologne, Germany. *Source*: ESA.

In this experiment to simulate the astronaut environment in orbit, the number six was considered ideal from the viewpoint of group dynamics on a Space Station. Women were excluded in the first experiment in order to reduce the number of parameters, but the "Emsinauts" were unanimous in recommending, "Next time we would like a mixed crew." Jacques Collet, head of ESA's Long Term Programme Office, said that Europe has to achieve autonomy in manned spaceflight because the next century will be centered upon human presence in space. He announced that ISEMSI will be followed by other experiments to study the logistics and operational requirement of ground/crew relations. Raymond Fife, head of Space Station and International Affairs at the Norwegian Space Centre, noted that NUTEC had acquired expertise and technology in offshore operations which could be applied to manned space missions. The medical computer system developed by Paul Sabatier of the University of Toulouse and other telemedicine innovations are already being applied to the offshore oil industry, and could be utilized aloft.

Subsequently, the European Space Agency sponsored a project **EXEMI '92**, a 60-day multidisciplinary scientific study in isolation and confinement (see Exhibit 56). A crew of four—a Swedish woman and three men from Austria, the Netherlands, and France—lived in the TITAN hyperbaric complex at the DLR aeronautical medicine institute near Cologne, Germany. ESA–LTPO then sponsored a symposium on the **Study of Human Behaviour in Extended Spaceflights** (UNESCO, Paris, Dec. 1–2, 1993) to report their results, inviting others who have researched similar objectives in underwater habitat, submarines, polar stations, etc. The assembled scholars exchanged on such aspects as group assessment and dynamics, individual perceptions and performance, psychological/physiological changes in isolation, and international experiences.[26] ESA/NUTEC's Raymonde Fife, observed:

> **Working and studying the extreme environments of inner and outer Space create synergy effects which benefit both environments while producing spinoffs for everyday life on Earth, such as improved equipment, medical procedures and instrumentation.**

But the European Space Agency has also discovered the limits to ground-based simulations, so they have expanded their in-orbit studies of ESA astronauts on board both American and Russian spacecraft. Projects to prepare for long manned missions, such as **Euromir**, will be discussed further in Chapter 8.

6.1.3 Biosphere 2—Private Enterprise

One of the most interesting, comprehensive, and longest experiments in isolated, confined environments was conducted under private auspices in Oracle, Arizona (see Exhibit 57). Located in the American southwest near the foothills of Catalina Mountains on SunSpace Ranch, just 50 km north of Tucson, the enterprise is funded by Texas billionaire Ed Bass, a co-designer of Disney's Epcot Center. Under the name Space Biosphere Ventures, scientists, architects, ecotechnicians, and entrepreneurs put together a $150 million project. The founders consider the Earth the first biosphere, and call their undertaking Biosphere 2.[27] On 12 000 m^2, they have built a glass-enclosed ecological system—with laboratories or "biomes" for rainforest, savannah, marine, marsh, desert, agriculture, and human habitat. In Spring 1991, this miniature world was tested for its ability to recycle and maintain environments—air, water, and nutrients—supporting four thousand inhabitant species of plants and animals, including a crew of eight people. Built supposedly for a one-hundred-year life span, the initial test period was for two years to determine if the inhabitants can live in harmony within their biosphere. Open to public tours, the facility encompasses the Biospheric Research and Development Center, as well as a conference center and gift shop. To cover escalating costs, the venture expects to make money on tourism, media, and spin-off technology.

Exhibit 57—Biosphere 2. Privately financed isolation studies have been carried out in this enclosed facility in Oracle, Arizona, U.S.A., to discover what may someday be applicable to life in space colonies. The scientific focus has shifted away from human subjects in 1990 to closed ecological systems with plants and animals. *Source:* Biosphere 2 Center, Inc.

The project base is considered a world of discovery, an excursion into the future, the genesis of a space colony. It contains miniature mountain, desert, forests, and a 9 million liter saltwater sea. The aim is to test the ability of life forms to survive and thrive in a sealed environment for the extended time period. The whole endeavor is envisioned as a prototype for human habitation in space, a learning laboratory for other biospheres on space stations, the Moon or Mars. As Dr Walter Adey, director of the Smithsonian Marine Systems Laboratory and a consultant on the program, commented that scientists try to make simple rules for complex processes, so:[28]

> Biosphere 2 is the first serious attempt to step beyond this and work with organisms and ecosystems as the complex systems they really are.

Mark Nelson, chairman of the Institute of Ecotechnics, a London-based ecological organization, which conceived and originally managed the research and development at Biosphere 2, concluded:[29]

> The technology for producing oxygen and other necessities of life are readily available for sustained human exploration of the heavens. The science of biospheres, understanding how biospheres operate on a planetary scale as well as our microscale at Biosphere 2, is one key to opening this frontier today in preparation for future frontiers on Earth and in space.

Prior to entry the first crew of eight—dubbed **biospherians**—trained together as a team for some months, including survival trips to remote areas. The four men and four women were sealed into Biosphere 2 by airlock on September 26, 1991, beginning continuous testing of the regenerative life-support systems in five biomes. Only one, Roy Walford, a pathologist from the University of California—Los Angeles, was a trained scientist. Two thousand electronic sensors linked to one of the most advanced artificial intelligence computing systems assisted biospherians in monitoring the ecosystem within the 90% closed environment. Inside the air-conditioned enclosure, the plan was for the crew to grow, harvest and eat their own food—about 2500 calories a day, selecting their own menus from the 26 crops to be raised, as well as to drink recycled water. They enjoyed television and videos, as well as other creative hobbies, and were supposed to exchange only information and energy with the outside world during their 24-month stay within this interior space. Project secrecy was such that little was known as to how the initial crew were selected and trained, as well as how they were to be monitored and supported. But by October 1991 one crew member had left the airlock for medical treatment and reportedly carried in extra supplies upon her return. By December 1991, project operators conceded that they needed to pump in extra air, and were coping with problems within the enclosure, such as the spread of plant rust disease. The first mission ended in September 1993, and the crew members emerged haggard, having lost an average of around 11 kg during their two-year isolation in an artificial environment. Among the serious problems discovered in "Biosphere 2" were: (1) the farm feeding did not live up to expectations, partially because the external El Niño effect reduced sunlight from the outside—the closed ecosystem could not supply 2500 calories per day for the biospherians; (2) declining oxygen levels within the biosphere itself—dropping below 14.5%.[30] The pioneer mission was also marked by controversy over management's: concealment of setbacks; differences with the scientific community over the experimental nature of the project. The special independent committee of ten scientists who were to oversee the simulated "space habitat" resigned because of lack of cooperation by the operators, who were not listening to the advisors. To address the Biosphere 2 scientific shortcomings, a panel of scientists under Tom Lovejoy, Smithsonian Institution's ecologist, studied the whole project, issuing a report and recommendations which are being acted upon (*The Scientist*, February 8, 1993).

A second mission was undertaken in February 1994 with a crew of seven, but the time enclosed was curtailed. Five men and two women stayed in isolation for only six months, but emerged from the sealed laboratory healthy and well-fed. By April 1994, six officers of the original operating company were relieved of duties. The new administration created a research consortium with scientists from Columbia University's Lamont-Doherty Earth

Observatory where the emphasis at Biosphere 2 is now on research in environmental sciences, especially global climate change, biodiversity, and sustainable agriculture. Jointly a non-profit institute has been created to open Biosphere 2 for research by universities, national laboratories, and private teams. Dr Bruno Martino, a biochemist from Harvard University, was appointed in August 1994 as its new scientific director. From both a tourism and an ecological perspective, Biosphere 2 was successful. The hope is that its experiments may eventually contribute to NASA's Controlled Ecological Life Support System, which is developing biogenerative life-support systems to produce, process, and recycle biomass. The Biosphere 2 hardware may someday be tested in an orbiting space biosphere. Some other positive byproducts of this venture are:

- A consortium of Japanese corporations and NASDA are planning **Biosphere J**—"j" is for junior since it would be a quarter the size of its Arizona counterpart. It is to be built in northern Japan.
- Biosphere 2 patron Edward Bass has donated $25 million to Yale University to form an **Institute for Biospheric Studies (YIBS)** to conduct further ecological research.
- Spin-off technologies that will have terrestrial applications, as well as in space.

6.1.4 Design for Extreme Environments Assembly

Prof. Larry Bell of the University of Houston's College of Architecture and other colleagues in the United States and Canada, Russia and Ukraine, Great Britain and Japan came up with a brilliant concept for professional cooperation relative to studies on exotic environments. They formulated an international organization that sponsors conferences every two years for representatives from academia, industry, commerce, government, and consultancies. Called the **International Design for Extreme Environments Assembly (IDEEA)**, it is open to those interested in research, planning, and operations. The focus is on sustainable development in extreme, harsh, and hazardous conditions, as well as the impact of extreme environments on environmental change and policy. An environment is defined as **extreme** when circumstances of the locality threaten, disrupt, or cannot normally sustain physical, social, institutional, and economic life. Such conditions may be caused by:

- extremes in climate, geographies, topographies;
- environmental change, hazards, or contamination;
- chronic poverty, natural disasters, political/civil conflict.

Extreme environments, such as Antarctica or outer space, present severe operating conditions that drive science and technology development, causing the invention of new designs, management methods and models which can be utilized elsewhere.

The **IDEEA** participants are also concerned about: (1) application of innovative, advanced and intermediate technologies in such settings; (2) transfer of knowledge, methodologies, approaches, materials, technologies, and experiences from one setting or location to another; (3) identification of positive initiatives for global security and environmental policy with a network of organizations, governmental bodies, and individual leaders. What is unique about this strategy is that it is multidisciplinary, bringing together a cross-section of both researchers and practitioners. Thus, there are exchanges on a central theme that draw together a wide variety of specialists, such as architects and construction experts, behavioral scientists and medical professionals, lawyers and political scientists, as well as members from various national space agencies.

IDEEA ONE was convened at the University of Houston in November 1991, attracting almost 500 attendees from a dozen countries. The exciting presentations resulted in a 958-page volume of proceedings ... . **IDEEA TWO** was held in October 1993 at McGill University in Montreal, Canada, and saw an expansion of networking and information sharing for problem solving in extreme environments ... . **IDEEA THREE** was cancelled in March 1996 by the Oxford School of Architecture, for Oxford, U.K.[31] The range of topics to be discussed is depicted in Exhibit 58.

REPRESENTATIVE TOPICS FOR IDEEA THREE

GLOBAL FUTURE•ENVIRONMENTAL EXTREMES•TECHNOLOGY TRANSFER
ENVIRONMENTAL CHANGE•GLOBAL SECURITY•PLANETARY EXPLORATION
FRONTIER TECHNOLOGY•KEY BIOMES•SUSTAINABLE DEVELOPMENT

ENVIRONMENTS	FOCI / EMPHASES
DESERTS	AUTOMATION / ROBOTICS
DENSELY-POPULATED CENTRES	DESIGN / CONSTRUCTION
DISASTER ZONES	DEVELOPMENT PRACTICES
MINES / TUNNELS	EMERGENCY PLANNING
MOUNTAINS	ENERGY SYSTEMS
OCEANS	ENVIRONMENTAL IMPACTS
POLAR	HUMAN FACTORS
SAVANNA	INFORMATION TECHNOLOGY
SPACE	LIFE SCIENCES
SUBTERRANEAN	OPERATIONS / LOGISTICS
TROPICAL RAINFORESTS	POLICY / LAW
URBAN CONGLOMERATIONS	PRACTICE / MANAGEMENT
WILDERNESS	WASTE / LIFE SUPPORT

" There is one thing stronger than all the armies of the world and that is an idea whose time has come "

Exhibit 58—**IDEEA** Conference Topics. The scope of themes and emphasis at such multidisciplinary meetings is illustrated above for the Third **International Design for Extreme Environments Assembly**. *Source:* **IDEEA UK** Secretariat, Oxford School of Architecture, Oxford OX3 OBP, U.K.

In concluding this opening section on terrestrial analogs, we have reviewed four professional activities that have great implications for the high frontier. But ground simulations, such as Biosphere 2 and ESA isolation studies, cannot test for all the effects of closed ecosystems within a microgravity environment—such as, the impact of radiation, ozone sunlight levels, severe cold, and soil nutrient conditions. However, there are important contributions to living and working in space which come from the accumulation of knowledge and experience about isolated, confined, or extreme environments. Learning to construct and manage habitat biospheres under such circumstances allows for the transfer of technology and insight aloft on the high ground. We have even more to learn from humankind's collective experience in orbit to date.

Personnel Flow To and From Orbit

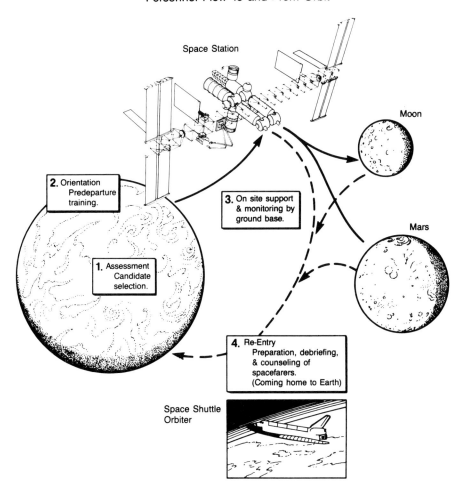

Exhibit 59—Space Personnel Deployment System*. The author's concept of a four-stage system for sending people to/from space is illustrated here by Dennis M. Davidson.

6.2 SPACE PERSONNEL DEPLOYMENT STRATEGIES

In sending humans offworld, especially for long durations, the author has hypothesized that a "systems" approach should be developed for such voyagers based upon the previous terrestrial analogs and our limited experience aloft.[32] Incorporated in this system would be careful evaluation and selection of every would-be spacefarer, as well as preparation in interpersonal and intercultural relations and team building. Such an approach should contribute to mission achievement while preventing or reducing depression, withdrawal, hostility, paranoia, and other mental health problems that may afflict orbital travelers.

A **space personnel deployment system** (SPDS) may be defined as a planned and orderly means for the exchange of people to/from Earth and the space frontier. The system described goes beyond what is currently in place for elite astronauts and cosmonauts on relatively short-duration missions. SPDS is intended for all "those others" who are going into orbit as contractors or settlers.[33] The essential components for this would consist of four stages as are illustrated in Exhibit 59—namely, assessment and orientation before launch, in-space support services, and re-entry policies and programs. We shall now examine each element in the model, attempting to answer four key questions within each phase—who, what, how, and why—as well as to consider some research issues involved with each stage of this deployment system:

When the Soviets selected their first cosmonauts, they recruited 20 who were to be brave, reliable, physically fit, not prone to panic, capable of mental endurance, and familiar with 'things up there.' Later selections, like their U.S. counterparts, put more emphasis on engineering backgrounds, persistence, and more stable personalities. A high rate of attrition was built into recruitment, so about three times the number required were selected. Gradually selection went beyond the military and pilots to include civilians, especially scientists, with a wide mix of age, experience, and background. When man's first home in space, the Salyut station was launched in April l971, crewing policy changed to an all-Soviet resident crew, and visiting Intercosmos crews with internationals and persons with more diverse specializations.

—*Race into Space: the Soviet Space Program* by Brian Harvey.
Chichester, U.K.: Ellis Horwood, 1988.

Stage One—ASSESSMENT

Assessment aims to recruit and select the most suitable people to go aloft, while screening out potential misfits who would jeopardize their own or others' wellbeing, as well as to ascertain the level of candidates' competencies and coping skills.

Space developments in the 21st century will find a variety of organizations, in addition to national space agencies, sending people aloft. In time, we can expect sponsors to range from aerospace contractors and commercial enterprises to various public agencies and the

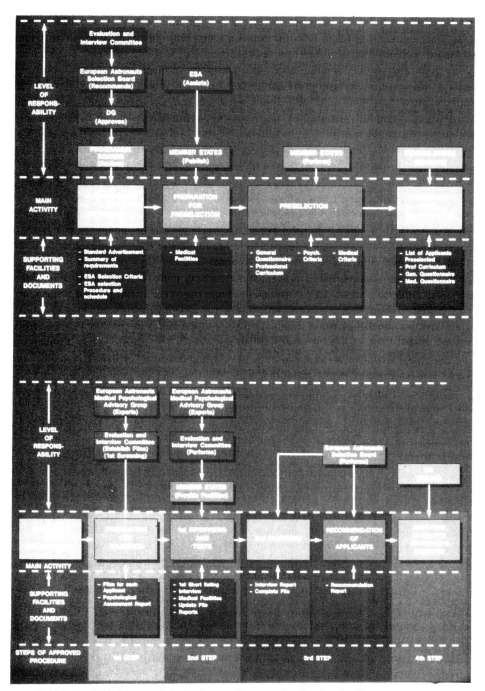

Exhibit 60—ESA Astronaut Selection Process. The procedure for choosing European astronauts is illustrated in this flow chart. *Source*: The ESA's European Astronaut Centre, Cologne, Germany.

military, media networks and even tourist firms. Eventually, non-profit associations with special interests may wish to underwrite their own space community, just as happened among the early American colonists. Whatever the institution that sends a human onto the high frontier, it should exercise a measure of responsibility in the spacefarer's selection, preparation, and wellbeing, particularly with reference to that person's impact on the space group of which he or she is a part. Pre-departure evaluation of candidates delimits costly care and rehabilitation aloft. In the next century, the selection and training issues may move beyond the control of individual sponsoring organizations. Some world entity may have to be established, such as a Global Space Administration or a **Space Metanation**, to regulate and monitor space assessment and access, as well as to resolve other multinational space issues.[34]

A possible model may exist in the European Astronaut Centre, which is responsible for defining the selection, qualification criteria, and coordinating the training of future European astronauts from many nations on that continent.[35] The Centre in Cologne, Germany, also selects candidates for flight or ground activities, such as equipment development, or for special training to liaise with astronauts in orbit, whether they fly on American or Russian spacecraft. Since the Centre is responsible for human safety aloft, it has the authority to define rules and ensure their observance. In its supervision of Europe's manned space programme, the EAC activities include crew assignments, instructor selection, management of training facilities and external relations. Under the ESA, this regional center represents 13 member countries in their participation on NASA's Shuttle flights (**Spacelab**) and the future **International Space Station**, especially with reference to the European-crewed orbiting laboratory, **Columbus**, and a planned spaceplane/mini-shuttle, **Hermes**. The same responsibility extends to Europeans who fly on the Russian's **Mir** space station. Exhibit 60 provides an overview of their selection procedure within the ESA management structure. According to André Ripoll, head of the European Astronaut Centre, the candidates for astronaut training must be citizens of ESA member states, healthy, efficient human beings who are both good professionals and operators with a technical or scientific background (*ESA Newsletter*, January 1991). Pre-selection begins in the member states, who put forward 3 to 5 candidates; the ESA selection board will then choose the persons to be trained for specific assignments, such as laboratory or spaceplane specialists. Other formal requirements for ESA astronauts are:

- preferred age (27–37); height (153–190 centimeters; ability to speak and read English and/or Russian; physically fit and psychologically apt, including a sound health history and normal weight;
- possess a university degree or equivalent; have professional experience in a relevant field (e.g., natural science, engineering, medicine, pilot).

The typical ESA **Columbus** astronaut will be expected to be able to stay up to six months in orbit, be a career person eligible to fly over twenty years. It is no wonder, then, that the German Aerospace Research Establishment (DLR) recommends that astronauts get both basic human factors training and special psychological preparation for specific crew missions.[36]

In attempting to evaluate personnel for long-term missions, the following **guidelines**, developed by the author from terrestrial experience with foreign deployment systems, may prove useful, possibly for testing the **International Space Station** or a lunar base in the next decade:[37]

• WHO—Screening requirements of space travelers and settlers would apply to all who utilize tax-supported shuttle or spaceplane transportation, whether space agency personnel, contract workers, visiting professionals, public servants, media representatives, military members, foreign dignitaries, and eventually tourists. When the time arrives that family members accompany such personnel into space, then they too should be included in the assessment process. For the 21st century, at least, the principle proposed here initially is **selectivity**—the time for experimentation with this process is in the immediate century ahead.

• WHAT—the aim of the evaluation should be to determine the suitability and adaptability of space candidates to deal effectively with the new environment and situation aloft; both physiological and psychological evaluations should determine the individual's capability to deal with differences and difficulties in long space travel, to live in outer space under constrained conditions and cope with its stressors. The latter include confinement and dependence on life-support systems; lack of privacy and sleep disturbances; high work load interspersed with monotony; reduction in leisure-time activities and social contacts. Essentially, the process seeks to identify proneness to "space culture shock," as well as areas for special training so as to improve coping, stress management, and human relations skills when in orbit.

• HOW—assessment might be conducted through a specialized center that would utilize a variety of disciplines and means for evaluation purposes, such as psychological interviews and group meeting, instrumentation and testing, and simulations (live gaming, computer cases, and virtual reality exercises). It is conceivable that expert systems could be designed for this purpose which are based upon accumulated research about the qualities that make for success within the isolated, confined environment. In the beginning, the objective might be to choose those with competencies that ensure survival and mission accomplishment, but eventually the aim may be to group spacefarers by personalities and interests that share values, goals, and concerns for a specific type of settlement.

• WHY—recruitment and selection should not only seek appropriate space "pioneers," but determine special counseling and training needs of those chosen, particularly with reference to the family that is left behind on Earth. Obviously, the first concern is to exclude from space communities those who would be unsatisfactory for such an assignment, subject to premature return symptoms, or likely to become a disruptive influence. Ultimately, the evaluation should choose people who can contribute the most to the emerging community in orbit. The United Nations long ago indicated that spacefarers were **Envoys of Mankind**; therefore we should choose people to go aloft who will be representative and diplomatic![38]

Initially, the sheer cost of transporting people to and from space should justify a careful assessment policy and program; certainly, the expenses associated with premature off-world return would warrant investment in a deployment system. Outside of space agency personnel, the mission specialists are the prototype of other civilian space workers to come. Preferences may, for example, be given to those with special qualifications, such as:

— veterans of some type of service (astronaut/cosmonaut, military or Peace Corps training, extreme environments, etc.) ...

— those committed to long-term space living and willing to assume inherent risks ...

— healthy, well-balanced married couples with dual competencies ...

— professional specialists whose expertise is required in orbit.

Exhibit 61—Crew Prototypes of the Future. In June 1995, there was a docking of the Shuttle **Atlantis** to the Russian space station. The STS-71 crew contained two cosmonauts when the link-up was made with the **Mir 19**, which had one NASA astronaut on board. The above photograph of the combined Russian/American crews, including two women, is symbolic of future diversity and synergy in mission personnel selection. *Source*: NASA/*Countdown*, May/June 1995, p. 54.

Exhibit 61 provides a view of heterogeneous crews of the future, a mixture of gender, nationalities, and professional backgrounds.

The assessment process also involves **screening out** certain types of candidates, such as those with personality problems. Regarding long-duration spaceflights, NASA has expressed concern about excluding those who: (a) lack social competencies or human relations skills; (b) lack communication skills for reliable interaction among the crew, and between the crew and ground control; (c) tend to be perfectionists and autonomous. Predeparture evaluation is difficult in predicting "wild cards" that may arise aloft in crew relationships, such as may be caused by jealousy or sex/love factors.

Currently, NASA selects for two types of astronauts: Pilots *who serve as Shuttle commanders (onboard responsibility for crew) and pilots who assist in controlling and operating the vehicle and in deploying/retrieving satellites. Such candidates must have a minimum of an undergraduate engineering or science degree and preferably advanced degrees; 1000 hours of pilot-in-command of jet aircraft with flight test experience considered desirable; ability to pass a Class 1 space physical; height 64–76 inches (limit is 6 foot 4) ... .*

Mission Specialist *who works with commander and pilot relative to crew planning and operational activities, consumables usage, conducting experiments and payload handling; persons with unique expertise who do not have to be agency personnel, but may be sent by a contractor or foreign government. Required to have undergraduate and preferably an advanced degree, especially in engineering or biological/physical science, or mathematics. Minimum of 3 years professional experience in speciality unless an advanced degree. Must be able to pass Class II space physical, and have height between 58.5 and 76 inches. Military may apply through their branch of the Armed Forces, but civilian teacher program is presently suspended. There are no gender or age requirements.*

Those selected for either program are Astronaut Candidates who must successfully complete one year of training, including interviews and orientation. Civilian candidates who are not finally selected as astronauts may continue in other positions.

—*Astronauts Selection and Training*, AHX Astronaut Selection Office, NASA Johnson Space Center, Houston, TX 77059, U.S.A.

Assessment Research Needs

The most immediate need is for information sharing among worldwide space agencies on current and future practices in crew selection and training. Because of the proclivity in the past to give preference to pilots with flight experience, many speculate as to whether this should remain a factor in future spacefaring candidates, such as a "civilian-in-space" program. The rationale was that such "flyers" come from a stressed background and are used to dealing with life-threatening circumstances, so therefore are better able to cope with the microgravity environment. Apparently, aviation provides a three-dimensional experience, teaches crew coordination, accustoming one to be concerned about displays and consumables. This is a matter for further study, along with the neuropsychological tests which the U.S. Air Force use to select their personnel for spaceflight. For the manned spaceflight engineer (MSE), the military sought qualified personnel with proven performance and initiative, self-confidence and perseverance, as well as technical competence. Are these still such valid criteria for other space travelers? The Russians have already entertained on their **Mir** station a variety of professionals from Japanese journalists and a British female chemist to a variety of international astronauts from ESA/NASA, as well as from other countries in Eastern Europe, Vietnam, India, etc. All received additional training at the cosmonaut center in Star City.[39] It is just a matter of time before national space agencies will no longer be able to control access to space, so research is necessary on some basic global criteria for selecting future **Envoys of Humankind.**

As mentioned previously *The Human Role in Space* project was an early attempt to examine human qualifications for space activities, including sensory/perceptual, psychomotor/motor, and intellectual competencies required.[40] The study was deficient in ignoring psychosocial factors. Harrison and Connors[4], previously cited, also reviewed the experience of humans living successfully in exotic environments, identifying some socially desirable traits among which have implications for selection of spacefarers: personal competence, emotional stability, and social versatility. The top performers in such situations manifest strong task or mission orientation; coping skills with depression, stress, and anxiety; appropriate group interaction skills, being neither too introverted or too extroverted.

Today's payload specialists, usually scientists, are more like tomorrow's average spacefarer—typically they are chosen by their companies or a foreign space agency; most women and internationals who have flown the Shuttle came aboard under this category. Their profile includes high levels of graduate education and scientific accomplishment; a generalist with multiple skills or expertise in a required specialized field; ability to pass NASA's medical and psychological screening, as well as to learn quickly in their intensive training programs. One example of an industry astronaut is Charles Walker of McDonnell-Douglas, who returned from orbit to become president of the National Space Society.

Once the criteria for spacefarer assessment have been agreed upon, a related issue is what astronauts are selected for specific missions and who does the selection? Dr W. K. Douglas (ref 16 in Chapter 5) interviewed astronauts on this subject of crew selection but received mixed suggestions as to who should make the evaluation and choice, ranging from an agency committee or management on the ground to the crew commander for a particular mission. For long-duration flights, these astronauts seemed to prefer peer review, so as to ensure compatability. They recommended that the astronauts should have had experience training and working together before going into orbit. These spacefarers thought a commander should have a veto over group membership, including mission specialists. They envisioned that space station crews would be selected for their skills, motivation, physical/mental health. One commented that what is needed for longer flights aloft are sensitive, intelligent people, not "macho-man" types.

But the real opportunity for further research is evaluation relative to groups who have passed the ground screening and are actually working together aloft. This would be a performance assessment that goes beyond their work productivity to their teamwork and morale. That is, how do they apply their interpersonal, intercultural, and team skill training as they actually interact as self-directed teams? In analyzing crew selection for a future Mars expedition of two years in orbit, Filbert and Keller suggested four categories of critical skills requirements:[41]

(1) competencies in navigation and piloting;

(2) competencies in equipment operation and maintenance;

(3) competencies in scientific study and analysis;

(4) competencies in medical/dental diagnosis and treatment.

Note the complete absence of psychosocial and team skills, so essential for a successful mission of this length and risk! Further, researchers indicate that social competency should be a dimension to assess in crew composition.

Although American/Russian/European/Canadian selection criteria and experience for spacefarers are valuable, more broad-based policies and procedures are required for larger, more diverse space populations. The potential size and number of sta- tions/bases/settlements in the next century demand a more sophisticated system for choos- ing and evaluating candidates, so expanded assessment research is in order. This is partic- ularly necessary on the matter of multicultural crews of which the STS-71/**Mir** crews is a forerunner, as shown in Exhibit 61.[42]

Stage Two—ORIENTATION

Once spacefarers have been carefully evaluated and selected, the next step is to effectively orient them before departure to space living, safety, and culture through both self-learning and training ... .

In preparing and educating space explorers and settlers, the curriculum and its scope would vary depending on the intended orbit, length of stay, particular mission and previ- ous spaceflight experience. Certainly, it would be based upon the training of global space agencies, but go beyond that with civilian space pioneers. Since the selection and training of NASA astronauts at the Johnson Space Center and the Russian cosmonauts in Star City are well documented, we feature here the emerging European program as outlined in Ex- hibit 62.[43] The basic training is a one-year, and whole program four years in length. It begins at the European Astronaut Centre, Cologne, Germany, where the principal crew training and medical facilities are located. Four branches specialize in additional prepara- tion of spacefarers from that continent: in France, Marseille hosts the External Servicing Facilities where EVA training takes place, while Toulouse is the site of the Hermes Sub- system Training Facilities; in Belgium, Brussels offers the Pilots Training Facility for the **Hermes** flight simulator and trainer aircraft; in the Netherlands, the Robotic Training Centre provides training on the **Hermes** robotic arm. The German Aerospace Research Establishment (DLR) has been the leader in the psychological training of scientists for space flight.[44]

As the European Space Agency enters spaceflight agreements with the American, Canadian, and Russian space agencies, their astronauts/cosmonauts undergo further crew training for specific multinational missions at either the Johnson Space Center in Texas or Star City near Moscow (e.g., for flights on the NASA Shuttle or Russian **Mir** station). ESA is involved with 16 other nations in preparations for construction and staffing of the **International Space Station** as previously explained, so it is anticipated that multi- national crew training may occur for that macroproject.

Next we continue with prognosis on the future orientation of spacefarers for long- duration missions and settlement.

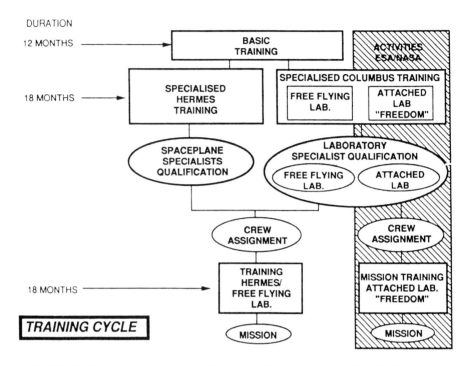

Exhibit 62—European Astronauts Training Cycle. *Source*: ESA Fact File #3—European Astronaut Centre, Cologne, Germany.

- WHO—A short-time visitor, for example, might not be required to undergo as extensive indoctrination as one who is expected to stay months or years aloft. Novice spacefarers, on the other hand, might undertake a longer training course than veterans of spaceflight. Families of the space voyagers should be involved, but their program would vary, depending on whether they will remain on the ground or go aloft. Those who are to become space settlers on the Moon or Mars, for instance, might be exposed to the most rigorous and lengthy preparation, possibly as family teams.

- WHAT—The pre-departure training should focus upon the challenges in extraterrestrial living, from survival and safety to cultural and interpersonal skills.[45] One area of learning has to do with achievement of mission objectives, including the technical and operational aspects of transport, equipment, ecosystems, and coping in a zero or low gravity environment. Another would be in the area of the behavioral sciences so that crewmembers and future settlers function effectively as a team within a space community. The latter preparation might include human behavior issues and skills related to motivation, communication, conflict resolution, negotiation, leadership, team management, and family relations. Eventually, as colonies evolve, the preparation in advance of lift-off will have to include matter of governance and administration. Consideration needs to be given to appropriate and relevant mission instruction for spacefarers in the following subject areas:

(a) space sciences, such as astronomy, resource observation (including mapping and photography);
(b) space technologies, such as automation and robotics, communications and computers, orbital construction and mining;
(c) space migration, such as wellness and creative use of leisure aloft, and eventually community development;
(d) foreign languages/cultural understanding for multicultural crews and missions, such as basic English, Russian, Japanese, or Chinese as required.

The human resource development (HRD) program designed for 21st century spacefarers must be cross-functional and enriched beyond pragmatic training of today's astronauts and cosmonauts.[46] Tomorrow, entirely new learning systems will have to be created, such as in **space ergonomics and ecology**, and **biocultural courses** in long-term human adaptation to microgravity. Such innovations will emerge eventually from long-term orbital experience of **living and working in space**.

Creative input will have to be developed for planning on new space careers which are only originating.[47] As missions expand in size and scope, diverse multicultural workforces will require group dynamics and intercultural training to become synergistic teams. Furthermore, crews must receive special education to undertake expanded roles—Dr David Baker reported that cosmonauts, for instance, have been given instruction in observing resource deposits, both gas and oil, within their country; have helped in the development of maps for remote areas of the former U.S.S.R. as well as in photography of civil engineering projects.[48] Similarly, Shuttle astronauts engaged in "Mission to the Planet Earth" have to acquire new competencies to fulfill that goal.

Cosmonaut training began with survival skills and expanded to include other competencies. Like NASA, the training takes places in various bases around the country, but is largely in Gagarian Training Center at Star City located near Moscow. Like the Johnson Space Center in Houston, it has facilities for both training and mission control. When Valentina Tereshkova was chosen to be the first woman in space in 1963, there were also four other women selected with her for training—though they never got into orbit, they worked in ground support jobs. It would be nineteen years before the Soviet system would permit the next Russian woman to fly in space.

—Dennis Newkirk, *Almanac of Soviet Manned Space Flight.*Houston, TX: Gulf Publishing, 1990

• HOW—innovation in pre-departure orientation of spacefarers must extend from the design of self-learning and intensive group learning, to new educational methodology and technology. The instructional possibilities for conveying the above subject content might include interactive video, video case studies, simulations (live and through computerized graphics), expert systems, programmed learning, instrumentation and tests, and a variety of group process, as well as field trips. Partnerships should be formed among government, corporations, and institutions of higher education not only to produce HRD instructional programs and materials for the deployment of spacefarers, but perhaps to plan for the establishment of space *academies and universities—first ground-based and later in orbit.*

• WHY—the costs, complexities, and risks presently inherent in long-duration space missions demand that initially access be limited to the competent, educated, and multi-skilled, as well as those who are well-balanced "copers" and "can-doers." In a space-based culture, knowledge and information ensure survival and success. As anthropologist Edward Hall remind us: any culture is primarily a system for creating, sending, storing, and processing information with communication underlying everything, whether in words, behavior, or material things. Thus, the preparation for space culture should be comprehensive.[49]

Orientation Research
Pre-launch training today for relatively short missions concentrates on technical skills needed and the biomedical demands of the orbital environment, including the housekeeping procedures that facilitate life on the Shuttle or **Mir** station. Science writer Alcestis Oberg has described the typical preparation of an astronaut, with its emphasis on interdisciplinary familiarization in case of emergencies.[50] Currently, classroom work focuses on basic sciences, particularly as this relates to shuttle systems: electronics, computers, aerodynamics, orbital mechanics, spaceflight physiology, Earth observations, astronomy, planetary science and star identification, spacecraft design and operations, ground support activities, and mission control procedures. The current methodology for this includes real and computer simulations, technical work assignments, flying experience, and public relations activities. The training differs somewhat relative to role as either pilot or mission specialist, but exceptional performance and enthusiasm in applying what is learned is a factor in mission assignment. NASA's head of flight operations makes a final crew selection based on this training process, plus demonstration of competence, dedication, willingness to sacrifice and take on disagreeable tasks, and cooperation. The need for further research centers on how this orientation is suitable not just for future astronauts, but "technauts"—technical specialists and contractors who will go aloft to build and maintain space infrastructures, as well as possibly settle there.

Because the **International Space Station** will create a different living situation aloft Douglas, in the previously cited THURSIS study, did ask experienced astronauts for recommendations relative to changes in the traditional NASA pre-launch training. Their pragmatic feedback included:

• **paramedic training so all crewmembers could assist flight surgeons on long-duration missions ...**
• **group dynamics and interactive training, so the crew may work together more effectively as a team ...**
• **family training or information programs as to what to expect and do, which would be supportive and helpful while their spacefarers are away on long missions ...**

- operational training within a mock-up of the actual Space Station with simulations of actual control room situations ...
- safety training on established procedures, dual warnings of problems, and how to cope with system failures and learn from them ...
- potential problem-solving training on what to do if a certain event or problem occurs.

The group interviewed had faced the hazards of spaceflight, so they advocated cross-training, procedure conditioning, and continuing familiarization with specialized activities (e.g., EVA underwater training). We should remind ourselves that even ground training can be risky, for both astronauts and cosmonauts have died during their pre-launch preparations.

Clearwater and Harrison also identified some of the differences in training for long-duration missions, such as Mars:[51]

- length of spaceflight precludes full mission simulations, and forces partial training built around critical activities—it is more important to learn how to acquire and apply general knowledge to specific situations not covered in simulations ...
- more emphasis on imparting maintenance skills, so critical on prolonged missions ...
- on-board training will be required to shorten mission preparation time, reducing time between acquiring and using skills ...
- expand training in interpersonal dynamics, human relations, and teamwork, so crew understands each other better.

Some of the recommendations in the above studies have been implemented, but additional research would be useful into those not yet acted upon. Further investigations need to address such questions as:

(1) Is contemporary pre-departure training adequate in terms of subject matter, methodology, and final selection of such personnel? What should be dropped, what should be added, especially with reference to long-duration spaceflight?
(2) What can be learned and adapted for spacefarer orientation from the pre-departure training practices of ground-based organizations for professionals on expatriate global assignments?
(3) What innovative educational technology can be utilized in preparing people to go aloft (e.g., laptop computers, virtual reality, expert systems)?

With regard to advanced HRD research with spacefarers, studies should be directed toward diversification both in population going aloft and in organizational sponsorship. The trend would seem to be toward less theoretical and more experimental education, with greater emphasis on psychosocial learning for group living and space culture. Behavioral science research to date on such matters has been inadequate. For example, the Canadian Space Agency and DARA in Germany have integrated psychological components into their human factors training, but have there also been follow-up studies on its effectiveness after their astronauts have returned from orbit? Other **Space Education Analogs** are provided on the next page.

SPACE EDUCATION ANALOGS:

Numerous other analogs exist for developing a more broad-based program for spacefarers at all levels of education and for all ages (some are listed in Appendix C). In the U.S.A., for example, these are wide-ranging from the NASA Astronaut Training Corps in Houston to the U.S. Space Camp in Huntsville, from the NASA National Space Grants and Fellowship Program to the Challenger Centers for Space Science Education, to consortia like the University Space Research Association and the University Corporation for Atmospheric Research.

Within worldwide universities and colleges, there are many space courses being offered without much central coordination or information sharing. In the United States, for instance, various university space centers and institutes offer majors in space sciences, astrophysics, and astronomy (e.g., University of California—San Diego, University of Arizona, and M.I.T. There are numerous schools of aerospace study (e.g., University of North Dakota). The space courses in higher education today may be the curriculum for tomorrow's space academy. For example in the social sciences, at the University of Hawaii, there is a course offered in space anthropology; at the University of California—Davis, in space psychology; at California State University—Northridge in space sociology and at Niagara University in space settlements; at the University of Houston—Clear Lake in space architecture, habitats, and food services. Space commercialization and policy courses are currently available at Stanford University, Harvard University Business School, and George Washington University in the District of Columbia. In Russia, there is the Moscow Aviation Institute, a technical university; the Space Research Institute of the Russian Academy of Science, the Russian Institute of Space Device Engineering, and other such specialized institutes; and the National Youth League which is oriented to space activities. In both countries, there are professional societies which offer in-service education to members, such as the American Institute of Aeronautics and Astronautics or the Association for the Advancement of Space Sciences and Technology in Russia. In other countries within Europe, Canada, Japan, and China there are numerous space educational institutions. The problem is that there is little coordination or exchanges of information among these resources on designing a learning program for spacefarers. The **International Space University**, previously mentioned, is in the best position to provide this leadership.

There have been proposals discussed with the U.S. Congress to use the "land grant university" concept which so successfully promoted public higher education to fund university space education on the ground and in orbit. For example, in 1987, Senator Lloyd Bentsen and Rep. Michael Andrews introduced a bill into both houses of Congress to establish and fund a system of space grant colleges and fellowships modeled upon the land grant agricultural system of the past. The proposal has yet to be translated into legislation and funded.

The orientation research should also evaluate the effectiveness of any counseling, group therapy, and social work provided to the space voyagers' families left behind on

Earth. Eventually when spouse and dependants are permitted to accompany the spacefarers aloft and participate in regular pre-departure training, studies should be directed to family performance in orbit.

Relative to specialized educational institutions for the preparation of space travelers, two prototypes are emerging which are worthy of further investigation by scholars:

(1) The Japanese are developing plans for a **Moon Park** that would combine research and education on space settlements (*Space World,* June 1988, p. 22). The primary focus of the facility will be to study space crew relationships and psychological problems caused by isolating people who are dependent upon closed ecological life-support systems. At a site yet to be determined, the Earth-bound base would use off-the-shelf equipment and closed-loop systems in water and air reclamation. The project sponsors are the Japanese Institute of Space and Aeronautical Science and the Obbayashi Corporation.

(2) The **International Space University** (Appendix C) would seem the ideal entity to coordinate the global orientation and education for space citizens of the future. Presently, ISU coordinates a multinational, multi-institutional, mobile educational program in space sciences and management at the graduate level. In different nations, it offers its instruction during the summer with a distinguished visiting faculty from many countries and institutions. It also has a permanent campus in Strasbourg, while developing satellite affiliate branches linked together by an electronic network. Eventually, ISU expects to develop campus extensions on the high frontier.

There may come a time when national governments or the international community will consider founding **space academies** for the preparation of a more diverse spacefarer in everything from astrolaw to astrobusiness! In a novel under way on lunar industrialization, the author envisioned such a Unispace Academy adjoining the proposed spaceport on the Big Island of Hawaii, and operating in conjunction with the federal government's Center for Cultural and Technical Interchange between East and West. In December 1985, a Japanese parliamentarian who headed up his nation's space committee, Tetsuo Kondos, was credited with circulating a proposal for the establishment of an international space center/university to be located in the Pacific area. Within the U.S.A., the late Senator Spark Matsunaga suggested the state of Hawaii for a space academy during a 1986 public forum held by the National Commission on Space. Later that same year when the NRC issued its famous report on *Pioneering the High Frontier*, acknowledgement was given to the plan for a "Pacific Space Center"—a potential international spaceport that would promote space-related educational and business activities for the Pacific Basin during the 21st century.[52]

Dr Bruce Cordell, a California space consultant and astronomer, has done extensive studies on all aspects of manned Mars missions, offering this profound observation on the role of education for such an accomplishment:[53]

Education and an elevated level of human consciousness are essential to human survival and prospects ... . In reality, education should involve not merely the acquisition of skills, but an increase in perspective and an elevation of the human spirit. The prospect of the evolution of civilization into the cosmos offers humankind a superb opportunity to achieve continuance, expansion, prosperity and wisdom.

Stage Three—IN-SPACE SERVICES

In-space services are the on-site support and monitoring, both physical and psychological, which is necessary for survival and mission accomplishment, as well as to facilitate human integration into emerging space cultures and communities ... .

• WHO—those who depart the Earth to live and work aloft deserve support from their sponsoring organizations until such time as a space settlement or colony can function somewhat independently. In addition to the families of the space voyagers, a host of professional personnel on the ground will be necessary to provide in-space services. Once at a space station or base, some specialists there should be designated responsibility for the management, distribution, and maintenance of such services. One proposal, for example, calls for contract corporation employees at NASA's planned Space Station to render separate services ranging from hotel accommodation and food, to heat and fuel (for both the station and spacecraft). Currently, the record for humans in space at the same time is only thirteen; imagine the challenges when those numbers jump to a hundred, a thousand, a million!

• WHAT—the support services aloft generally are likely to be of three types. The first would be operational—ensuring the technical or mechanical aspects of a station or base so that it remains in orbit or is functional ... . Another dimension involves physical and biological support to maintain a group of people in outer space. Obviously, this involves transport, habitat, communication, nourishment, equipment and supplies to accomplish the mission. More specifically, it involves provision of "biospheres" for a total Earth-like environment, a closed ecological system for recycled air, water, food, and waste management (CELSS). Maurice Averner, a manager for this program at NASA headquarters, says that the Agency is examining regenerable life support for Moon and Mars' outposts, possibly a physical/chemical system based on growing plants to provide food, water, and oxygen (*Space News*, March 4–10, 1991, p. 6). NASA–Ames has contracted for a **Life Science Centrifuge Facility and Support System** for the space station, which will evaluate the effects of different levels of gravity on living organisms, including plants and small animals. Life-support systems can be built not only into habitats and spacesuits, but also in trailer-like vehicles that can be transported around the surface of another planet. On the Moon, for example, "life pods" that supply astronauts with emergency shelter, oxygen, water, food, first-aid supplies, and spacesuit repair kits could be strategically located around the lunar surface.[54] Human engineering is improving the orbital work environment by new designs for workstations, habitability systems, human–computer interface, and extravehicular activities ... .

The final dimension of in-space services is psychosocial, helping to delimit "space culture shock" on long-term flights, while furthering learning in space safety, survival, and knowledge. Such activities contribute to morale, productivity, and high performance of the spacefarers, as well as their integration into the space environment and culture. The author has been involved in several recent support proposals within the NASA Small Business Innovation Research Program. An example of one such educational/recreational service has been proposed by Nadja and Michael O'Hagan of Orincon Inc. in San Diego, California. These researchers seek to develop an interactive expert system for space personnel to assess and maintain psychological wellbeing onboard a spacecraft. The strategy would involve self-reporting interactive evaluation and supportive treatment of stress, coupled with computerized game playing based on adaptations of fuzzy set logic. Two other proposals engendered by NETROLOGIC Inc., also of San Diego, were under Dr James Grier Miller as principal investigator. One was a comparison of an expert system and neural network as a decision-aid for optimizing human performance in space. The objective was development of intellectual systems that would enhance the human/automation interface, group interactions and choices, and assist other cognitive processes. The other proposal for the Department of Defense involved creation of portable infrared transponders and wall-mounted receivers to locate and monitor shipboard (or space) personnel, permitting commanders to maintain real-time assessment of the whereabouts and status of key personnel; the on-board system would enhance mission safety and efficiency. The whole field of artificial intelligence and expert systems has vast potential for space frontier applications as a means for improving human productivity aloft.

Throughout spaceflight, there have to be mechanisms and procedures in place for those on the ground and in orbit to evaluate human need, progress, and performance. Ground training can never quite prepare spacefarers for the **unknowns** confronted aloft, especially when the equipment fails in some spectacular way. For example, in 1992, the rescue of **Intelsat 6** required extensive improvisation before Shuttle astronauts could salvage the wayward satellite.

GROUND BASE MANUFACTURING FOR EVA

To construct an orbiting station or colony requires extra vehicular activity. Astronauts will have to work 300 or more kilometers above the Earth for long hours in temperatures as low as -90°C. For that purpose ground manufacturers must design and produce spacesuits that will protect humans aloft from the cold and support their other needs. Two spacewalkers from Shuttle *Discovery* discovered in February 1995 that their hands and feet were getting too cold, so their 5 hour test outside had to be cut short. Dr Bernard Harris, an African-American physician, and Dr Michael Foale, the first British-born spacewalker, performed well, but the insulated suit and other special clothing provided by NASA proved to be inadequate.

• HOW—To illustrate the enormities of these human support challenges, consider these two reports. Since 1961 the Russians have been preparing for a two-and-a-half year manned mission to Mars within a few decades. Their Academy of Sciences have, therefore, conducted extensive research for that purpose on long periods of confinement, and self-contained or "closed-loop" ecosystems; currently, these findings are being applied to the **Mir** station where crewmembers have already remained aloft for over a year. To place humans on Mars, the Russians estimate that a crew of 3 would need 31.5 tonnes of logistics, whereas a 10-person expedition would require 105 tonnes. Thus, to transport such weight from the Earth's gravity well into orbit, the Russians, in addition to their large **Energia**, are experimenting with a heavy-lift rocket (capacity 110 tonnes) and cryogenic rocket propellants. Support services to Mars require a multinational joint venture ... .

In-space support services for Russian space stations have included morale boosts for psychological support, such as:

(a) **special radio and video programs, plus two-way television communication every Sunday between cosmonauts aloft and their families on the ground, and on special occasions (e.g., birthday of a daughter);**

(b) **visiting cosmonauts who stay one week, bringing mail and personal packages, as well as official supplies;**

(c) **partnerships with ground-based scientists through participation in their orbital experiments, especially with plants, gardening, or even animals;**

(d) **periodic television broadcasts to the homeland on their work and accomplishments for the Moscow evening news;**

(e) **opportunities for spacewalks and experiencing the majestic panorama outside, such as time for looking out windows and imagining;**

(f) **maintaining an "Earthlike" work schedule, including at least a day or two off weekly to rest and absorb the space experience.**

To support a person in space with air, food, water, propellants, replacement parts, and other supplies, the Russians developed an unmanned cargo ship called **Progress** that visits their orbiting station at least every three months.

In the Bluth and Helppie analysis for NASA's analysis of Soviet space station analogs, previously cited, these sociologists examined the cosmonauts' environment, technology, organization systems, personality systems, and physical condition. Regarding support services aloft, these findings are appropriate to note here:

- **Salyut/Mir stations are designed for a 7–10 year lifespan with improving life-support systems, windows, air/temperature/noise controls, washable leather walls, easy maintenance with replaceable parts, brighter lighting, light color orientation, computer and automated systems, adequate living quarters with private crew quarters ...**
- **combinations of light and music to relax and stimulate, as well as to give a sense of Earth's seasons ...**
- **disposable clothing designed for ventilation and avoiding wrinkles, as well as provisions for individual preferences to colors and pockets ...**
- **variety in food selection, resupply, and eating arrangements, as well as use of vitamins, artificial stimulants, tranquilizers and sleeping pills ...**
- **structured and surprise leisure activities and exercises, including now two week-**

end days off with a 5-day work week (typically 8 a.m. to ll p.m.) and no split work shifts ...

- psychological support programs through a ground-based team, monitoring system of psychological reactions and stress, and periodic visits of psychologists, sociologists, and physicians.

This report underscores broader human factors in the Russian program which contribute to high performance on extended missions. It notes details about crew quarters, personal hygiene, living and social environment that contribute to crew morale and productivity. This remarkable research records matters of cosmonaut safety training, physical conditioning, sleep habits, role relationships, group management skills, and other station strategies to ensure alertness and performance. Bluth and her research associate, Martha Helppie, have described a space deployment system worthy of some emulation.

THE CASE OF THE PSYCHOLOGICALLY UNPREPARED ASTRONAUT ON MIR

Norm Thagard is a Marine aviator who flew 163 combat missions in Vietnam ... a graduate engineer and medical doctor ... a teacher and researcher ... a husband and father of two children. He also is a NASA astronaut who successfully completed a year of evaluation and training and became a mission specialist. But despite four short duration missions on the Shuttle, at 52 years of age he was not prepared for the isolation, confinement, and cultural differences of 115 days of weightlessness on the **Mir** station. As the first American to come aboard their orbiting home, his four cosmonaut colleagues gave him a welcoming celebration in March 1995. In fact, Elena Kondakova gave him a big hug and a kiss, before giving Norm the customary newcomer gift of bread and salt. At the start Thagard found **Mir** nice and roomy, though he was disappointed that his experimental equipment was delayed by Russian customs; it was weeks into the mission before it finally arrived. Occasionally, he reported to ground control that he felt isolated, unable to speak enough with family and colleagues in Houston. He found the food strange and complained about not being able to speak English enough. Yet in a conversation with the White House, he told the President, "The Russians took great care of me. We're great friends, so I think if what we did on the interpersonal level is any indication, there won't be any problems with this on the intergovernmental level as well."

And after 3½ months aloft, the Shuttle **Atlantis** arrived to deliver a fresh Russian crew and to take Norm and two cosmonaut comrades back to Earth. On arrival home, Thagard was by far the hardiest of the three—the Russians were carried off on stretchers. Dr Thagard walked off without help, hugged his wife of 30 years and their three sons. All three spacefarers then underwent the medical tests and debriefing. Norm observed that if the mission had stretched to six months, he might not have made it ... .Then to his credit, the new NASA administrator apologized to this American who had been the longest in space among his countrymen, compared to the Russian record of 438 days aloft. Dan Goldin admitted, "We put all our emphasis on the physical well-being of the astronauts and the success of the mission. We neglected the psychological well-being, and Dr Thagard made it clear to us."

Now NASA is expanding its life sciences and human factors efforts on behalf of astronauts. Among its plans for offterrestrial manned missions, the Agency has been engaged in long-term research to circumvent the high transportation costs of taking food to space, a critical factor for future colonies in orbit. Under the "Controlled Ecology Life Support System" (CELSS), the agency is experimenting with "astrofarming" so that food can be grown by spacefarers. At their Ames Research Center, biochemists are experimenting with "astrocrops" which are sealed in lucite cylinders used for growing wheat, corn, tomatoes, and soybeans on porous platforms at ten times the density of ordinary farming! The light, air, and wind are controlled artificially, while automatic spraying of roots takes place with nutrient mist—sensors monitor everything and a computer adjusts the mixes of the variables. Before the end of this decade, they plan a Variable G Research Facility of canisters tethered together and to the Space Station which grow crops in an completely controlled orbiting garden. Other scientists at universities and private corporations have been contracted to test astrofarming on a real-life scale, ranging from fish farming, animal raising, and robotic farmers.

In-space services begin to function with lift-off, as experiences with Skylab and the Shuttle so well demonstrate. Upon reaching a space station or base, more life-support systems are utilized in unique ways. Perhaps the new arrival might be paired off with an experienced "buddy" and participate in some form of indoctrination briefings on the local situation aloft. Communication procedures are established to link the individual not just with "mission control," but with families and friends on the home planet. Medical and mental health services are made available, particularly to cope with stressors related to gravity-free living and space sickness, circadian rhythms and sleep patterns, lack of privacy, and anxieties. The work schedule, as well as social and recreational opportunities afforded spacefarers, is important for enhancing the quality of space life. Education and training aloft can be continued by means of team development and technology-assisted learning.

• WHY—Until space settlements become self-sufficient well into the next century, space voyagers will be dependent upon the home planet. In their seminal book, *Envoys of Mankind*, Robinson and White, previously cited, suggest that two mutually dependent types of people and societies are emerging—Earthkind and Spacekind. In the synergistic relations created between these two kinds, the former provides support and service to those who live and work in space, so that its resources may be utilized to the benefit of all (see Appendices A and D). The situation is somewhat analogous to that of 17th and 18th century colonists to the New World who depended upon Old World institutions and people until they could eventually function on their own, contributing in time much more back to Europe than they initially received.

Perhaps the scope of ground support for human spaceflight can best be appreciated in the macromanagement required for the manned Apollo missions—it took at one point 300 000 people in 20 000 companies and NASA to lift the Saturn V rocket and spacecraft (2867 tonnes) for 17 missions which put 12 astronauts on the Moon!

As the third element in a deployment system, in-space services requires the building of an infrastructure on Earth to support space-based activities. This necessitates more than the construction of space vehicles and the providing of life support. It includes on this planet, for instance, regional bases, spaceports, space training centers, communication and tracking systems, etc. Similarly, other infrastructures have to be financed and erected in orbit so that humans can live and work there, as well as explore and settle the Solar System. For example, Exhibit 63 illustrates one type of hardware that can provide human support both on the Moon, and on Mars. The International Academy of Astronauts proposed in 1990 that a multinational lunar base inhabited by 12 persons be established by 2010 (see Chapter 10). Co-author of the proposal, H. H. Koelle of the Technical University of Berlin, said the base would cost $600 billion to build, $1.5 billion to operate each year, envisioning an expenditure of $75 billion over a 15-year period. Previously, in an *Ad Astra* feature column (January 1989, p. 48), Timothy Morgan made the case for such a base by reminding readers that the Moon's resources, just 3 days away, are essential to further human expansion. He recommended that a Lunar Settlement Simulator be constructed on Earth for applicants to spend a year in preparation before being permitted to settle on the high ground.

Exhibit 63—Lunar/Mars Excursion Vehicle. Pat Rawlings of SAIC provides an artistic rendition of an exploration vehicle that could be first tested on the Moon, along with other relevant systems and technologies, with a view to later use on Mars. Operating on the lunar surface in this manner requires a whole new type of long-duration mission training.

In-Space Services Research

Homo sapiens has only accumulated a little over thirty-five years' experience on how to sustain human life in space. The references cited in this book confirm the accomplishments and the setbacks, especially in terms of the U.S.A. **Skylab** station and the Shuttle fleet, but particularly the Russian **Salyut** and **Soyuz/Mir** missions. On-going habitat re-search for the forthcoming **International Space Station**, as well as an initial lunar out-post, should provide more support and comfort for their inhabitants, and improve prospects for Moon and Mars dwellers. R&D must be more aggressively directed toward the infrastructure essential to support space outposts, then settlements.

In the U.S.A, the Shuttle Orbiter is the current prototype of not only future transporta-tion, but also habitats—it is a model of "biospherics." Presently, it provides a home and work laboratory 300 kilometers aloft featuring a closed ecological system and complete life support for seven people for up to six days, or possibly more. Its food services offer a galley with microwave that can supply a diet programmed for 3000 calories per day to compensate for the body loss in orbit of potassium, calcium, nitrogen, and other minerals. On board, fitness equipment can slow the negative effects of weightlessness—loss of mus-cle tone, bone mass, concentration, and disposition. Apart from ground analogs and simu-lations, until it actually docks at a space station, the Orbiter is a real laboratory for ex-panded human factor and life science research related to in-space support services. Mis-sion length there has been extended up to 16 days aloft for behavioral and biological, as well as other material science, investigations on this fleet of four Orbiters—**Columbia, Discovery, Atlantis**, and **Endeavor**. Within them, there are unique research capabilities emerging that hold promise for researchers. One proven example is the European Space Agency's **Spacelab**—a 4 m × 7 m cylindrical laboratory that is also fitted into the Orbiter's cargo bay (see Exhibit 64). This pressurized lab permits ESA researchers to work with a number of experiments, including using external pallets that can be exposed to the vacuum of space. Another orbiting laboratory for studies up to a year or more is the Russian **Mir** station, so that joint human factor research is expanding involving agreements by the RKA with NASA, ESA, CSA, and NASDA.

Private enterprise has many proposals for orbiting research facilities, such as that of **SPACEHAB, Inc.** (600 Maryland Ave. SW, Washington, DC 20024, U.S.A.) in conjunc-tion with the McDonnell Douglas Astronautics Company and international partners in Eu-rope and Japan. This would be a manned experimental module that fits into the Orbiter's payload area and is connected with the maindeck crew compartment. Originally conceived to carry researchers as passengers, current suggestions are to use it primarily for industrial microgravity experiments. It offers opportunities for relatively low-cost access to a manned space environment for the conduct of short-term experiments with reduced docu-mentation and protection of proprietary rights. Its 21 lockers also provide possibilities for Space Station test-bed experiments; even in life sciences and human factors, the schemes exist to take advantages of the pressurized volume of space for research ... . Another promising commercial prospect was described by John Cassanto of Instrumentation Tech-nology Associates, Malvern, Pennsylvania. In an article entitled, "A University Among the Stars" (*International Space Business Review*, Spring 1986), he proposed a **Material Dispersion Apparatus**—MDA is a minilab that weighs less than 2.3 kg and occupies a volume less than 0.01 m^3. Because of its small size, sample experiments can be sent aloft not only on the Shuttle and Space Station, but aboard expendable launch vehicles (ELV). Those seeking to do basic research on materials behavior in space can use the MDA on NASA's Hitchiker program. Now if such devices can only be adapted for life sciences and human factors experiments!

Exhibit 64—Artist's impression of Shuttle Orbiter Carrying **Spacelab**. A high-angle front view of the Shuttle Orbiter carrying **Spacelab** as its primary cargo. Flying since 1983 under management of NASA Marshall Space Flight Center in Huntsville, Alabama, the facility has hosted American, European, and Japanese astronauts engaged in utilizing a variety of technologies. *Source*: NASA/Johnson Space Center.

Although much attention has been paid to the psychophysiological aspects of space-flight, much research yet remains to be undertaken relative to the psychosocial dimensions, particularly with regard to long-duration missions. On-site support will have to satisfy psychological and sociological needs contributing to effective group and community life in the orbital environment. Let us review three ground-based analogs which provide insights for further investigations related to life aloft:

(1)   In examining studies of group living in exotic environments, Harrison and Connors hypothesized that people have to be provided with latitude to respond to changing conditions in order to satisfy human needs—sharply defining roles can discourage functional, as well as dysfunctional, behaviors. What is the optimal balance aloft between prescribed and discretionary roles? How can effective leadership and management practice avoid role overload?

(2) Military serving in NATO's northern flanks operate during the winter in near total darkness. Sources in Oslo, Norway, have reported a suicide epidemic recently that took

four times as many young soldiers lives as during the previous ten years; some killed themselves while on duty; others on home leave did so as a result of suffering from severe depression and alcoholism. Apart from the impact of prolonged darkness and long separation from girlfriends, a fundamental factor contributing to the problem was the attitude of the defense establishment, which regarded this posting to the far north as punishment (*Insight*, November 25, 1985 ... . A lunar personnel deployment system must consider the effects of two weeks of darkness after two weeks of light. What type of support services will Moon dwellers require to counteract this natural condition there? How can the latest research in human circadian rhythm and chronobiology (study of periodic changes in biological patterns) be utilized for the benefit of **Selenians**, the lunar settlers? For example, could a spacefarer's adjustment be facilitated by resetting their biological clocks during dark lunar periods through increased exposure to artificial light?

(3) Research should also be directed to sleeping problems in orbit. While one-third of terrestrial adults are estimated to have such difficulties, approximately 75% of astronauts on Shuttle missions experienced trouble with sleep. Former astronaut Bruce McCandless II of Martin Marietta Astronautics admits that he had the problem, and attributes it to a combination of intense emotional and environmental problems, such as lack of gravity, odd sleeping times, noise and frequent sunrises. Again, the matter is related to disruption in circadian rhythm, the body's daily cycle of sleep and wakefulness. To deal with this issue of concern, Dr Jeffrey Davis, director of Preventative Medicine at the University of Texas, Galveston, reports that NASA is experimenting with astronaut behavior modifica-tion, such as bright light therapy, adjustment of sleeping schedules, sleeping masks, re-straining devices, and opaque windows. 50% of the astronauts are utilizing sleep medica-tion that works quickly with minimal side-effects to impair performance aloft. One of the newest is **Ambien** which is manufactured by G. D. Searles & Company of Skokie, Illinois (for information call the 24-hour hotline, 1-800-SHUTEYE).

(4)   Research literature on living overseas abounds with descriptions of the phenomenon of culture shock and how to minimize it. Such transitional experiences can cause an identity crisis in an individual:

> **Culture shock is a trauma experience in a new or different culture or environment because of having to learn and cope with a vast array of new cultural cues and expectations, while discovering that old ones probably do not fit or work. It is precipitated by anxiety that results from losing all our familiar signs and symbols of social intercourse which orient one to the situations of daily life, contributing to peace of mind and performance efficiency, even when not consciously aware of them.**
> (Kalervo Oberg in Harris & Moran, *Managing Cultural Differences*. Houston, TX: Gulf Publications, 1996, Ch. 7)

For long-duration missions aloft, how can on-site support services counter the dysfunctional effects of **space culture shock**? The latter is psychological disorientation, exacerbated by lack of understanding and knowledge of an alien orbital environment. How can research lead to program provisions and facilities that lessen personal rigidity in responding to the challenges of this new offterrestrial living situation? That is, counseling and group and/or drug therapy to reduce excessive fears, hypochondria, psychosomatic illness; severe confusion, loneliness and depression; withdrawal, hostility, and paranoia; substance abuse or excessive interpersonal conflict. Space living, culture, and climate is different, but people can be educated and helped to cope positively, while creating a cultural synergy with their unusual circumstances.

Other key areas for immediate spacefarer research which will affect in-space services are:

Workload and Stress
Astronauts on short missions described space as a place without stress, but cosmonauts on longer flights (140 or more days aloft) showed evident signs of stress and fatigue. One hypothesis about long space voyages is that the body's immune system will respond more to psychological, rather than physical, conditions. Proper attention to habitability factors, such as esthetics, proxemics, and privacy, can contribute to lessening maladjustment, while positive programs can foster stress reduction.[55] Thoughtful scheduling can also counter excessive workload stress. Excessive workload aloft has been shown to affect cognitive performance by narrowing the attention space and interfering with concentration leading to forgetting of proper task sequence, incorrect evaluations, slowness in decisions or failure to carry out decisions made. Related to this, further investigations are desirable in the emerging field of **cognitive space ergonomics** with reference to mental workload, man–machine interface, and adaptation of ground-based stress training to weightless environs. Pre-departure orientation that deals with issues of emergencies, illness, accidents, and death in orbit may facilitate stress management aloft when such events do occur.

Douglas (ref. 16 in Chapter 5) confronted astronauts with the potential problem of mortuary affairs on the forthcoming Space Station—does one preserve the body, or dispose of it on-site? He also found that astronauts thought there would be less stress if the crew received psychological support aloft, and knew that their families had access to similar services on the ground.

Dr Peter Hancock, when at the University of Southern California's Institute of Safety Science & Systems, engaged in research for modeling and measuring the interaction between physical and psychological stress. His unobtrusive measures for evaluating physiological stress levels might be adapted for a station or base aloft. As part of the Living Systems Applications to Space Habitation project, Hancock proposes to collect and analyze data on an "envelope of life" in orbit, so as to determine a point beyond which stressors from the interacting physical, biological, and social variables may not go without terminating survival (refer to Chapter 3).

Monitoring Performance
Another area for study identified by scientists providing input to the National Commission Space report (1986). The issue was described in *The Space Settlement Papers* as:[56]

> The experimental and demanding nature of human settlements in space will require constant, non-intrusive monitoring of experience, as well as analysis of the social forces in an isolated environment that is without precedent. Progress will depend heavily on learning from this process of feedback.

The Living Systems proposal of Drs James Grier Miller and Philip R. Harris described in Chapter 3 envisions a computer/sensor technology for such non-intrusive monitoring of inhabitants of a space station or base. Besides improved management and safety which could result from such procedures, a continuing learning cycle could be established for the benefit of future spacefarers.[57] Experiments could begin now with such a system aboard the Orbiter fleet and be extended to a lunar base!

Just as NASA today monitors its astronauts aloft as to vital functions and physical well-being, someday organizational sponsors may wish to have means for evaluating wellness and acculturation of their space travelers. In addition to the above-mentioned methods for physical evaluation, psychological monitoring might also be accomplished aloft through computerized needs adjustment surveys, individual and group counseling for the development of case studies, as well as other forms of performance data analysis. The purpose of any monitoring and data collection with spacefarers should not be to invade their privacy, but to ensure their wellbeing, and through such research to improve the performance and productivity of those who follow them in orbit.[58]

• *Play and Performance in Space*—Long-duration space living must have provisions for relaxation and "recreation" among spacefarers. Interviews with astronauts by Douglas (1984) revealed the issue of "humanness"—what a crew might sacrifice during a ten-day mission, they are unwilling to renounce on a ninety-day stay at a space station! These space veterans warned of overscheduling and the need for compensations for demands made on crewmembers. For the sake of mental health, they recommended opportunities for private time to contemplate the grandeur of the space experience, to socialize and foster interpersonal relations.

When the **Skylab IV** crew overworked an average of sixteen hours a day with no day off, a sense of malaise set in. That crew found that they performed best when they worked and dined together, even going to bed at the same time. Both the American and the Russian experience aloft confirms these favorite pastimes—looking out the window, gardening, visitations from Earth with mail deliveries, packages, even grab bags. Among those who kept diaries or logs of their space experience, Hoffman detailed what a play day on the Shuttle **Discovery** was like—running on the treadmill, zero-g games, night observation out flight-deck windows, last-meal-in-orbit feasting, listening to music with headphones while floating free, and sleep trances.[59]

In their seminal *Living Aloft* study, Connors, Harrison, and Akins (ref. 8 in Chapter 5) did examine leisure-time activities aloft and discovered that they ranged from passive (media watching, reading, and meditation) to active (group singing and dancing, playing musical instruments, games and contests, and letter writing). On **Soyuz 9**, for instance, the cosmonauts watched concerts on video, listened to news and music piped up from their homeland, held two-way communication with family and friends in Russia, as well as with scientists and celebrities. One Indian cosmonaut practiced yoga.

The narrow human factors research to date has generally ignored the play phenomenon aloft, as well as interpersonal issues relating to sex, drugs, and alcohol. We have yet to focus investigations on advanced communication technologies in orbit which could facilitate education and recreation, such as better use of multimedia and expert systems. With more permanent habitats aloft, planners will have to expand the range of in-space studies and services implied in settlements. The focus should be upon development of positive, "space wellness" strategies and programs.

Human biological and psychosocial evolution will be accelerated in the centuries ahead by extraterrestrial living! Survival adaptations in orbit may eventually lead to species mutations. No wonder that the head of Soviet biomedicine, Dr Oleg Gazenko, believes that the limitation to our living in space is psychological, not physiological; the Russian space program has always given psychological support a high priority—woman cosmonaut Svetlana Savitskaya goes further, to suggest that psychologists be included on long-duration missions. In engaging in space human factors or life science studies, researchers would be well advised to consult NASA's *Man–Systems Integration Standards*, which contains much useful information, such as on human performance, orbital environment and safety, health management, workstations, and extravehicular activities.[60]

To close the space deployment loop, we now consider issues related to the return of the spacefarer to this planet. As in all people migrations, eventually some colonists may prefer to remain aloft for the remainder of their lifetimes, unwilling or unable to come home again to Earth!

Stage Four—RE-ENTRY POLICIES AND PROGRAMS
These are required to deal with spacefarers' relocation back to Earth, including reacculturation problems. Living in a low or gravity-free environment, especially for lengthy stays, obviously affects body, mind, and spirit. Information should be systematically gathered and analyzed from space "expatriates" to improve future recruitment, selection, and training of spacefarers as is currently done with foreign deployment on this planet. Re-entry preparation, counseling, and guidance, as well as debriefings, complete the space personnel deployment system.

• WHO—just as with pre-departure orientation, the space voyager and his or her family should be involved, whether the latter accompanies that individual into space or remains back on Earth. The extraterrestrial experience, especially one that may involve many months or years aloft, impacts family and close friends, affecting both personal and organizational relations. In the 21st century when space settlers volunteer permanently for the new orbiting colonies, re-entry procedures will be necessary for their possible visits to this planet. Perhaps astronauts and cosmonauts who have been in space are the best persons to act as consultants in this and other aspects of the deployment system. Who better understands the "overview effect and re-entry shock"?

• HOW—policies need to be established on pragmatic matters as to

(a) the length of space assignments and sabbaticals from such outposts as the Space Station or Lunar Base ...
(b) when, where, and how the reconditioning is to take place ...
(c) what the length and process of debriefing before leaves may begin or re-assignments to orbit made ...
(d) what type of data-gathering and follow-up procedures are to be undertaken upon return to Earth ...

(e) what rights, benefits, and assignments the spacefarers may expect upon return to the sponsoring organization.

Before any rotation back to this planet from long-duration missions, the re-entry program should likely be inaugurated while still in orbit—extensive conditioning will be required to ready people again for the gravity environment below.

• WHAT—the re-entry process must deal with physical, psychological, social, and organizational aspects of return to gravity living. Research in this area will have to deal with programs that recondition the body, foster reacculturation to home planet conditions and societies, as well as guidance for further career development or retirement. The programs will differ for those who remain in orbit for a limited duration, in contrast to those who are away for years.

• WHY—spaceflight for most people will involve a **transitional experience**—from a new perspective of seeing our planet as an interconnected ecosystem to ground-based views that are quite myopic and provincial. Upon return, some may complain of severe adjustment problems or "re-entry shock." In his book, *The Overview Effect*, Frank Oberg reported on this homegoing phenomenon among astronauts brought on by the fiery re-entry, adulation by well wishers, sense of gratitude for safe return, and even letdown. Constructs and perceptions may be so altered as to require psychological counseling so as to lead a normal Earthlife again. Some space expatriates bring back a global consciousness that inspires them to a leadership role; others get so depressed on the ground that may verge toward a nervous breakdown.

The whole space experience—even for those who never go aloft but learn of it through media, be it photographs, paintings, television, or films—changes our imagery and imagination, our field of perception or life space. For millions world-wide the live television broadcasts from the Moon in 1969 had an impact comparable to viewing the 1995 film, *Apollo 13*, which was but a re-enactment. Think what the actual experience meant for the twelve men who actually landed on the lunar surface, the first among billions of their species!

Re-entry Research
Ideally, under the leadership of global space agencies, a data or knowledge bank should be developed with a pattern recognition and retrieval system relative to the broad human role, performance, and problems in space habitation. In addition to recording findings of terrestrial analogs as described previously, this information system would be a comprehensive storehouse on the on-going human experience aloft.[61] Perhaps the Living Systems template which Dr James Grier Miller describes in Chapter 3 would serve as a framework for this cross-disciplinary analysis. Although NASA has been forthright about medical and biological insights gained from previous spaceflights, the agency has been circumspect on studying or releasing information on the psychosocial experience of its personnel in space. In the past, NASA has limited access to astronauts by social science researchers, even by its own psychiatrists and psychologists. According to psychologist Dr. Albert A. Harrison the agency has failed to capitalize on the data it has collected which could improve spaceflight and living for others to follow. For example, there are indications that transcripts of

crew communications going back as far as the Mercury flights, that data have not been analyzed from a behavioral science context so as to obtain clues to improve future missions and avoid tragedies.

There does not appear to have been any global, systematic data gathering surveys of past and present astronauts/cosmonauts who have actually flown in space. This expert group could provide valuable input to better human performance in space if behavioral scientists were given the opportunity to construct an appropriate questionnaire, and analyze the results. For Neil Armstrong, Buzz Aldrin and the other ten who visited the Moon, their lives and viewpoints were altered forever. A limited number of astronauts and cosmonauts have reported after their return—sometimes in books, poetry, and paintings—about their offworld lives. They experienced everything from vertigo, claustrophobia, and depression, to a sense of renewal, euphoria, and ecstasy. Veterans of numerous spaceflights seem to learn from prior missions and get better at their jobs aloft. Documenting the space explorer and settlers' experience on the high frontier should be a scholarly priority.

Early in the space program, the Space Science Board (1966) became concerned about long-duration mission effects, decrying NASA's neglect of psychological and behavioral factors in such spaceflights. In an interview on his Skylab and Spacelab experience, former astronaut Owen Garriott observed that it was human presence which ensured the success of those missions, and that there would be no functional limits to human performance on longer flights if radiation levels are contained and a daily exercise routine of one hour is maintained. Then he significantly added, "Neuromuscular weakness is not evident in space; it is only noticed after return to Earth" (Ref. 40, p. 210).

Bluth and Helppie[15] in recording Soviet space station experience, include this pithy quotation from O. G. Gazenko: "All people who have been in orbit experience certain difficulties, often significant ones ... after returning to Earth." These researchers described what cosmonauts noticed about themselves after landing—weakness, fatigue, feelings of increased body and object weight, gravitation shift in vector and vertigo, skin paleness, face puffiness, limited locomotor function, orthostatic stability decrease, perspiration increase, faintness, and other bodily difficulties. Consider this diary account of a cosmonaut:[62]

One Russian, Lebedev, wrote in his diary of his 1981 station experience that the reality is a story of loneliness, of fear, and homesickness, but the sheer routine of operating their Salyut 7 (meaning "salute") diverted these feelings somewhat. But he and his crew member, Berezovoi, were getting so fatigued, they required 12 hours of sleep each night. When they expressed a strong wish to be with their families by the New Year, the doctors and flight director approved of an early return ahead of schedule. After 211 days aloft, they came back to their families by New Year's Eve ... .Yet, they could be satisfied with their high performance when in orbit—they produced 20 000 photographic plates, helped to locate oil and gas fields, guided the route for a transcontinental pipeline and railway, and saved a billion roubles a year by weather forecasts giving their countrymen advance warnings.

This is the type of anecdotal data that researchers need to be collecting worldwide on the spacefarer experience, both in orbit and upon return!

Since the Russians parachute their capsule upon re-entry to land, and normally not water, their cosmonauts have had many adventures—from crash landings in remote areas to encounters with blizzards and wolves. Again we quote from Brian Harvey's first edition of that wonderful volume now extensively revised and updated and published by Wiley–Praxis as *The New Russian Space Programme:*[39]

> **When cosmonauts Kovalynok and Ivanchenkov plunged through the atmosphere and floated down to Earth after almost 140 days aloft in 1978, the former climbed out unaided, picked up a handful of soil and promptly fell back into the doctor's arms. Ivanchenkov said they just loved the intoxicating fresh air of mother Earth and the breeze of wind in their faces! Within two days they were up and walking to the canteen, trying to make up for weight loss—Kovalynok had lost 1 kg, while Ivanchenkov lost 1.8 kg.**
>
> (Harvey, 1988, pp. 281–2)

The same source reported that the former Soviets often sent their cosmonauts to a health resort in the Caucasus to recover and readapt after prolonged spaceflight. No small wonder when we consider how their endurance is tested as is evident in this 1987 report of a return from the **Mir/Kvant** complex:

> **Romanenko achieved a space endurance record of being in space over 326 days, flown 5149 orbits, and covered 135 million miles. After the search team found him and his companions, Alexandrov and Levchenko, in their snow-covered capsule amidst a ferocious blizzard, they were whisked away by helicopter to their Baikonour base. Once there, Anatoli Levchenko's work was far from over—he was put behind the controls of a waiting TU-134 airliner and required to fly it to Moscow and back, so as to simulate the ability of shuttle pilots to fly airplanes in level flight after a period of weightlessness!**
>
> (Harvey, 1988, p. 322)

Aboard **Mir**, the Russians discovered that cosmonauts started getting on one another's nerves after 8 or 9 months aloft, but recover dramatically within two weeks when told they are coming home. This type of realistic account is used here to illustrate the possibilities such recordings, especially on videotape, need to be recaptured by researchers, then utilized in orientation of future space voyagers.

Since the longer the duration of the space experience, the more likely the spacefarer will be subject to "re-entry-shock," it is thus necessary also to conduct studies on the physical and mental debilitation of returns which require rehabilitation through therapy or other treatment. As masses begin to move beyond terrestrial boundaries, there are questions to be answered that will further human absorption back into the gravity environment, such as:

(1) How much has the space experience raised consciousness, altered perception, and actualized potential?

(2) Are top performers in space equally successful on the ground after their return?
(3) What can be learned from spacefarers who have been high performers aloft or well adjusted since their re-entry?
(4) What is the role in the readjustment process of specialized services ranging from social work to occupational placement?
(5) Does post-flight psychological counseling facilitate re-integration into family, society, and organizations?

Exhibit 65—A Mars Lander. A painting by artist Robert McCall of a **Mars Lander** arriving and departing from a Mars Spaceport in the 21st century. *Source*: National Commission on Space, *Pioneering the Space Frontier*. New York, NY: Bantam Books, 1986, p. 69.

6.3 CONCLUSIONS ON SPACE PERSONNEL DEPLOYMENT

If space habitation is to advance rapidly, as suggested in Exhibit 65, then a more comprehensive systems approach to personnel deployment would seem to be appropriate. The model described in this presentation offers components which could be utilized in this decade for analysis of "wintering over" experience in Antarctica. Furthermore, the learnings from an Antarctic Personnel Deployment System would be directly transferable in the next decades for the creation of comparable deployment systems for a Space Station and lunar base. At this preliminary stage in the human "passover" from Earth's boundaries, it is my contention that ground-based research of this type helps to further our transition to a space-based culture.

There is increasing Russian–American exchange on the human experience in space-flight, but certainly such cross-cultural communication should be expanded. In fact, all international interchanges about human behavior in isolated, confined, and extreme environments can improve survivability, habitats, and life style in orbit. The range of data exchanged, whether about living in arctic or desert regions, or in space, should be multidimensional (e.g., the requirements for food and shelter, health and wellness, group dynamics and interpersonal skills, transport and ecosystems, as well as play, performance and productivity). Coping successfully with alien environments contributes to self-discovery and the actualization of human potential.

Among the deployment issues, there are three worthy of immediate study which can be undertaken equally in Antarctica or in orbit. The first has been mentioned above—the creation of **new cultures** by scientists and the military, by construction workers and colonists as they explore and develop unknown territories. Space exploration represents an unusual opportunity to employ methodologies of cultural anthropology so as to consciously plan for such cultural community experiments on the Moon and Mars.

The second research opportunity has to do with the creation of **new space careers**, whether on the ground or aloft. The high frontier will need people with a variety of occupational skills—"technauts"—to build and service spacecraft and facilities, as well as their inhabitants. A recent U.S. Presidential commission anticipated such emerging vocational endeavors:

> Many people from non-aerospace fields are turning to space as a career. In the 21st century, your professionals will view space as a new arena in which to develop their careers. Space doctors and medical researchers will be challenged by the physiological effects of prolonged weightlessness on the human body. Researchers in the interactive human sciences will study human adaptation in alien worlds and environments. Space architects, environmental engineers, and human factor engineers will join together to design remote living and working quarters. Virtually every trade and discipline will be involved in space endeavors, from obstetrics to insurance.
>
> (National Commission on Space, 1986, p. 78).

With such a diverse multiskilled, multicultural workforce leaving the home planet, broader personnel deployment systems, such as discussed in this chapter, will be essential!

The third opportunity for personnel deployment research is for **behavioral scientists in orbit**, whether such studies are formal or informal. Psychologists, psychiatrists, sociologists, and other social scientists can be cross-trained to perform other functions at space settlements, in addition to applications of their own professional expertise.

During the next fifty years, the initial stations and bases aloft are the "laboratories" for making later interstellar migration of our species possible. These outposts permit us to experiment with offworld transfer of our human and material resources. The next few thousand space pioneers could make a remarkable contribution to a fund of knowledge for the establishment of spacefaring civilizations. But as William MacDaniel, Niagara University professor emeritus of sociology, wrote to me in 1986, the real migration challenges are downrange—**selecting permanent human populations** in orbit or on celestial bodies. Maj. MacDaniel believes that systems research will then be directed toward selection of the most appropriate groups (optimum size, age, sex, occupation, and social distribution) to support necessary social institutions on the new frontier.

The late James Fletcher, former NASA administrator, stated that one of the most intriguing options ahead is the transforming of the environments of other worlds to make them habitable for human beings (*Space World*, July 1987, p. 36). Presently, there is no convincing evidence of life in our galaxy other than here on Earth. Seemingly, the human mammals from this planet have the opportunity to extend life into our Solar System and beyond!

STEPPING OUT FOR A SPACE WALK

I was not in a strange or forbidding place at all, but in a place where I—"I" in the sense of being a member of the human race—was meant to be. I thought of those who claim that man should not fly because he was not given wings, and have conjured up, I am sure, similar analogies for space travel. As I looked at my space suit and the MMU, however, I knew that we are meant to travel away from the Earth because we have been given curiosity, the intelligence, and the will to devise the means and the wonderful machines—such as those that now enclosed me—to permit such adventures.

Astronaut Dale Gardner in *The Jetpack: Flight Without Wings* by
Nelson Lewis Olivo, 1995.

SPACEKIND

We may talk about exploring the Solar System, but Earthmen aren't going to do it. It may be done by lunar colonists or other space settlers. They are going to become another breed of human beings. They are going to live a totally different life. They are going to be a different kind of people psychologically and they are going to be the cutting edge of humanity. They will be the ones who get industry out into space. They are the ones who will build the solar powered ships. They are the ones who will create a whole new kind of technology, different from anything we've got here.

(From a symposium address of the futurist writer, Isaac Asimov, quoted in *Lunar Development Council Newsletter,* January 1987, No. 1, p. 1)

7

Macrothinking in Global Space Planning

To open up and develop the space frontier, a new type of **macrothinking** is required. Such reasoning and action were evident during the Apollo mission period when the human family was "turned on" to the daring achievement of rocketing Man to the Moon! Unfortunately, since that decade of the '60s, space programs throughout the world have become the exclusive concern of national space agencies, aerospace contractors, planetary scientists, and government legislators. Their somewhat myopic focus on special interest missions has garnered only the support of the direct beneficiaries and committed space activists. As a result, the public interest has waned over technical accomplishments and failures which do not capture the imagination of *Everyman*. Because would-be space planners fail to provide a 2lst century vision of humankind's future within our Solar System, they do not inspire the masses to financially back risk-taking endeavors in the orbital environment. Further, because too many mission planners and managers do not control escalating costs and provide adequate information about the return on investment, they alienate suffering taxpayers and potential investors.

7.1 UNDERSTANDING MACROTHINKING AND PLANNING

After more than a quarter of a century since humanity's lunar landings, we need again to think **big** and think **systems** if we are to permanently return to the Moon and beyond for the New Millennium. The case for investment in space enterprise has to reflect **macrothinking** if public support is to be garnered.

Macro, in its Greek origin, means large, great, or long. In human endeavors, the macro concept has many applications. When intellectual disciplines or academic specialities devote themselves to issues and activities of the **larger society**, they are termed macrosociology, macroeconomics, or even macromanagement (a topic discussed in Chapter 8). The late professor Dalton McFarland explained that macrosocial phenomena include a variety of forces that integrate the elements of a society, such as, consensus.[1] Since society is itself a system, there must be linkages among subsystems that give legitimacy to the central institutional system, so that the larger community's purpose can be achieved and resources/rewards allocated. Macrothinkers—foresighted leaders and visionaries—give expression to goals and plans in which a citizenry concur, mobilizing themselves to achieve what is envisioned. President John F. Kennedy did just that in 1961 when he set a goal for the United States within a decade to land a man on the Moon and safely return him. National conditions were such that the system supported and achieved that ambition. On July 20, 1989, while commemorating the 20th anniversary of that feat, another U.S. President, George Bush, attempted to set forth the next major space goal for the nation. As explained in Chapter 1, the **Space Exploration Initiative**'s objective included construction of a space station, a permanent return to the Moon, and a manned mission to Mars by 2019. While SEI showed elements of macrothinking, it failed to engender sufficient consensus, backing, and resources. So the President's ambitions were aborted—the whole proposal lacked macroplanning, including a rationale that Congress would finance.

Enterprise has been defined as a project of some purpose and importance which is undertaken with energy and boldness by resolute participants. People are said to be enterprising when they engage in ventures of great difficulty with imagination and initiative. When individual enterprise combines with collective action, then big thinking and things happen! **Macroenterprises**, whether on the ground or aloft, are by their very nature **macroengineering** programs which pose a combination of technical, managerial, legal, financial and financial challenges. Macroengineering requires macrothinking to study, plan, and manage comprehensive technological ventures integrating people, hardware, and software for major accomplishments. While macroengineering builds upon present knowledge and technology, it strains the outer limits of the "state of the art," stretching the competencies of the macroengineer. A macroproject, for instance, has been described as a large, complex, long-term undertaking, usually involving expenditures of $100 million plus—the case with most space endeavors, such as the Space Shuttle or Hubble Space Telescope. Macromanagement refers to a new type of 21st century, interdisciplinary management for such large-scale and complex programs, as in constructing and maintaining the international space station, a lunar base, or other proposals for utilizing space resources.[2] In macroengineering projects, significant quantities of public and private resources must be committed and put at risk within extended time frames. They tend to have profound sociocultural impacts and environmental implications.[3]

Unfortunately, today's professional schools, whether in engineering, science, business or whatever, do not prepare their graduates to engage in either macroprojects or macromanagement where practitioners must move beyond narrow academic disciplines and specialities to conceive "holistically" in terms of interacting systems. Complex space enterprises today, and more so tomorrow, are both multinational/multidisciplinary, requiring multicultural skills. Macrothinkers manifest multifaceted capacities to effectively coordinate human, material and technical human resources, as well as to deal with sociocultural, political, financial, and legal dimensions, plus ecological and environmental concerns. Too much contemporary space planning is largely concerned with hardware and technical matters, neglecting broader human factors, so essential to long-duration missions and space settlement (refer to Exhibit 15, Chapter 2). Most space planning is narrow and mission-oriented, rather than placed in a larger macrothinking context.

As the "prophet of macroengineering," MIT's Frank Davidson has so aptly stated in a score of books and articles, we need a clear vision of how technology should shape our future. In addition, this farsighted lawyer argues that to benefit from modern science and technology, we have to transform our thinking, behavior, institutions and vocabulary, so that "we abandon the sterile political rhetoric of public versus private sector, and learn instead the neglected art of intersectoral harmony and collaboration."[4]

In Exhibit 66, macroengineers C. Lawrence Meader and Arthur C. Parthe, suggest eight aspects to include in any macroplanning.

Emphasizes:

- problems critical to large-scale developments
- database management systems and development tools
- analytical tools
- integration of available program management facilities
- computer graphics and report generation
- exploratory data analysis
- natural language and knowledge representation.

Exhibit 66—Macroplanning Characteristics. *Source*: Meader, C. L. and Parthe, A. C. "Managing Macro-Development: Policy, Planning and Control Implications," in *How Big and Still Beautiful? Macroengineering Revisited*, edited by Davidson, Meader, and Salkeld. AIAA Selected Symposium Volume 40, 1980, p. 90.

From the vantage point of thirty years of system consulting experience, I have concluded that those in the scientific and aerospace communities who offer proposals for space investment would benefit by studying the field of strategic planning and management. One lesson I learned from this speciality is to "think big and think ahead," at least a decade or more. As a consultant in NASA headquarters before the **Apollo 11** landing, I experienced the opposite view. When I asked what plans these top administrators had for space exploration and development after putting the first persons on the Moon's surface, they expressed uncertainty. Finally, reflective of an engineering mindset, they replied, "We're waiting for the Congress to tell us." Perhaps such thinking may help to explain why the Congress and Administration were unwilling then to capitalize upon the momentous feat of placing American astronauts on the Moon, subsequently often criticizing the Agency's leaders for lacking vision. A global macro space plan has the capacity to capture humanity's sense of destiny and spirit of exploration, even when construction of an offterrestrial infrastructure is described in incremental steps and features miniaturization over costly "big science" programs.

To conclude this introduction, Exhibit 67 synthesizes the requirements for **macrothinking and planning.**

A convergence in thinking and planning for large-scale enterprise which is:

- **Multidimensional**
- **Transnational in scope and leadership**
- **Interdisciplinary**
- **Market-oriented**
- **Engaging in preliminary studies, demonstration models, and prototype testing**
- **Able to mobilize resources (technical, material and human)**
- **Sensitive to the ecology and environment**
- **Synergistic and able to promote collaboration.**

Exhibit 67—Requirements for Macrothinking and Macroplanning. *Source*: P. R. Harris, "Macrothinking in Global Space Planning," *Earth Space Review*, Vol. 3:1, 1994, pp. 5–8.

Finally, Exhibit 68 offers an overview of the characteristics for successful **macrothinking and planning.**

- **Demonstrates an imaginative mindset that practices the art of the longview.**
- **Promotes coalitions and consortia which match purpose to means.**
- **Advances integrative education and practice of advanced technologies.**
- **Facilitates joint technological venturing across sectors, both public and private.**
- **Conceptualizes holistically in terms of systems and human needs.**
- **Ensures multicultural participation while considering broad impacts.**
- **Transforms attitudes and behavior to stimulate innovation.**

Exhibit 68—Attributes of Macrothinking and Macroplanning. *Source*: Inspired by Frank B. Davidson's *MACRO—A Clear Vision of How Science and Technology will Shape our Future.* New York, NY: William Morrow and Company, 1993.

7.2 MACROTHINKING ILLUSTRATIONS

Macrothinking has been carried out by **Homo sapiens** since our ancestors climbed out of trees and walked upright. Hunters and gatherers demonstrated macrothinking in the way they migrated across the planet. The story of civilization contains numerous examples of large-scale engineering projects, beginning with the agricultural period of human development and the need for infrastructure. Over thousands of years, history records macrothinking feats—building the walls of Jericho, great cities like Babylon, aqueducts of the Roman

Empire, the great wall of China, medieval castles and cathedrals, Mayan temples and metropolises. But it has been in this millennium that humankind distinguished itself by macrothinking which resulted in huge system engineering of roadways, dams, canals, bridges, waterways, pipelines, transcontinental railways, ship/air ports and many a mega-lopolis—the very infrastructure of a functioning society. Thinking big happens every-where, and is most evident on Earth in the Suez and Panama Canals, the Manhattan Pro-ject, the St. Lawrence Seaway and Power Project, the Aswan High Dam, the Osaka island redevelopment, the Saudi Arabian Jubal city, and more recently the English–French Chan-nel tunnel or Chunnel. As Davidson and Meador indicate, macrothinking demonstrates that a society has "both the competence and compassion to address human needs on a scale commensurate with our problems."[5]

Examples of individual **macrothinkers** abound in past and present—from ancient builders of the Seven Wonders of the World and the Egyptian Pharaoh, Sesosteris I, who connected the River Nile to the Red Sea in 2000 B.C., to the 16th century Dutch dike builders such as engineer-scientist, Andries Vierlingh. They are found in every century, country, culture and career—including artists such as the Italian, Michelangelo; printers and statesmen such as the American, Ben Franklin; architects such as the Englishman, John Rennie, who built the original London Bridge; soldier-builders such as French em-peror, Napoleon Bonaparte; writers such as the Frenchman, Jules Verne; engineer-managers such as the French-English, Isambard Kingdom Brunel; pragmatic philosophers such as the American, William James; theoretical scientists such as Europe's Albert Ein-stein; politicians such as U.S. President, Franklin D. Roosevelt, who launched massive social programs and projects. Many entrepreneurs illustrate macrothinking at its best, from Pierre-Paul Riquet in the 17th century, who designed French canals, to Ted Turner in the 20th century, who created a global television network.

However, it is in the Space Age, that macrothinking became both a science and a neces-sity. It is evident in the thinking of Russians, such as mathematician, Konstantin Tsi-olkovsky, who wrote philosophical and scientific space books in the 1880s, while design-ing pre-revolutionary atomic spaceships; Yuri Kondratyuk, who described orbiting space stations in the 1920s; Friedrich Tsander, who outlined space shuttles in the 1930s. Despite the secrecy and mystery in the former Soviet Union, the names of its brilliant 20th century space macrothinkers emerged, such as Valentin Glushko, Vladimir Chelomei, and espe-cially chief grand designer, Sergei Korolev, a Ukrainian.[6] With their Soviet colleagues, they launched a spectacular series of macroengineering "firsts," both unmanned and manned, including the first orbiting satellite, **Sputnik 1** (1957); the first man in space, Yuri Gagarin, in spacecraft, **Vostok 1** (1961); the first woman in space, Valentina Tereshkova (1963); the first spacewalk with cosmonaut Alexei Leonov in **Voskhod 2** (1965); the first soft-landing on the Moon with automated spacecraft, **Luna 9** (1966); the first automated docking of spacecraft with **Cosmos 186** (1967); the first unmanned lunar flyby/return with **Zond 5** (1968); the first unmanned lunar 'crawler', **Lunakhod 1** (1970); the first manned space station with **Salyut 1** (1971).[7]

Of course, there were many manifestations elsewhere of space macrothinkers, such as Romanian-born Hermann Oberth, an inspiring space writer; Germany's rocket engineers, Wernher von Braun and Krafft A. Ehricke; the United States' Robert Goddard, pioneering rocket scientist, and James E. Webb, second NASA administrator; British scientists, Fred

Hoyle, the astronomer, and Arthur C. Clarke, proposer of global communication satellites and science fiction writer. During the Cold War, the Americans joined the space race with their own demonstrations of macrothinking at its best, including: first successful planetary probe with **Mariner 2** to Venus (l962); first successful communications satellite with **Telstar** (1962); first manned lunar landing with **Apollo 11** (1969); first successfully manned space station of long duration with **Skylab** (1973); first successful unmanned landing on Mars with **Viking** (1975); the world's first planetary Grand Tour, **Voyager 2** (1977–1989); the world's first reusable Space Shuttle, **Columbia** (1981). Both the Luna and the Apollo mission series not only reflected exceptional macrothinking, but gave impetus to the whole field of macroengineering.[8] In both instances, space planners coordinated and managed cumulative complexity. Throughout this opening stage of space exploration in the last quarter of the 20th century, there are numerous examples of macrothinking and -engineering by both leading spacefaring nations,[9] as Exhibit 69 demonstrates.

But humanity has yet to reveal macrothinking in terms of offworld **settlement and industrialization**. Here are seminal illustrations pointing in that direction within the U.S.A. Each shows prospects for nearterm payback which would engender broader public and industry support, while gaining investment by entrepreneurs, venture capitalists and consortia:

- the 1976 publication of physics professor and macrothinker, Gerard K. O'Neill's *The High Frontier—Human Colonies in Space*, and the founding the next year of his Space Studies Institute to promote and research his ideas. In 1977, O'Neill directed a NASA summer study and its proceedings on *Space Resource and Space Settlements* (NASA SP-428, 1979) ...
- the 1979 General Dynamics Space Systems study on *Lunar Resources for Space Construction* that envisioned putting l600 people in space, including up to 400 on the Moon ...
- the 1981–1993 conferences of the **Mars Underground** for space experts and activists interested in developing Mars as a place for travel, science, and settlement. Co-sponsored by various professional space organizations, it has produced four volumes to date on *The Case for Mars*, published by the American Astronautical Society (available from UNIVELT, PO Box 28130, San Diego, CA 92128) ...
- the 1984 scholarly summer study and resulting proceedings by NASA with the California Space Institute—*Space Resources* (NASA SP-509, 1992); with the National Academy of Science—*Lunar Bases and Space Activities of the 21st Century* (Lunar Planetary Institute, 1985) ...
- the 1986 report and recommendations of the National Space Commission for *Pioneering the Space Frontier*, chaired by macrothinker, Dr Thomas O. Paine ...
- the l986 and l987 Smithsonian conferences at the National Air and Space Museum, leading to a **Declaration of First Principles for the Governance of Outer Space Societies** (refer to Appendix A, and George S. Robinson and Harold M. White's *Envoys of Mankind.* Washington, DC: Smithsonian Institution Press, 1986) ...
- the l988 amendment of Rep. George E. Brown, Jr. to NASA's charter through a Space Settlement Act, subsequently passed in part by the 100th U.S. Congress as an attachment to the Agency's appropriations; this legislation calls for strategies and programs that promote colonization and permanent presence aloft ...

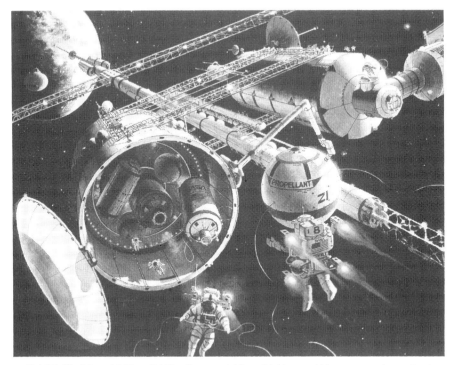

Exhibit 69—Macrothinking: Orbiting Spaceport. Macrothinking would be necessary to construct an **L1 Spaceport** as a jumping off point for spaceships cycling between Earth and Mars. Transfer vehicles, like those pictured in this painting by artist Robert McCall, would be serviced by astronauts through EVAs. *Source:* National Commission on Space, *Pioneering the Space Frontier.* New York, NY: Bantam Books, 1986, p. 135.

There is evidence of macrothinking among the many current space proposals and plans, such as:

- for Russian/American space synergy with a variety of other national sponsors, beginning with possible combined missions for (a) **International Space Station**; (b) a lunar base; (c) shared research on utilizing space-based energy; (d) automated and manned Mars exploration ...
- the 1993 announcement of an ISLEA joint venture between America's International Space Enterprises and Russia's Lavochkin Association using Russian lunar experience and hardware to return to the Moon before the end of this decade with government and commercial payloads: (a) to deliver scientific instruments and communication satellites into lunar orbit; (b) to conduct seven automated landing missions ...
- the 1994 Summer issue of *The Journal of Practical Applications in Space* with its editorial by publisher General Daniel O. Graham on **Space Policy 2000—Double Prime** ...
- the on-going program for Value-Added Transfer of Space Technology to NASA's commercial field and incubation centers under the leadership of macrothinker, Dr George Kozmetsky at the University of Texas' Institute of Constructive Capitalism ...

- the restructuring of the aerospace industry worldwide to changing global market conditions, particularly by mergers and acquisitions such as in the synergistic consolidation of Lockheed Corporation, purchaser of General Dynamics fighter airplane division, with Martin Marietta, acquirer of General Electric Aerospace and General Dynamics Space Systems ...
- the development of the International Space University, founded in the U.S.A., but now with a permanent campus in Strasbourg, France, plus satellite academic centers and electronic networks around the globe, and eventually perhaps in orbit.

In the near future, macrothinking will be evident if and when the U.S. Departments of Energy and Defense: (a) fund R&D demonstration power-beaming studies either for Solar Power Satellites, such as proposed by Dr Peter Glaser,[10] or a Lunar Solar Power System as advocated by Drs David Criswell and Robert Waldron (refer to Appendix B); (b) DOE's designation of one national laboratory as the research center for space solar energy. Exhibit 70 visually depicts how solar power could be collected on the Moon and then transmitted from there to Earth, thus providing for rising global energy needs.

Exhibit 70—Space-based Energy from the Moon. Artist Paul DiMare's conception of a Lunar Solar Power System (see Appendix B). *Source*: T. A. Sullivan and D. S. McKay *Using Space Resources*. Houston, TX: NASA Johnson Space Center, 1991, p. 12.

7.3 MINI CASE STUDIES OF MACROTHINKING

The following three cases illustrate **contemporary space macrothinking** that will impact our future—one from a space agency in the public sector, and two from non-profit enterprises in the private sector. Each is aimed toward 21st century undertakings that would involve the practice of macroplanning, macroengineering, and macromanagement. Each demonstrates vision and innovation toward space development.

A. The European Moon Program

The European Space Agency (ESA) has studies under way, planning, and reports for a four-phase lunar program early in the next century.[11] R. M. Bonnet, their director of scientific programmes, indicates that Europeans envision the Moon as a natural space platform or station (38 million square kilometers) that only requires infrastructure to exploit its scientific potential and its resources. Although they are aware of problems inherent in a lunar base, ESA sees it as a macroarena for developing high technology and undertaking challenging projects in exobiology, radiation biology, ecology, and human physiology. Further, the ESA planners see this Earth satellite as a test bed for learning more about our universe by astronomical observations from the lunar surface, and by developing skills there for interplanetary travel and exploration. They are well aware of a range of lunar resources to be utilized from energy and water to mining and manufacturing.

Such lunar development provides the European member states with a common long-term vision and an independent peaceful goal for scientific and cultural knowledge which would also have political and sociological benefits. This European Moon program has four steps that build upon initial experience through gradual progress and interrelated missions:

(A.1) *Unmanned, automated investigations* including possibly MORO (Moon Orbiting Observatory) to characterize the lunar surface in terms of geology, morphology, geochemistry, mineralogy, topography and heat flow, as well as to study the interior in terms of geodesy and gravimetry (anticipated launch date is 2003 on **Ariane 5**).

(A.2) *Permanent robotic presences* including possibly LEDA (Lunar European Demonstration Approach), a lunar lander carrying a payload of robotic and scientific instruments (a rover, soil-processing facility, and a robotic arm) which will land close to the lunar south pole.

(A.3) *Robotic deployment of scientific instruments* to gather information to establish the human outpost, such as: micrometeorite flux, seismic noise, soil mechanics, thermal properties of the surface, soil characterization, stereo imaging of the surface, measurements (gases, dust particles, sky background, and radiation doses), and the search for water.

(A.4) *Establishing a human outpost* including building facilities and an oxygen production plant (refer to Exhibit 71).

In a letter to the author (January 7, 1994), Dr Bonnet indicated that a second-phase report is under way to define a European global view for both lunar exploration and exploitation. However, he wrote that ESA sees lunar industrialization as a longer-term goal which they are not addressing at this stage. This writer believes that true **macrothinking** would necessitate that aspect now if European taxpayers and investors are to support this macroproject—they expect to see some return on the huge expenditures required to carry out these lunar goals!

Exhibit 71—Exploitation of the Moon's Resources. The European Space Agency envisions humans at a lunar outpost engaged in power generation by thermonuclear fusion. *Source*: ESA-SP 1150, *Mission to the Moon*, Paris, FR: European Space Agency, 1992, p. 130.

Wisely, the ESA proposes international involvement—a worldwide initiative aimed at creating consensus among scientists, politicians, and space organizations. They suggest a start-up on January 1, 2000, with an International Lunar Quinquennium (ILQ), seemingly the mechanism for transnational participation every five years. The assumption here is that the Moon is already part of human territory, and should be made accessible to all humanity through telepresence and virtual reality. (For further information on ESA, refer to Chapter 1.3 case study.)

B. United Societies in Space, Inc.

In harmony with the UN Outer Space Treaty (1967), this private initiative confirms that outer space is indeed the "common heritage of humankind," and its goal is to settle and populate the high frontier as an "interplanetary commons" for the human species (see Appendix D). **United Societies in Space (USIS)**, recognized as a non-profit corporation by the U.S. Internal Revenue Service, fosters **space** legal, financial, and governance systems which encourage investment, industrialization, and settlement aloft. **USIS** operates on the concept of **metaspace**, which views space as humanity's common territory. It coordinates individual and institutional efforts promoting freedom and peaceful enterprise in our solar system. Its motto is **space is a place for synergy**, its official journal is entitled *Space Governance*, and it sponsors an annual Countdown Conference, as well as a Global Space Essay Contest, to encourage solutions to these challenges. (Refer to Exhibit 72 and Appendix D.)

Incorporated in Colorado, U.S.A. (1993) by attorney, Declan J. O'Donnell, J.D., and others, organization planners demonstrate macrothinking in these ways:[12]

Exhibit 72—Peace Symbol of United Societies in Space, Inc. Designed for **USIS** by the Royal Publishing Company of Denver, Colorado, the dove of peace flies over our Solar System (see Appendix D).

(B.1) *Space Metanation* to be established on behalf of all humanity, possibly under the UN Charter provision of the Trusteeship Council for New Territories with the existing space-faring nations as the trustees. This innovative entity would then have the legal, financial and governance capability to further exploration, utilization, and colonization of the space frontier. With the adoption of a new Declaration of Interdependence between **earthkind** and **spacekind,** as well as a Constitution for the space nation, a transportation system and infrastructure can be undertaken to promote offworld development. To this end a Constitutional Convention is being convened by **USIS** for August 2000 at the Denver Milehigh Stadium in Colorado.

(B.2) *Space authorities* are seen as the mechanism for Metanation to encourage orbital investment and development. These quasi-governmental bodies could coordinate transnational endeavors for specific planets, such as Mars. Such multijurisdictional authorities would be modeled upon the legislation and administration of the Tennessee Valley Authority and the New York Port Authority. Initially, the plan calls for creation within ten years of a **Lunar Economic Development Authority (LEDA)** to issue bonds, leases, and insurance, as well as provide public information, regulation, and management for development projects and industrial parks on the Moon, or its vicinity (see Chapter 10). **LEDA** might be founded privately, under an Act of the U.S. Congress, under the United Nations, or a combination thereof. Certainly, it could provide the legal framework for the projects of the above European Moon Program. In 1995, the second National Space Society's **International Lunar Exploration Conference** endorsed the formation of an Advisory Consortia to the Lunar Economic Development Authority. Subsequently, O'Donnell, a tax/space lawyer prepared articles of incorporation for the state of Colorado, U.S.A. In 1996 while preparing the **Omnibus Space Commercialization** bill for the U.S. House of Representatives, Congressman Robert Walker, chairman of the House Science Committee, requested a proposal from **USIS** on the **LEDA** strategy. In his role as president of the World Bar Association, the legal lobbying entity for **USIS**, O'Donnell submitted a document with these unique features regarding the proposed **Lunar Economic Development Authority**:

- **Lunar Venue**—extending the legal scope of space resource management to the Moon based upon the precedent of **Intelsat**, as a means for worldwide participation ...
- **Triple Tax Exemption**—legislative provision for three income tax exemptions to investors in Authority bonds intended to promote space commerce and lunar infrastructure ...
- **International Participation**—provisions for both private/public sectors involvement (e.g., Advisory consortia, nation states, and national space agencies)—as reported in *Space Governance*, the **USIS** journal, June 1996, Vol. 3:1.

(B.3) *Space macroprojects* could be underwritten through creative funding alternatives, such as the above bonds, or through a metabank that would issue its own currency. In addition to funding lower cost, mass access to space, beginning with a Single-Stage-to-Orbit Vehicle, other relevant income-producing proposals might be sponsored, such as:

(a) space-based energy via power-beaming and fusion energy;
(b) lunar industrial park and base;
(c) lunar campus of the International Space University;
(d) a space transportation system capable of interplanetary travel, beginning with Mars;
(e) research and development for constructing orbital "city ships" comparable to Gerard O'Neill's "Stanford torus."

In these ways, Metanation and the space authorities would contract with global consortia of entrepreneurs, agencies, corporations, and universities to advance free enterprise aloft, with the purpose of facilitating mass human migration and colonization of the Cosmos.

C. First Millennium Foundation
Macroplanning at its best is evident in the work of another Colorado visionary, Marshall T. Savage, and his colleagues. This non-profit foundation of macrothinkers believes that there is a cosmic imperative to colonize space which will lead to a dramatic explosion in economic growth and creation of unlimited sources of new wealth. Founded in 1994, the **First Millennial Foundation (FMF)** has core members who have evolved a step-by-step strategy for this purpose, publish *First Foundation News* and *New Millennium* magazine, and hold an annual core conclave. Among their programs are:[13]

(C.1) *The Millennial Project*, a remarkable 515-page, well-documented volume which spells out in detail the FMF strategy for eight stages of colonizing the galaxy during the next millennium. These include: (a) **Aquarius**, a prototype space colony under the sea; (b) **Bi-frost**, a 21st century space launch system; (c) **Asgard**, the first space colony in orbit for 100 000 residents; (d) **Avallon**, ecospheres on the Moon; (e) **Elysium,** the terraforming of Mars; (f) **Solaria**, the colonizing of the solar system; **Galactia**, the colonizing of the galaxy; (g) **Foundation**, the Millennial Movement which has already been started.

(C.2) *Aquarius Rising in St. Croix* —the U.S. Virgin Islands site for the FMF prototype sea colony to prepare for space living. Plans are under way to develop a 520-acre parcel with a half mile of beach near the town of Frederikstead. The sea/space facility is also intended as a tourist attraction connected to the cruise industry, which will educate the public about space opportunities through this land/undersea enterprise.

(C.3) *Ocean Thermal Energy Conversion (OTEC)*—the deep water plant for producing thermal energy, drinking water, and air conditioning for the sea/space colony at St. Croix. Closed-cycle OTEC uses the temperature difference between the ocean's warm surface and cold depths to boil a working fluid in an evaporator. Energy is then extracted from the hot gas in a turbine and gas recondensed in a turbine. An open-cycle OTEC directly evaporates warm seawater in a large vacuum container. The epicenter for this OTEC research is in the Natural Energy Lab in Hawaii and the Solar Energy Research Institute in Colorado.[14] GENOTEC of Washington, DC, has already designed a shore-based, closed-cycle plant for St. Croix near the proposed FMF site. It is this type of imaginative technological development that is needed in space colonies, possibly using fusion, nuclear, or a combination of such power. Whether under the sea or out in space, Marshall Savage and his cohorts **think big and far ahead** about a fantastically rich extraterrestrial world!

7.4 CONCLUSIONS

The type of **macrothinking** described in this chapter has the following advantages if applied to development of the space frontier:

(a) captures public imagination by its boldness, scope, and timeframe, for it provides supportable goals ...
(b) permits the incorporation of incremental stages for accomplishment, putting individual space missions into a larger, longer context;
(c) encourages synergy and interdependent actions, ideal for international cooperation and financing;
(d) gives the younger generation a sense of their future, and possible orbital career opportunities.

Macrothinking is in consonance with exploring and pioneering the "final" frontier, offering solutions to the challenge of creating a cosmic civilization. For macrothinking first opens up the frontiers of the mind, stretching our capacities through long-term planning. In his book, *The Pale Blue Dot*, astronomer Carl Sagan places our planet in the context of the universe.[15] That is the context of macrothinking!
    Macrothinker Krafft Ehricke said it best:[16]

> Nobody and nothing under the natural laws of this universe impose any limitations on man except man himself ... . [This law] of astronautics' challenge man

to write his own declaration of independence from *a priori* thinking, from uncritically accepted conditions; in other words, from a past and principally different, pre-technological world clinging to him. This can be done.

It is our **thinking** which may constrain us from developing the richness of outer space, just as it did with those inadequate planners in 1969 who failed to follow up after the last Apollo mission with a program of lunar science and industrialization. Since space exploration and exploitation are imperatives for our species, only macrothinking can adequately deal with mass migration aloft, providing the rationale for human spaceflight and interstellar travel. Only macrothinking can translate the excitement of that journey *en masse* by means of cyberspace. Only macrothinking can transfer the resources of space to protect this planet, and counter poverty among its inhabitants. Only macrothinking will enable us to voyage to the stars!

In his classic work, *MACRO*, previously cited, Frank Davidson concludes with a discussion of the neo-industrial paradigm that links technology to articulated purposes for improving the human condition. Macrothinking provides space technology with such purpose. Futurist Alvin Toffler said it best:[17]

> Space will not only shape our descendants' view of our time; it also shapes our view of the future. It increases the gravity of tomorrow, the consciousness of the future, in our culture today and in a world of exploding change.

This macrothinker suggests that space exploration and space-derived knowledge is defining humanity's place in the universe!

SCENARIO FOR MASS SPACE MIGRATION—2050

Humanity's access to outer space is presently constrained by high transportation costs and limited spacecraft capacity. Mass migration to lower Earth or geosynchronous orbits may become feasible in the 21st century when transportation costs of space travel are reduced as low as $9 to $26 per kilogram of payload (compared to current estimates of $4000 to $9000 per kilogram via today's shuttle and expendable launch vehicles).

Presently, laser propulsion scientists have research under way that may dramatically lower such costs, while boosting the amount of people able togo aloft. Arthur Kantrowitz of Dartmouth University and Avco Everett Research Laboratories, as well as Leik Myrabo of Rennsselaer Polytechnic Institute, forecast that by the middle of the next century, air breathing/laser propulsion technologies may result in small private spacecraft (5 meters in diameter) for 3–5 persons, capable of a round trip to and from LEO at a transportation cost of $8.40 per kilogram!

Before powerful lasers began to lift inexpensive payloads into orbit, including tens of thousands of people, more study is also needed on personnel deployment and habitation systems for long-duration missions ... . The media fans of Gene Roddenberry's *Star Trek* already know that the ultimate answer to transmigration and space travel will be, "Beam me up, Mr Spock" and STARSHIPS!

8

Macromanagement of Space Enterprises

The prototypes of 21st century management, particularly with reference to managing large-scale enterprises, may be partially found in the last half of this century within the global space program.[1] The new space era inaugurated a number of macroprojects in both the U.S.A. and the former U.S.S.R. of such scope and magnitude that another type of management had to be created to ensure successful achievement.[2] The pushing back of the high frontier is a powerful catalyst not only to the development of new technologies, but also to the advancement of fields such as macroengineering and **macromanagement**.

8.1 MANAGEMENT CHALLENGES FROM A NEW SPACE ERA

Further extension of human presence into outer space during the decades ahead offers opportunities for management innovations by those with responsibilities for projects aloft, whether in the public or private sector, whether manned or unmanned missions. In the past, NASA spaceflights, such as **Mercury**, **Gemini**, **Apollo** and **Skylab**, as well as building and operating the Space Shuttle Transportation System or the Hubble Space Telescope, are examples of large-scale enterprises requiring an inventive type of administration and contracting. While a recent study of *Living Aloft* anticipates the expanded human requirements for long-duration missions, it devotes only a chapter to organization and management, limiting discussion to spacecrew structures, motivation, and external relations.[3] In fact, space and management literature is sparse relative to the how of **space management**—the exception is books and journals that deal with the subject of macroengineering.[4] Yet, currently, as well as in the immediate future, post-industrial technical and management challenges exist relative to satellite expansion, planetary scientific exploration, an aerospace plane, heavy-lift launch vehicles, an international space station, a lunar outpost, and possibly a Mars mission.

Today reduced funding and rising costs compel the world's space agencies and companies to collaborate and share, especially with reference to information and management systems. Two examples confirm how the space superpowers have been financially constricted: (1) the American **Space Exploration Initiative** proposed by President George Bush was postponed because the U.S. Congress failed to finance the program; (2) cash-short Russia has several times kept cosmonauts aloft for extra stays on station **Mir** because of financial problems. In October 1995, the Russian Space Agency (RKA) reported that construction of a booster rocket for crew replacement was delayed—this was caused by a shortage of funds that caused two Russians and one German to stay in orbit for an extra month. Thus, it is obvious that for both NASA and RKA, there will be no manned Moon or Mars missions without multinational partnerships and financing. Whether for scientific, commerce, settlement, or defense purposes, international space activities feature more complex missions, involving multiple locations, plus greater numbers and varieties of personnel. Furthermore, such major global space projects require an integration and synthesis of differing technologies and management systems/styles. Even within one nation, such as the U.S.A., the sponsorship of most space operations increasingly will come from multiple sources—not just the National Space and Aeronautics Administration, the Department of Defense or military branches, but also other governmental agencies, such as the Department of Transportation or Commerce, as well as the private sector—from the aerospace industry to new commercial space enterprises.

The management of space enterprises is in transition—a passage from the way short-duration spaceflights have been *managed* for the past forty years, to high-technology, computer-based project management and leadership, which is quite flexible and multinational in scope. The new global space approach may possibly create an *ad hoc* United Nations consortium as the project sponsor, comparable to the coalition of allies in the Gulf War. Or it may require the creation of a world space administration. One harbinger of the future may be the contemporary **Space Agency Forum**, which presently meets annually in conjunction with the International Astronautical Federation. There, **SAF** brings together the heads of 30 space agencies from around the world with representatives of numerous international space organizations. At the moment, it is simply a useful exchange of information and views on issues facing them in undertaking upcoming space activities ... . Another model for tomorrow's multinational space management is today's European Space Agency. It is managed by an ESA Council of Ministers representing fourteen nations (13 member states and a cooperating state of Canada). As the governing body, the Council approves or disapproves the Agency's space plans for Europe. The Agency then manages ESA's research, operations and facilities. Exhibit 73 provides the locations of some of the latter which are spread across the continent and the Americas, as well as in the Russian Federation ... . Eventually, the private sector may lead in confronting the interwoven issues of space governance and management, as was suggested in the last chapter relative to establishment of space authorities within a metanation (see Appendix D).

The inherent orbital challenges go far beyond space technology and management, encompassing human resource and cultural dimensions. Ten years ago, William MacDaniel, a professor emeritus of sociology, forewarned of such multiple difficulties in terms of just one project being planned for this decade—an **International Space Station (ISS)**:[5]

> **Any way that we look at it ... NASA will be confronted with management problems that will be totally unique. Space station management is going to be an entirely new ball game, requiring new and imaginative approaches if serious problems are to be resolved and conflict avoided.**

MacDianiel, a retired Air Force pilot who co-founded Space Settlement Studies Project (3SP) at Niagara University, then analyzed one people-management dimension that results from the sociocultural mix of international scientific and engineering teams, along with the

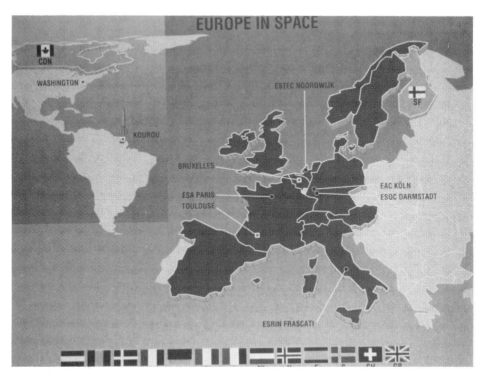

Exhibit 73—ESA Ground-based and Global Establishment. This illustration of Europe's space facilities and offices indicates the global scope of ESA in carrying out its mission: **to provide for and to promote, for exclusively peaceful purposes, cooperation among the European states in space research and technology, and their space applications**. The flags of the 14 countries participating in this international organization are shown. *Source: Europe in Space* published in 1993 by the European Space Agency, Paris, France.

on-board space crews. The multicultural inhabitants of the space station will have to cope with many practical aspects of their cultural differences which alter their perceptions and ways of functioning relative to everything from communication and problem-solving to spatial needs and diet. Because of these management challenges, a special case study on the **International Space Station** is provided later in this chapter.

Whether the orbiting of greater numbers of people for longer periods of time is done by the U.S.A., Russia, Europe, or a combined consortium, project leaders will also have to set the management of cultural synergy as a priority, while creating new systems capabilities.[6] Unmanned probes of spacecraft into far corners of the universe will also demand increased inter-institutional and international cooperative management. In any event, futurists, students of management, and those concerned with technological administration, would do well to review the limited literature and experience of emerging *space management,* for its wider implications. Despite current difficulties associated with the management of NASA programs, that Agency does offer scholars both experiential history and a possible demonstrated model of future management practice.

8.2 THE APOLLO HERITAGE OF INNOVATIVE MANAGEMENT

In worldwide management, a transformation is under way from industrial organizational designs and leadership styles to those suitable for a new work culture.[7] In an A.T. &T. report on emerging issues, the term *metaindustrial* was used to designate the evolving management approach to human systems. One driving force for this transition may very well have been the innovations within the space program by NASA in the 1950s and 1960s. That agency in conjunction with its partners in the aerospace business, contributed to the emergence of the high-technology industry and its flexible management practices. Because of the very complexity of the **Apollo** lunar mission, space administrators and contractors also invented new ways of organizing and managing in order to achieve agency goals.[8] As Seamans and Ordway observed:[9]

> **The Apollo project which landed a team of American astronauts on the Moon is generally considered as one of the greatest technological endeavors in the history of mankind. but in order to achieve this, a managerial effort, no less prodigious than the technological one, was required.**

It is this writer's contention that much of what is currently being characterized as the *new management* is partially the heritage of that space effort. This is especially pertinent with the building of large-scale technological undertakings, whether on this planet or in orbit. Those engaged in macro endeavors that involve many systems, disciplines, institutions, and even nations will have to apply in even more creative ways the legacy which the **Apollo** program gave to management. Not only is more research needed in this regard by academics, but studies and conferences should be directed to what constitutes **macromanagement**, which the late Dalton McFarland described as the managerial imperative.[10] Currently, the term has a double, but complementary meaning—macromanagement represents a post- or metaindustrial management approach, but it can also be employed to describe the complex management of macroprojects (such as required for the building of the Anglo-French Channel Tunnel or in a Mission to Mars).

In the 1983 inaugural issue of the periodical *New Management*, the editor, James O'Toole, provided ten guidelines of what contributes today toward organizational greatness or excellence;

(1) **oriented toward tomorrow—attuned to long-term future;**
(2) **oriented toward people—development of human resources;**
(3) **oriented toward product—commitment to consumer market;**
(4) **oriented toward technology—employment of most advanced tools and technical process;**
(5) **oriented toward quality—emphasis upon excellence, service and competence;**
(6) **oriented toward external environment—concern for all stakeholders;**
(7) **oriented toward free-market competition—imbued with spirit of risk-taking capitalism;**
(8) **oriented toward continuing examination and revision of organizational values, compensation, rewards, and incentives;**
(9) **oriented toward basic management concerns—making, selling or providing services;**

(10) oriented toward innovation/openness to new idea—nurtures and encourages those who question organizational assumptions and propose bold changes.

Dr O'Toole was later to elaborate on this theme through a book entitled *Vanguard Management.*[11]

An examination of the **Apollo** program history will confirm that leaders then in NASA and among its aerospace contractors followed such principles, with the possible exception of the third item, which does not quite apply to public enterprise. But their aerospace civilian counterparts had this concern for the customer—in this case the agency and its personnel. Otherwise, the Moon mission would not have been so successful. NASA, over two decades ago, almost anticipated post-industrial management. The very scope and complicatedness of putting humans on the lunar surface forced such creative alterations in administration.

Among the primary management innovations coming out of the space program of that **Apollo** period was the matrix organization, with its emphasis on dual reporting and team management. The adoption of this management approach was necessitated because of the complexity of mission tasks, and the inadequacy of traditional industrial-age management. Among the space contractors, for instance, TRW Systems of Redondo Beach, California, was a leader in a matrix-type management that two decades later would be a common feature of "new" management operations. Their vice president then, Sheldon Davis, pioneered team building as a means of helping technical people to work together better with those from other disciplines. Other aerospace contractors used project and team management as a form of a*d hoc* organization with new start-up activities. General Dynamics, for instance, could quickly assemble experienced team members for its later Shuttle-Centaur project because of experienced work groups who had developed the Atlas-Centaur rocket. Among these contractors, Hughes Aircraft was a principal exponent of the new matrix management. One of its executives, Jack Baugh, did a doctoral dissertation on how their decision-making was accomplished through this means.[12] His thesis was that matrix is essential in some aerospace projects *when*:

- simultaneous dual decisions are needed in situations of great uncertainty generated by high information processing requirements; when strong financial and human resource constraints are present;
- decisions must be speeded up;
- quantity of data, products, and services to be managed demand it

Today, and more frequently tomorrow, these are the conditions demanding macromanagement and high-performance teams. A contemporary profile of a metaindustrial organization would include these characteristics:[13]

- **use of state-of-the-art technology, ranging from microcomputers to robotics;**
- **flexibility in management policies, procedures, and priorities, continuously adapting to the market—a norm of ultrastability;**
- **autonomy and decentralization, so that people have more control over their own work space and are responsible for decisions, but with integrating controls present;**

- open, circular communication with emphasis on rapid feedback, relevant information exchange at all locations, networking, and multimedia use;
- participation and involvement of personnel encouraged, especially through team, project, or matrix management;
- work relations that are informal and interdependent, cooperative and mutually respectful, adaptive and cross-functional;
- organizational norms that support competence, high performance, professionalism, innovation, and risk-taking, even to making allowance for failure occasionally;
- creative work environment that energizes people and enhances the quality of work life, so that it is more meaningful;
- research and development orientation that continually seeks to identify the best people, processes, products, markets, services, so as to better achieve mission objectives.

Today's high-tech firms operate on such principles. It is interesting that many of these qualities were also identified over thirty years ago as essential to the interdisciplinary management of large-scale endeavors.[14] These were also the characteristics, to a great extent, practiced by NASA management in the **Apollo** era. They are considered essential for organizational excellence now and in the future, particularly for large-scale programs such as renewing the American infrastructure or developing a permanent infrastructure for the space frontier.[15] Space agencies and aerospace corporations which have developed huge bureaucratic management must also transform or reengineer themselves into such entities to prosper in the next century.[16]

Since those in R&D management, especially those coming from engineering and technological fields, may have some differing perceptions about the management process, Exhibit 74 is included here. This classic model by Professor R. Alec MacKenzie illustrates the comprehensive art and science of managing both human and material resources effectively. Its central activities are managing ideas/information, materials, and people through conceptual thinking, administration and leadership. This involves analyzing problems, making decisions, and communicating. Management's task is to plan, organize, staff, direct, and control. The paradigm highlights among its central facets, the management of change (stimulation of innovation and creativity in achieving goals), as well as differences (encouraging interdependent thought and resolution conflict). It still seems a relevant conceptualization for managing large-scale undertakings, whether on the ground or in orbit. From the viewpoint of this researcher, a management psychologist who has served as a NASA consultant, it would appear that the main difficulties facing space leaders in the future may be found on the right side of the model—the people dimension calls for behavioral science management solutions.

The former UN Chief of the Outer Space Affairs Division, Lubos Perek, suggests we think of outer space as a resource to be managed.[17] Thus, in that context, "management is everything that improves safety, efficiency, and economy of space activities for our own and future generations." In this conceptualization of Perek, space management preserves the offworld environment; prevents interference in space communications; controls spacecraft traffic and debris aloft; delimits human error in spacecraft manufacture and opera-

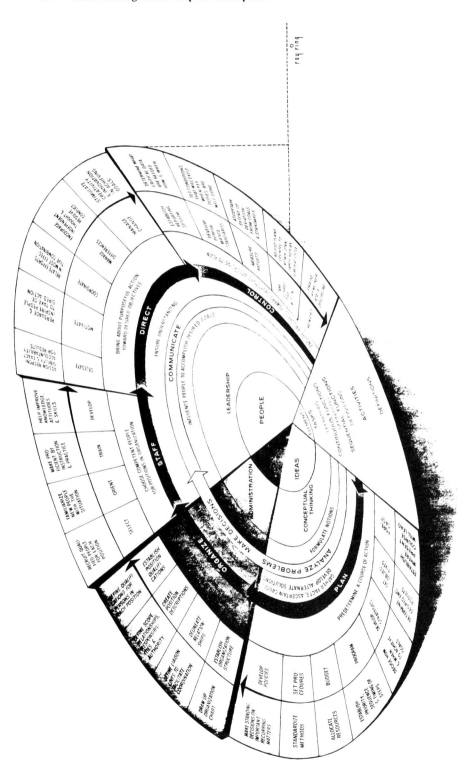

Exhibit 74—The Management Process. *Source:* R. Alec MacKenzie, "The Management Process in 3-D," *Harvard Business Review,* Nov.–Dec. 1969. Reproduced with permission.

tion; prevents and controls technical malfunctions; creates organizational structures to ensure that space is used for reasonable and beneficial purposes.

8.3 THE IMPACT OF ORGANIZATIONAL CULTURE

The work culture affects organizational planning, decisions, and behavior. Edgar Schein, MIT professor, maintains that this is the mechanism for conveying—explicitly, ambiguously, or implicitly the values, norms, and assumptions of the institution.[18] Organizational culture is embedded and transmitted through:

- formal statements of philosophy or mission, charters, creeds, published materials for recruitment or personnel;
- design of physical spaces, facades, buildings;
- leader role modeling, training, coaching, or assessing;
- explicit reward and status system, promotion criteria;
- organizational fit—recruitment, selection, career development, retirements or "excommunication;"
- stories, legends, myths, parables about key people and events;
- leader reactions or coping with organizational crises and critical situations;
- organizational design, structure, and systems;
- organizational policies, procedures, and processes.

   In Chapter 4, the case was made for the influence of **culture** on space developments, including the organizational culture of space agencies and the aerospace industry. (Refer back to Exhibit 40 on AeroSpace Organizational Culture, which illustrates the many other dimensions of a system;s expression of identity.) Since research indicates that excellent organizations manifest strong functional cultures, NASA obviously did this during its **Apollo** period. (In a 1995 lecture on this subject, U.S. House of Representatives Speaker Newt Gingrich observed that perhaps the Agency should have been disbanded after the Moon landings—"you build a project team, you get the job done, then you close it down and start a new project." *Reuters News Service*.) Now in the mid 1990s, NASA is a case in point about changing organizational culture. It is attempting to reinvent itself, to excel again in its shuttle/space station phase of its organizational development under the leadership of its present administrator, Dan Goldin, but it is being done in a political and economic climate that demands downsizing and cut back in expenditures. Testifying before the House Appropriations Subcommittee in 1995, Goldin outlined these plans for restructuring NASA management:[19]

- Cut out duplications/overlaps by consolidation.
- Transfer functions that need not be done by NASA to universities and/or private sector.
- Change relationships with prime contractors, so NASA does less and contractors do more.
- Privatize and commercialize where appropriate.
- Shift the operations activities to the private sectors, while focusing on "cutting edge" research and development.
- Emphasize objective contracting that enables rather than directs industry and academia.

Space management was an issue examined at the California Space Institute when the author was a faculty fellow during a NASA Summer Study.[20] To ensure success with future space planning, a team of scholars and space experts at that time recommended that a survey and analysis of NASA organizational culture be conducted from its headquarters to various field centers. Although this would facilitate planned change and renewal within the Agency, the proposal has yet to be implemented. Furthermore, these professionals concluded that if post-space-station plans for a lunar base are to be effectively implemented, then it is likely that NASA must literally transform its management attitudes, styles, strategies and operations. Whether it is NASA as an organizational system, or one of its top aerospace contracting partners, the aerospace work culture must shift from the industrial or bureaucratic mode to that of enterprises characterized as *metaindustrial*. It may even require some structural changes in this Agency to give that organization more autonomy and mission focus. Then, NASA would be positioned to live up to its Apollo heritage, enabling it to remain a principal actor in global space business that takes advantage of the vast resources on the high frontier well through the next century.

Management consultants envision organizations as energy exchange systems. Institutional culture can energize the use of psychic and physical energies of its people in achieving organizational goals, or it can undermine and dissipate those efforts. this is the lesson of the **Apollo** Moon project for all engaged in space enterprises. To achieve comparable success in large-scale technological ventures, here is a sampling of **cultural** issues in space management to be confronted and altered:

(1) **The mindset of the engineer and technologist requires expansion to more of a generalist—too often present approaches exclude consideration of human issues, and the contributions of the management and behavioral sciences to planning and decision-making are downplayed.**

(2) **The need for more synergistic relationships in space endeavors should replace obsolete competitive postures by individual companies. Tasks related to exploitation of space resources are so immense that national space entities need to collaborate more effectively. Perhaps the time is coming for a Global Space Administration to be formulated to coordinate the space endeavors of various countries involved in exploration and development?**

(3) **In the U.S.A., despite archaic antitrust regulations, legal mechanisms must be created that enable the aerospace industry to work together in solving common problems, be it matters of quality control on launch pads and space vehicles, or greater sharing of research and development knowledge, such as in the creation of the next generation aerospace plane. The great space corporations like Lockheed Martin, McDonnell-Douglas, Rockwell International, Boeing, and others, can do more for the nations's space program by joint venturing and sharing, than by competitive duplication. Furthermore, new ways for cooperative inclusion of university and government research laboratories should be explore—again as in the Apollo era. Perhaps the consortium model being currently developed by the European Space Agency in this regard is worthy of increased emulation in North America? Canada, for instance, has entered into numerous NASA and ESA agreements, as well as with NASDA and RKA.**

(4) As space endeavors reach out to include more business participation beyond the traditional aerospace companies, attitudes and regulations regarding contractors deserve revision. Perhaps the NASA tradition of partnership with its suppliers is more appropriate than the DOD mentality of seeing such as "users". Space enterprises would benefit from marketplace concerns for satisfying clients and customers. The commercial challenge changes in space management go beyond the opening of an office for this purpose in the U.S. Department of Transportation or Commerce. Today NASA is moving in this direction with its Centers for Commercial Development.

(5) Technology development time spans have been increased, rather than condensed, because those in the space arena have become more bureaucratic, less entrepreneurial and innovative. From goal-setting to implementation, the Apollo's mission was accomplished in less than a decade. Now NASA planners use a 12–15-year time frame from inception to completion of a new technology, while just the opposite experience occurs in the growing high-technology industry. Perhaps the time has come to re-examine the cultural assumptions upon which practices of redundancy, over-design, over-preparation, over-study, and excessive timidity become imbedded habits and traditions? Such matters are worsened when these approaches go beyond aerospace design and get applied even to non-technical areas, like conference management and reporting.

(6) Organizational renewal implies a continuing process for clarification of roles, relationships and missions. It requires tradeoffs from the ways we always did it to the adaptations and inventions necessary to remain in the emerging 21st century "space game." Perhaps space stations and lunar outposts are better designed as habitat modules by architects and hotel chains than by traditional aerospace vehicle designers? Possibly the functions of such space activities should be privatized, so that the space agency personnel can be given a more supervisory rather than operational role, thus freeing them for more basic space research and development!

These are some issues and questions that deserve consideration by management leaders in the global space community who would revitalize their organizational cultures and design management strategy attune to future demands.

Unfortunately, the events connected with the space shuttle disaster on January 28, 1986 brought new meaning to the above space management observations. In the aftermath of the demise of **Challenger** and its crew, the Presidential Commission (1986) investigating the tragedy revealed that its source lay not just in technological flaws, but also in management failures by NASA and its contractors. These ranged from *inadequate* quality control and decision-making to organizational communications and information flow. The *Los Angeles Times* (March 1, 1986, p. 1/26) used a telling headline to report this story: "Humans vs. Technology: How NASA System Failed." The magazine *Research & Development* (March 1986, p. 44) succinctly entitled its feature on the subject, "Challenger Disaster will Change the Face of NASA." One of the positive outcomes was improved space management and scheduling; the agency has already instituted a new super computer and management information system. In 1990, The Advisory Committee on the Future of the U.S.

Space Program also provided an objective analysis of NASA's management turbulence, institutional aging, personnel, and technology base. Its report to both the agency and the National Space Council included a section on management with detailed recommendations for internal reorganization, such as:[21]

- **That a Systems Concepts and Analysis Group be formed in the NASA headquarters to serve the Administrators.**
- **That NASA adopt as standard for the management of multi-center programs, a headquarters project manager and staff located at or near the "Primary Center" in the undertaking.**
- **That NASA management review the mission of each (field) center, consolidating and refocusing centers of excellence in currently relevant fields of science and technology with minimum overlap between centers; an appropriate balance between in-house and external activity should also be developed.**
- **That NASA should concentrate its "hands on" expertise in those areas unique to its mission ... . Contract monitoring is best accomplished by a cadre of professional systems managers with appropriate experience. Increased use of performance requirements, rather than design specifications, will increase the effectiveness of this approach.**

Were one other recommendation in that Advisory Council report adopted, it could profoundly change the management of NASA's nine field centers. The proposal called for the agency to selectively convert some of these into a university-operated and government-owned organization, managed like the successful prototype model, the Jet Propulsion Laboratory in Pasadena, California, which operates in conjunction with the California Institute of Technology. It would appear from the Congressional testimony of Administrator Goldin that NASA is heeding these recommendations. Other space agencies are also in the process of organizational renewal for varying reasons. With the European Space Agency, the alterations are being propelled by creation of the European Union and recent launch failures with their **Ariane** rocket. For the Russian Space Agency, the reforms have been prompted by the collapse of the U.S.S.R. and shift toward a market-economy, as well as the need for international agreements to finance existing space programs. In the private sector, space management is being transformed by economic forces which have caused many aerospace companies to disappear, such as General Dynamics Space Systems, in the process of acquisition and merger by more viable industry leaders, such as Lockeed Martin.

8.4 NEW SPACE ROLES FOR EARTH-BASED MANAGERS

Many management issues identified in the previous section are basically cultural and point up the need for planned changes within organizations which manage space projects. At the above-mentioned California Space Institute/NASA study on "Technological Springboard to the 21st Century," resource speakers addressed such issues and provided numerous insights related to the strategic planning of a lunar base. Five significant volumes of proceedings were edited by NASA's Mary Fae and David McKay with Michael Duke, and

Exhibit 75—Managing Unmanned Missions: **Galileo**. This unmanned spacecraft was managed by teams of JPL scientists of different generations during its six-year journey across the Solar System until July 1995, when **Galileo** automatically released a probe to descend into Jupiter's atmosphere at a speed of 170 600 kilometres per hour. The probe, pictured going through the Jovian atmosphere, collected the first sample of **Jupiter's** chemical make-up, winds, and clouds, and radioed back data to Earth for 57 minutes. If all goes as planned, the two-year automated Galileo Orbiter mission will study the Jovian system's moons, rings, and magnetic fields. *Source*: Artwork, JPL/NASA in *Ad Astra*, September /October 1995, p. 17 ... . In a *Los Angeles Times* editorial (December 7, 1995), Tom Harris, a Canadian space writer, observed: **Today's Galileo spacecraft encounter with Jupiter provides exactly the kind of cultural influence society needs today ... Although Galileo is one of the last of the large planetary explorers, a new series of smaller, less costly spacecraft will soon be flying, giving us the thrill of extraterrestrial exploration and discovery on a regular basis.**

have been previously cited under the title *Space Resources* (SP-509). Several of the more telling comments from that report are relevant here to our management topic:

- William E. Wright, Defence Advanced Research Projects Agency—

The aerospace industry culture is extraordinarily conservative. It suffers from a syndrome, "If it has not been done for the last twenty years, forget it." The industry and NASA are not bold enough in their planning and requests for funding. Major programs come into being because someone champions them (puts his "butt" on the line and helps bring it into being).

- Vajk, SAI, and Michael Simon, then of General Dynamics—

Shared a document in the form of stock prospectus for the establishment of a fictional corporation, "Consolidated Space Enterprises." It envisioned nine companies that could profit by serving customer needs and functions on the space station. Four were in the category of space service providers engaged in space transport, repair, research, and products; three were housekeeping companies engaged in providing at the station, hotel, power, communications services; two were support companies providing special space services and fuel. The concept of commercial operations on the station, each "feeding" upon the other's needs, is not only stimulating to thought, but changes the role and relationship between public and private sector participation in space undertakings ... . Vakj, now an independent space consultant, also cited examples of new, more sophisticated management information systems that can alter the role of space project managers. New computer tools, such as relational database management systems, give managers better capability for literature search.

- Ronald Maehl, RCA—

Pointed out that management issues related to a space station and lunar base represent a departure from traditional NASA management practices. First, there is the matter of managing the development of such projects, and precursor missions; then there is the issue of operational management of a space facility when it is functioning. There are precedents with meteorological and communication satellites gained by NOAA and commercial operators. There are new challenges relative to man–machine interactions, operational cost containment, and private participation in such space activities.

These three samples of inputs from resource experts are indications for tomorrow's management of space enterprises that warrant innovative study and policy changes by space agencies and their contractors. University schools of business and management have yet to produce significant research on the challenges in managing offworld macroprojects.

Analysis needs to be devoted to the expertise and skills necessary for *Earth-based managers* of projects occurring hundreds or thousands of miles away from them. We have much to learn in this regard from previous project managers of unmanned probes by spacecraft, such as **Mariner, Viking, Pioneer, Voyager** and **Galileo** missions, as well as from those who managed **Luna, Soyuz, Mir**, and **Energia** programs. That book on space project management has yet to be written—the issues range from limited controls to teleoperations or the control on Earth by an operator of a machine that is at a remote location in space. New management problems are likely to be experienced related to "queuing time" (signal delays between operator and command and machine response and back). Some of the management alterations relate to the use of automation and robotics in space. The role of ground-based managers of space enterprises will be further altered by other emerging technologies, particularly in the field of information and communications, as this quotation illustrates for the new Mission Control Center at Houston's Johnson Space Center:

Replacing the huge custom-built mainframe computer that drove monochrome monitors is a high speed fiber optic cable network of 200 Digital Equipment Corporation workstations with full-color graphic capability. The upgrade will enable the center to control a space shuttle and the proposed space station simultaneously, but at $30 million less each year than the old mission control.

—Air & Space, Smithsonian Institution, October 1995, p. 14.

As more manned space operations occur at multiple locations and for longer durations, new infrastructure will be needed on this planet to support such activities, often international in scope. In addition to integrated worldwide mission control systems, there may be regional support centers, some under government or military auspices, and some under private corporations. As spaceports both on the ground and in orbit proliferate, unique management combinations will be invented which bring both the privat and public sectors together in global joint ventures. For the next fifty years, we are likely to experiment with a variety of Earth-based management practices for orbital undertakings. As spacefaring becomes more complex and permanent, one trend seems evident—less control by those on the ground, and more control by those aloft. Ground management's role is likely to become more supportive and consultative, while orbital managers take charge of programs that involve human survival and performance in the microgravity environment. Innovative **management practices in orbit** will occur through creative human responses to challenges posed by that unique environment, as illustrated in Exhibit 76.

Thus, we stand in the late 20th Century, on the threshold of extending old civilizations into space, perhaps even creating new ones, in which our own sons and daughters may become extraterrestrials from every point of view ... . Not only are our own sons and daughters pioneers in the firmament, they could also become biologically, if not taxonomically different.

What is the evidence that our original human space envoys are in any way different from their Earth-sitting counterparts or that both earthkind and spacekind might become different in the course of generations? Are we really suggesting the possibility of a new subspecies in space ... ?

—George S. Robinson & Harold M. White, Jr.[22]

Exhibit 76—Managing a New World: **Mars**. Human settlement on other planets raises significant management challenges and hazards! One is **environmental**, so the National Research Council has a committee concerned with protecting Mars from tainting by exotic Earth life. Thus, NASA has environmental standards for **Pathfinder** expected to conduct an unmanned mission to Mars in 1997. *Source: Planetary Exploration—The Immigrant Trail*, 1988, p. 478. Published by the former General Dynamics Space Systems Division, now Lockheed Martin.

8.5 MANAGEMENT IN ORBIT

Diverse people on-site at an **International Space Station**, and more especially at a Moon or Mars outpost, will require more freedom for decision-making and creative problem-solving than the astronauts and cosmonauts currently enjoy wherever a mission control center exists on Earth. Decentralized, space and team management aloft will come into prominence with the erection of 21st century space infrastructures—the creation of a spacefaring civilization. Now is the time to begin planning for the practical matters to be faced by orbiting station and outpost managers, especially when the personnel and organizational components come from varied sources. With international station partners, there are bound to be matters of cross-cultural management arising.[23] Relative to immediate macroprojects (e.g., an operational lunar base with advanced transportation vehicles), Earth/Moon systems research is needed now on management concerns, such as: communications, habitation, life support, emergencies, autonomy, leadership, and other such issues contributing to survival and quality of life on the lunar surface.[24] If a manned mission to Mars were to be undertaken, especially with Russian participation then such management matters are crucial, magnified by the cultural differences of Americans and Russians even regarding the nature of management. Use Exhibit 76 to visualize this New Martian World to be managed—the conditions, the processes, the products, the people on the Moon will require something quite different from terrestrial management.

Mixed crews aloft (men and women, military and civilian, private sector and public service workers, scientists and other professionals, diverse nationalities and cultures) will pose more complex management challenges and responses. The people in increasing numbers who visit a space station or lunar outpost by 2025 will include more than astronauts/cosmonauts, or even "technauts" (contractor technicians); they will involve a broader segment of Earth's society from politicians to tourists. In past colonial explorations, trading companies were formed to recruit and sustain colonists in new, remote environs—perhaps some of these previous exploration strategies could be replicated by a Space Trading Company? The commercialization of the high frontier will be a profound force in altering the management of space projects.[25]

As tomorrow's populations in orbiting habitations increase in size and heterogeneity, as well as in length of their stay away from this planet, Ben Finney, a University of Hawaii anthropologist, reminds planners to expect more stress and strain and to provide inhabitants with more autonomy. To maximize safe, congenial, effective performance by such pioneers in space living and work, behavioral science research and applications should be instituted related to team development and group dynamics, new leadership training and responsibilities, and even wellness programs in outer space communities (refer to Chapter 3). Space managers on the ground and in orbit will have to concern themselves with such deployment issues, so as to facilitate a spacefarer's acculturation in an alien, isolated, sometime hostile, environment as was discussed in Chapter 6. Managing multicultural crews in space on long-rotation participants who can deal with isolation and remoteness, are self-sufficient and multi-skilled, are sensitive to human relations issues, and operate by the norm of competence. It will call for new applications of leadership and followership. Whether on an orbiting platform in both LEO and GEO, or residing at a Moon or Martian base, the realities require costly, risky, and long-term programs involving new management procedures which provide continuity and consistency, quite apart from changes in personnel and annual budget constraints.

Another management concern to be addressed more vigorously is that of multipurpose missions, which combine civilian and military personnel and payloads. Economies of scale, piggybacking to constrain costs are arguments for such endeavors. Technical and management complexity, foreign policy and international cooperation considerations may provide stronger cases for not keeping separate the commercial and defense space activities. Perhaps the biggest challenge may be in the education and training of scientists and technologists in space management, or the reverse, educating business managers in science and technology. In the 21st century, diverse management competencies will be required in managing large-scale enterprises as Exhibit 77 portrays.

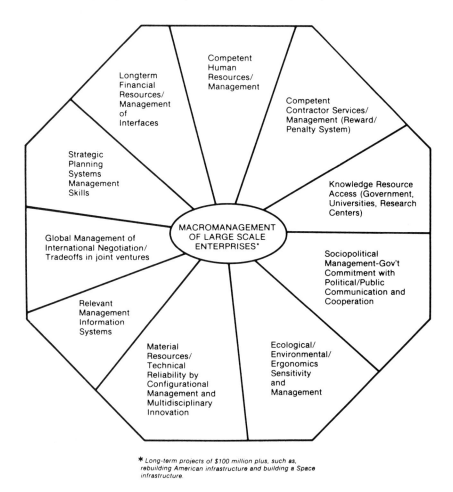

Exhibit 77—Macromanagement for Space Enterprises*. Astronautical artist, Dennis M. Davidson, illustrates the author's conceptualization for ten dimensions of **macromanagement** whether it is practiced on the ground or in orbit. Most space macroprojects require this type of new multidisciplinary, integrated management.

8.6 MACROMANAGEMENT FOR AND IN OUTER SPACE

As we have already indicated, large-scale and complex technical programs require a new type of **macromanagement**, whether to rebuild this planet's infrastructure or to create a space infrastructure. Exhibit 77 depicts my conceptual model in this regard, defining the scope of this term from a management perspective. Long-term projects upwards of a $100 million or more require administrative skills across a range of activities that begin with strategic planning and extend to global or interplanetary management of material and human resources.[26]

Macroengineering projects have occurred for centuries, and may well shape our future.[27] Space programs, like **Apollo** and the **Mir** station, advanced this field, furthering an allied but emerging discipline—**macromanagement**. Most space programs are macroengineering in scope because they usually share these characteristics:[27]

(1) **involve difficult, complex engineering and management problems which must be resolved before program completion;**
(2) **require significant public and private sector resources that must be committed over long time frames;**
(3) **include scientific and technical problems of complexity, unusual size or circumstances, and often unknown technologies or resources;**
(4) **have profound socio-cultural impacts on societies which develop them relative to environmental, legal and regulatory factors, economic and political effects, etc.**

(Davidson, Meador, and Salkeld, 1980)

To these four, Neal Carter of Battelle Institute would add these dimensions of macroengineering and management:[28]

(5) **require massive funding;**
(6) **have dimensions of large scale;**
(7) **extend technology's state of the art;**
(8) **involve long-construction periods;**
(9) **take place in difficulty and sometimes hostile environments;**
(10) **have multinational implications;**
(11) **require sophisticated project management techniques.**

In the past few decades, project management of large-scale enterprises has benefitted from creative tools, such as PERT (Program Evaluation and Review Technique), CPM (Critical Path Method), PMSS (Project Management State Space), Operations Management, systems modeling, and other such techniques. Similarly, developments in the supercomputer and management information systems have made macro projects more feasible and manageable. Many of these management innovations owe their origins and refinement to the U.S. Department of Defense and NASA.

Macromanagement may very well become a dominant theme in management theory and practice of the next century. According to the late Dalton McFarland, previously cited, it represents post-industrial management. This distinguished professor maintained:[29]

> **Macromanagement theory will not only serve to guide and interpret research on management processes in society at large, but will also play a vital role in the determination of social, political, and economic policy.**

As NASA seeks to redesign the shuttle or build a new supersonic spaceplane, to implement plans for a functioning, manned Space Station by 1999 and lunar outpost by 2010, it will not only have to use macromanagement strategies, it may also pioneer in the process. With more corporations outside the aerospace industry participating in space ventures, all space agencies will face a unique set of interface challenges with these new stakeholders. Already space entrepreneurs expect to launch small space satellites and a variety of other commercial ventures that require creating synergy with governmental entities that control space access. Burgeoning space technology enterprises will necessitate the adoption of macromanagement methods, as will the involvement of multiple countries in a single space macroproject.

Research funding should be expanded into matters of macromanagement by a global consortium made up of space agencies, global corporations, universities, and others because it influences the future of management, demanding a new type of managerial thinking, style and skills. Presently, only the World Development Council is addressing issues of planning/financing/managing global super projects.[30] Much must be done beyond the present endeavors of the Global Infrastructure Fund and the American Society of Macro-engineering. Such organizations are in a position to develop a curriculum to educate macromanagers. Another positive step was taken by IBM's support to Cornell University for a Large Scale Scientific Computer Project.

Exhibit 77 depicts macromanagement, whether on this planet or in space, as necessitating leadership capable of:

* *synergy*—facilitating cooperation and collaboration in bringing together diverse elements so as to produce more than the sums of the parts;
* *intercultural and interpersonal skill*—managing diversity by overcoming differences between and among peoples, groups, and nations, particularly through effective cross-cultural communication, negotiation and consensus-building;
* *political savvy*—gaining agreement and support for project goals from the various political or governmental actors/entities, as well as from the public if their support is essential;
* *financial competence*—understanding of economic realities of the long-term project, and capable of putting together the necessary business plan and funding to complete the undertaking, while containing excessive expenditures;
* *managing interfaces*—leadership in bringing together and on time the various resources required to achieve project goals (human, informational, technical, material).
* *cosmopolitan*—sensitivity to global and interplanetary issues affecting the project, such as legal, ecological, environmental, and human, as well as able to cope with such international over national concerns.

These are but a sampling of qualities desirable in tomorrow's macromanagers. In fact, no one person may possess all of these, but a management team may together exercise such competencies. If NASA engineers and administrators had possessed such competencies, their planned space station might have been in orbit in 1995. Certainly a traditionally educated engineer is not likely to possess such skills. At MIT, research has been under way by Frank Davidson on a more comprehensive, multidisciplinary education of macro-engineers. The American Society of Macro-Engineering, especially through its conferences and publications, such as *Technology in Review*, are also beginning to address these concerns.[31] However, schools of businesses and the Academy of Management are lagging behind in synergistic education of engineering managers. Technology or R&D managers need to become more management generalists, more open to new ideas outside their own fields and industry, more competent in emerging skills of strategic planning and management. For this to happen, schools of engineering and business will have to design comprehensive, integrated curricula which include instructors of law, political science, finance, entrepreneurialism or technological venturing, It may be on of the central issues of 21st century management, explaining why some scholar/practitioners, like Dr George Kozmetsky of the University of Texas—Austin, call for *transformational management* strategies.[32] There at his Institute for Constructive Capitalism, Kozmetsky has two programs under way to this chapter's theme:

(1) Co-sponsored with the Society for Design and Process Science, a 1995 conference and publication on **Integrated Design and Process Technology. IDPT** seeks to use both technologies to create a symbiotic relationship among scientists, engineers, decision-makers, and other critical thinkers.
(2) Developed through its Center for Commercialization and Enterprise, two NASA Technology Commercialization Centers at the Johnson Space Center in Houston and the Ames Research Center satellite in Sunnyvale, California. These are entrepreneurial, incubator programs to transfer space research and technology into commercial enterprises.

During the 1984 previously mentioned NASA summer study at the California Space Institute, two resource speakers pointed out existing management models worthy of further analysis by space planners. To create the necessary infrastructures for tomorrow's space programs, New York consultant Kathleen Murphy proposed that we could learn from macrodevelopment projects around the world as another terrestrial analog for space. Such major "greensite" projects in the Third World have already resolved problems of owners and contractors, conflict resolution and negotiations, reward and penalty provisions, as well as tested arrangements that might prove feasible for space development.[33] These include new financing models, joint ventures, consortiums, shared R&D between government and industry, national bank syndicate investments, and other macromanagement strategies. the other input came from consulting engineer Peter Vajk, who observed that many global projects related to new terrestrial materials (NTMs) may offer insights for the exploration and exploitation of space resources. Like NASA projects, they are high risk, capital intensive, and involve very large R&D, start-up, and operating costs. Companies developing NTMs are beginning to use a macromanagement approach through a corporate headquarters that sets general policy, negotiates major contracts, and keeps accounting/systems records, but follow distributed or semi-autonomous management with regard to subsidiary operations or facilities. NTM projects are also technology/multi-skill intensive, and involve macroengineering. Such project sponsors own, lease or high their transportation; they operate distribution centers, retail outlets, and sales/marketing offices. Their programs are extended in time and space with reference to R&D, deployment and operations phases. Their activities are transnational and utilize sophisticated computer information networks involving high-rate data transfer. Vajk believes that macroprojects involving extraterrestrial resources (ETM) can operate like these Earth-based analogs—"space is just a different place to do the same kinds of things we do on this planet."

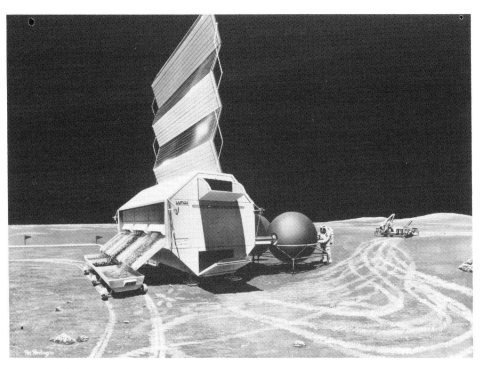

Exhibit 78—Lunar Power Plant. Macroprojects on the Moon might include building a compact oxygen plant like this one illustrated by Pat Rawlings of SAIC. It will process **lunar regolith**, using a hydrogen reduction treatment technique to refine oxygen for propulsion. *Source*: NASA Johnson Space Center.

8.7 SPACE MANAGEMENT IN THE FUTURE

But space is also a place for technological venturing and large-scale endeavors of a peaceful and commercial nature, such as envisioned in Exhibit 78.[34] It opens opportunities on the part of human institutions and governments to produce synergy, not war. It not only requires a change in mindsets, but new management, multilateral action, and collaborative research. A quarter century ago, a classic work provided us with the charter for that purpose. In *Managing Large Systems: Organization for the Future*, the authors reminded us that such enterprises are interdisciplinary in character, and integrate an array of scientific, technological, social, political and other personalities and resources.[35] That describes the large-scale programs of most national space agencies, and was well understood by the key administrators of the **Apollo** program. Sayles and Chandler's book is worthy of further restudy and application relative to the macromanagement of space enterprises, such as the Lunar Power System described in Appendix B.

As noted in this book, the National Commission on Space appointed by the President of the United States recommended investing $700 billion on the space frontier over the next 50 years. The macro space undertakings proposed in their 1996 report are of such scope as to need more than bold vision; they require a new systems approach for managing continuity over long time periods, despite fluctuations in personnel, policies, governmental administrations, and finances. Three such macroprojects examples are presented in Exhibit 79.

Exhibit 79—Managing Moon Macroprojects. Artist Robert McCall envisions activities of the next century on the Moon, which may include astronomy with both optical and radio waves; use of **lunar materials** to **manufacture** products in space; and a mass driver (foreground) that propels baseball-sized lunar materials into space. *Source*: National Commission on Space, *Pioneering the Space Frontier*. New York, NY: Bantam Books, 1986, p. 64.

Gaining a national consensus to support such space visions and ventures is a cultural problem. Implementing plans for the purpose implies innovative approaches to space management, such as discussed here. Why we are so timid in carrying out the Commission's bold recommendations has been well expressed by *Washington Post* writer, Charles Krauthammer (April 9, 1995):

> **Ours was the generation that first escaped gravity, walked the Moon, visited Saturn—and then, overtaken by inexpiable lassitude and narrowness of vision, turned its cathedrals of flight into wind tunnels.**

For existing space organizations, whether global space agencies or their aerospace partners, re-education of personnel is in order to prepare for the 21st century demands of space management in general, and macromanagement in particular. Whether for oldtimers or newcomers into space enterprises, new executive/management development programs should be redesigned to deal with the above considerations. In transportational partnerships, corporate HRD (human resource and development) specialists will have to cooperate with R&D professionals to create more appropriate in-house training of macromanagement skills.

Space management in the future will necessitate crossing traditional academic disciplines and industrial differentiations, as this quotation of Frank Davidson's book *Macro* so aptly describes:

> **Space development is a critical case-in-point, because it will test the ability of our diverse, rather relaxed society to set long-range goals, to hue the line despite disappointments and setbacks, and to devise institutional arrangements that will assure continuity ... . Low-cost approaches are indispensable, because an increasingly educated public will rightly insist on return on investment ... Now is the time, for the aerospace community to reachout to the mining industry, the heavy construction industry, and the ground transportational industry, so that joint ventures on land and sea, as well as "up there," may set a pattern of partnership and a network of personal relationships which will benefit all systems engineering programs that are so necessary for the future health, safety, and prosperity of the Republic.**

A prototype of future **macromanagement education** may be in the making at the International Space University in France in their Master's Degree Program in Space Science.[36] This graduate program is available during the regular academia year, as well as in a ten-week summer session. Its departments are multidisciplinary in **Space Systems Architecture and Mission Design; Space Business and Management; Space Engineering; Space Life Sciences; Space Policy and Law; Space Resources, Robotics and Manufacturing; Satellite Applications; Space Physical Sciences; Space and Society**. In the second department mentioned, the business and management aspects of space projects are examined. The core lectures cover a broad range of topics: management techniques, financing, costing, business planning and structure. Principles are illustrated with examples from space enterprises, while management tools are used in terms of the managing and economics of space activities. Special lectures deal with space failures and quality management, space project management, and space business case studies. The core curriculum is taught by an international faculty, supplemented by lecturers from around the world who are experts in their subject matter. This author sees this emerging model as most promising for the preparation of tomorrow's macromanagers. My hope someday is that this innovative approach will be spread to other universities electronically, and eventually in orbit!

8.8 SPACE STATION CASE STUDY

To appreciate why macromanagement is so vital to space developments, the following mini case study is presented next. As suggested in the last chapter on **macrothinking**, this summarizes why project managers must utilize a combination of political, financial, communication, and managerial skills to achieve their goals. In this instance, the macroproject is the construction of an orbital platform and laboratory as envisioned in Exhibit 80. Its proponents have had a difficult time putting together a case for investment that sufficiently convinces both the public and the legislators to support that venture whose future is still uncertain as of this writing.[37]

Exhibit 80—International Space Station Cooperation. As the 20th century closes, this macroproject represents the most challenging attempt at synergy in space among nations. The often-changed design for this orbiting joint venture by the U.S.A., Europe, Canada, Japan and Russia may face further alterations with the introduction of the last-mentioned federation into the allied partnership. See case study for the evolving story. *Source*: NASA/Johnson Space Center.

CASE: EVOLUTION OF THE INTERNATIONAL SPACE STATION

Visionaries have dreamed and written of housing people on islands in the sky for centuries. Wernher von Braun, who began the Space Age in Peenemunde, Germany, on October 3, 1942 with the launching of the V-2 rocket, wrote over forty years ago in *Colliers* magazine of future orbiting stations and platforms. This great rocket scientist actually designed a demonstration model—a 77-meter diameter wheel which was to become immortalized in the film, *2001: A Space Odyssey*. Von Braun envisioned the station as the starting point for expeditions to the Moon—instead the U.S. went there directly, postponing the orbiting station.

Background

After many abortive proposals and debates, NASA actually launched the first Earth-orbiting station in **Skylab**, flying three missions there in 1973 with crews staying up to three months aboard while performing 270 multidisciplinary in-vestigations.[38] The spacecraft contained a house-size workshop, airlock, docking adapter, solar observatory, and command-service module. Weighing a 100 tonnes, the orbiting facility provided 350 m^3 of workspace and was 54 meters in length. A walk through the *Skylab* mock-up in the Smithsonian Air & Space Museum will confirm the size and amenities which its astronauts en-joyed while aloft. Much was learned from this project about human capacity to live and work in the microgravity environment. With insufficient public support and devastating budget cuts, its lifetime was limited until *Skylab* plunged back into the atmosphere six years later.

After a year of experimenting with an orbiting space station, the former So-viets began to establish semi-permanent manned facilities in 1977 with the launch of **Salyut 6**. The Russian station evolved as modules were added to subsequent *Salyut* and *Cosmos* craft until the **Mir** superstation appeared in 1986. Political and social turmoil within the then U.S.S.R. in the 1990s post-poned their plans for **Mir 2** and Kosmograd, a space city in the sky. Within the new Commonwealth of Independent States, the Russian Federation has de-veloped agreements with NASA, ESA, CSA, and NASDA for their personnel to conduct life and material science experiments aboard **Mir**, and now both astronauts and cosmonauts have used each other's spacecraft for this pur-pose. In 1995, the U.S. Shuttle actually docked with the aging Russian station.

In the 1970s, NASA began the search for a permanently manned space station design, the configuration of which has been subject to numerous de-bates among politicians, space scientists, engineers, and the public. When President Ronald Reagan in 1981 requested from the Congress, $8 billion to actually build it, Hans Mark (1987) attempted to record the history of these furious debates which are still going on. An NCB network broadcaster re-ported on April 6, 1991 that NASA has now *spent* $5 billion on a space station that is not likely to be permanently occupied until 1999, and then by only four astronauts. This commentator said the price tag for the finished product is likely to be $30 billion, and some call the station an "orbiting pork barrel." By 1995, some insiders claim that the Agency has actually spent $12 billion on station *planning*, estimating that 70 cents of every $1 so expended goes to "paper work;" to complete the project, critics now say $40 billion more will be required ... . Each year when NASA seeks its annual funding authorization for this macroproject, the debate heats up again—the current Republican Party leadership in the U.S. Congress, according to Rep. Robert Walker, promise to set a multi-year authorization for the funding of the oft-redesigned station. Meanwhile, NASA engineers and bureaucrats have been severely criticized for overpromising what might be done elsewhere more cheaply. The basic problem with this macroproject is that the agency failed to make a substantial and convincing *case for investment* to the public, the Congress, the scientific and industrial communities. Over the years, the cost estimates and purposes of the program have altered many times. In effect, the Agency did not properly *macromanage* the program! Despite the efforts of well-meaning space ac-tivists to promote this large-scale technological venture, a political base and consensus for the station has yet to be developed. Many rationales have been offered by NASA and its aerospace partners as to why the station needs to be constructed at such huge cost, plus even larger sums that will be required to keep it in orbit in terms of maintenance and human occupancy. The justifications range from a place for the Shuttle fleet to dock and a platform for exploration of the universe, to a laboratory for microgravity manufacturing, for general scientific experimentation and observations, for life science research on effects of weightlessness, and even for defense purposes. The arguments-in-favor maintain the station is needed for learning about orbital living and mechanics; for development of new space-based vehicles, and technologies; for missions to return to the Moon and lunar industrialization; for going to Mars and beyond. Some expert study panels and commissions agree, such as the National Commission on Space (1986). Others did not, especially among planetary scientists whose astronomers propose putting the money into unmanned space explorations. The University Space Research Association sponsored a study by 57 universities recommended using the Shuttle's external tank, now discarded, to build the station; NASA showed no interest in the proposal. Given that the design life of the station is only projected for a dozen years, some observers, like your author (Harris, 1995), wonder if it would not be better to invest in the Moon itself as a permanent space station and launch pad into the universe? By mid-decade of the 90s, the project has been scaled backed in scope, life sciences laboratory is now given as the principal reason for building it, and international cooperation with Russian participation is the main thrust in the program. At this writing, there is no absolute certainty that the U.S. Congress will finally provide the funds to complete the undertaking.

NASA did try to win support for the enterprise—it produced beautiful brochures as to why the station named **Freedom** was necessary as a orbital foothold to the future. One of them by Martin and Finn (1987) said the station was a new national laboratory needed for science and technology, for exploration, for U.S. competitiveness and technology leadership, and for humankind. The agency did enlist **Spacecause** to lobby for the project; it did sign up four major contractors and numerous subcontractors who would also seek to "win friends" for the program. But its designs for a habitation, laboratory, and logistics modules were considered by some experts to be grandiose. What started as a U.S. Space Station **Freedom**, for a time called **Alpha**, had to be transformed today into an "international partnership" to reduce expenses by obtain funding and technical assistance from the space agencies of Europe, Japan, Canada, and now Russia. These same partners became increasingly disillusioned at: (a) the slow pace and scope of negotiations, even after agreements were signed; (b) the numerous redesigns and budget reductions which caused scheduling delays caused by U.S. politicians, which impacted their own programs to provide separate laboratory/habitat modules.

The U.S.A., which is to provide the primary funding, is distracted from underwriting such a massive space undertaking because of a tremendous national budget deficit, recession and unexpected costs of the Gulf War, compounded by ever growing demands for domestic spending to remedy problems in the homeland, from decaying infrastructure, to homelessness and substance abuse. In the minds of many Americans, these concerns seemed to outweigh the scientific, economic, social, and political advantages claimed by NASA for the station. Thus, budget cuts consistently erode original plans to house a crew of 8 permanently in a low inclination orbit of 28.5 degrees to the equator. A double-keeled station weighing 227 tonnes, 13.6 meters in length and 4.5 meters in diameter had to be scaled back on orders from the U.S. Congress (a $6 billion curtailment over five years). The complicated management support system had to be reorganized—the headquarters' Space Station Program Office was trying to coordinate station activities in 9 of the agency's field centers, with 3 international partners and numerous private contractors. Over the past decade, the station macroproject has produced monumental management problems—such as difficult negotiations to obtain agreement from all station partners to keep the facility design and equipment compatible; counteracting criticism of the station designs from engineers who boasted of superior, less expensive designs; critical reports from other governmental agencies, such as the White House's Office of Science and Technology. Meanwhile *the incredibly shrinking station* has been redesigned from originally 122 meters long, with one sixth the transmitting capability originally conceived; the $1.8 billion annually operating cost has been scaled down by 80% because of Congressional mandate permitting only $2.1 billion a year to be spent on the while endeavor. In 1995, NASA is promising to build the station by June 2002 for a mere $17.4 billion!

Current Reality
The **INTERNATIONAL SPACE STATION (ISS)** project is coordinated by the NASA Office of Space Flight for some seventeen international partners. NASA Office of Space Access and Technology has been designated to spearhead new ways for the Agency to do business, transferring space technology to commercial applications, including what is learned from the **ISS** operations. OSAT is currently building a Payload Traffic Model, a long-range operations plan determining which experiments will be sent to the station when it is actually in orbit. NASA states that the **ISS** goal now is to provide at 445 km above the Earth, a "long-term research platform in a near zero gravity environment to conduct basic and applied science, technology, and commercial driven research, as well as provide an observation platform for Earth and space science." The principal contractors Boeing, Rocketdyne, and McDonnell-Douglas are building hardware to launch in 1997. The new ISS configuration calls for eight pressurized modules, a long metal connecting truss, a Canadian robot arm for exterior work, a solar power system, and numerous smaller systems to maintain the "laboratory"—the latest is that it will be 88 meters long, with a 110 meters wingspan, for 1230 m^3 of volume with a total of 110 kW of power. The new orbital inclination of ISS will be 51.6 degrees, permitting launches from Russia. The start-up is November '97, inaugurating forty-four spaceflights by participating countries until June 2002, when assembly is complete

with crew installed. The Technical and Management Information System (TMIS) is the computerized means for assembling and distributing data about the project. The elaborate expenditures for ground support facilities and programs also had to be reduced.

According to NASA publications, the **ISS** assembly schedule as of July-August 1995 is as follows;[39]

Space station assembly schedule

Schedule	Date	Payload
First U.S. Element Launch	November 1997	FGB
First Russian Element Launch	April 1998	Service Module
Continuous Human Presence	May 1998	Soyuz
U.S. Lab Launch	November 1998	Pressurized Lab
Japanese Lab Launch	March 2000	Pressurized Lab
ESA Lab Launch	February 2001	Attached Pressurized Module
Centrifuge Launch	October 2001	Centrifuge Accommodation Module
Habitation Module Launch	February 2002	U.S. Habitation Module
Assembly Complete/ ContinuousFull Crew	June 2002	Crew Transfer Vehicle

It would seem to this writer that the program is in four activity phases requiring 73 launches to take place in a sequence over 55 months, involving possibly 888 hours of both American/Russian EVAs at great risk:

Phase One—June 1995–1997 to use existing capability of an adapted U.S. Shuttle and Russian orbital station for initial activity together comparable to what will take place on **ISS** ... . The next phase involves fully functioning vehicles which can be parked unattended in orbit, allowing for human presence to begin in 1998.

Phase Two—late 1997 to March 2000, actual deployment of core elements for ISS construction:
- from U.S.A.—two nodes for storage and electrical power equipment; a U.S. Lab with payload equipment/experiments) ...
- from Russia—Functional Energy Block (FGB, contingency fuel storage, propulsion and multiple docking points (purchased by U.S.A.)—self-powered cargo vehicle to be launched first on a **Proton** rocket; Service Module with environmental control/life-support system derived from **Mir 1** and also launched by **Proton** vehicle; Science Power Platform to provide

20kW power and heat rejection launched on **Zenit** rocket; and **Progress** Cargo Vehicles to carry reboost propellant to **ISS** four times a year ...
- from Canada —Mobile Servicing System (16.7-meter robot arm and mobile transporter) ... . The transportation to and from orbit will be provided by the NASA Shuttle Fleet, RSA's 57 launchers, and ESA's **Ariane 5** rockets ... .

Phase Three—March 2000 to September 2001, adding modules:
- from Europe—Columbus Orbital Facility, a pressurized laboratory, logistic services by **Ariane 5** rockets, an Italian and Automated Transfer Vehicle) ...
- from U.S.A—Centrifuge Facility illustrated in previous exhibit ...
- from Japan—JEM laboratory is an exposed platform or "back porch" for space experiments with small robotic arm for payload operations ... .

Phase Four—2002:
- from U.S.A.—November launch of Habitation Module with galley, toilet, shower, sleep stations, and medical facility; June **ISS** is fully operational with crew of up to 6 persons, living in spartan quarters ...
- from Russia—two crew Transfer Vehicles, a redesigned **Soyuz** capsule for emergency rescue for two crew members.

Station Management

Based on the Intergovernmental Agreement (IGA) previously signed and revised by Canada, ESA members, and Japan, as well as Russia in 1995, the **partners'** roles and responsibilities have been spelled out, along with their financial contributions and allocated on-board space ... . **Users** are of several types—facility-class payloads for microgravity/life science research; technology and commercial payloads, express and external payloads (latter are attached to four outside ports which may be plugged into station resources). The user community has greater voice in program preparation through a special representative; users now sit directly on various product teams developing the station ... . The seven science facilities available aboard station are: laboratory/2.5 m centrifuge, a furnace, and facilities for gravitational biology, human research, fluids and convection, and biotechnology.

At NASA, the sixth station chief is now Wilbur Trafton (Deputy Administrator, Office of Space Flight), who heads up a streamlined management team, most of whose members are located at JSC in Houston, Texas. The new design team includes the Americans Douglas Cook and David Mobley, with Russian Leonid Gorshkov, chief engineers for NPO **Energia**. The management approach of Total Quality Management has been incorporated into the *Integrated Product Team*. IPT is organized along product lines, instead of function and disciplines. Hundreds of these multidisciplinary teams will have resident product experts, including contractor, business manager, and customer. McDonnell-Douglas, for instance, has five IPTs for each segment of the main truss it must deliver to NASA under its contract. Within that company there is a Vice President for the Space Station, Gale Schulter, who thinks the IPT process "encourages management to get its act together and plan ... to make sure roles and responsibilities are well defined."

Current Situation

As of this writing in 1996, the **ISS** problems and debate go on. The Russian booster stockpile goes down just when experts question whether four Shuttles are enough to get the main parts of the station in orbit starting next year. Russian enthusiasm for the **ISS** macroproject diminishes as attempts to extend the ten-year life of **Mir** were rejected by the Americans and the Russians press to have U.S. taxpayers pay for their 10–12 launchings per year beginning in 1998. Meanwhile their conglomerate partners are becoming aware that the Russian segment of **ISS** could be unlatched and operated independently at lower cost as a stand-alone, orbit outpost—possibly a **Mir-2** (Jim Oberg, **Space Report 1** "Russia's Space Program," *IEEE Spectrum*, December 1995, p. 35). Furthermore, the head of Germany's largest aerospace company proposed to NASA's Dan Goldin that planning be directed now to privatizing the **International Space Station**. Since Daimler-Benz Aerospace is building the ESA's laboratory module for the station, the Administrator said he would take their CEO Manfred Bischoff's suggestion into consideration for possible later implementation. Meanwhile American political analysts, like Mark Whittington, have been writing in *Space News* (May 6–12, 1996, p. 13) that the whole **ISS** venture is becoming "NASA's Vietnam." He maintains that the original purpose of the station has long been forgotten in Agency rhetoric, and that planners should heed the warnings of the International Space Station Independent Assessment Report (January 30, 1996). That panel cautioned that ISS performance schedules continue to erode, budget reserves are inadequate, power demands of station operation will exceed generation capacity, and question Russia's ability to meet its promised contributions.

"Stay tuned, folks," to see if these well-honed, but often changed, plans actually get implemented and the **International Space Station** actually flies so that humans can live and work there for the NEW MILLENNIUM! It may yet prove to be a monument to the practice of synergy in space. Given this space station's history as described in the above case, there must be a better way to manage space macroprojects. We suggest it will require macromanagement skills if numerous spaceports are to be constructed in orbit during the next century. Relative to the so-called U.S. leadership in this process, this author agrees with the "Commentary" of S. Fred Singer, president of AIAA's Science and Environment Policy (*Aerospace America*, February 1995, page 3):

> **Our national goal ought to be a landing on the Moon, and eventually, Mars ... . We still have neither a supportable goal beyond space station nor a logical plan to reach that goal—25 years after the magnificent achievement of the Apollo project. I submit that we lack a credible, generally accepted rationale for planetary explorations by humans. It does exist—most clearly for Mars—but it must be spelled out!**

8.9 CONCLUSIONS ON SPACE MACROMANAGEMENT

The plans which global space agencies have for space developments for the next fifty years will have to include private enterprise if they are to be realized. Even at the minimum level of an **International Space Station** and a lunar outpost, construction and maintenance costs will be billions of dollars, requiring a new generation of technological advances. To accomplish such minimal goals demand: (1) more international technological partnerships involving private sector participation with the public sector; (2) a different type of space management and leadership from that exercised during the past fifty years which inaugurated the Space Age but only with short-duration missions; (3) a new type of financing for space infrastructure

Building upon the **Apollo** heritage of technological/administrative innovation, **macromanagement** has been proposed for 21st century space macroprojects. These latter might range from manned missions to Mars, to mining the asteroids, and actually building space settlements aloft. For that to happen, this chapter has highlighted some of the issues that need to be addressed in the strategic planning/management of space endeavors, such as:

(1) large-scale technological undertakings involving humans dictate multidisciplinary approaches to their financing and management;

(2) increases in orbital human populations for longer periods of time will necessitate a decrease in the influence and management of ground control from Earth;

(3) expanding space-based programs whether manned or unmanned entail greater management by spacefarers;

(4) **space management**, both on the ground an in orbit, will be the laboratory for advancing the art and science of all management in the third millennium.

With a \$4 trillion (10^{12}) gross national product, one would expect the United States of America to allocate more than 1% of its budget to the next stage of space exploration and exploitation. However, as the 20th century draws to a close, American **space leadership** wavers, seemingly lacking in clear **national space ethos and goals**, as suggested in our opening chapter. Instead Europeans, through both the ESA and the RSA, are beginning to provide the lead both in human spaceflight and in lunar development. After a decade in orbit, the Russian **Mir** space station is a proven success in terms of international participation and application. Perhaps further study of that program may offer another prototype of tomorrow's macromanagement. As a result of International Space Year, this decade has proven the value and the potential of **space synergy**.[40]

DIVERSITY IN SPACE

With the globalization of the space program, managers, trainers, and commanders in orbit will have to possess multiple management and diversity skills. Cross-cultural management competencies will be essential for missions that are more international in scope, in terms both of operations and personnel composition. Space macroprojects will bring together people from many nations and cultures, all conditioned in a particular way of managing and operating. Crew diversity will take on added signifcance in terms of the **International Space Station**, a lunar base, or manned mission to Mars. The New Millennium will be the focus of space settlement and industrialization, pointing up the interdependence of both Earthkind and Spacekind. Diversity begins as terrestrials migrate offworld, but real diversity will evolve when our species settles and adapts to the varied environments of our Solar Systems. And, if we should meet life from other worlds and universes ... .

—Philip R. Harris, Ph.D., Space Psychologist and Author

9

Challenges in Space Industrialization and Settlement—Commercial, Legal, Political

Although the human species required over three and a half billion years to evolve, George Robinson, a space philosopher and attorney, observed that in the past thirty years we have moved beyond Earth to penetrate near space, deep space, and other planets.[1] In the process of transforming our perceptions of humanity, space law scholars speculate that *Homo spatialis* or **spacekind** will develop as a new species—altered in time from *Homo sapiens*, physically, psychologically, and socially.[2] In contemplating the human occupation of outer space, issues related to its industrialization and settlement may be viewed as problems or challenges.[3] Preferring the latter approach, there are indeed numerous multidimensional challenges—technological, biological, and financial to start. However, in this chapter, we will delimit analysis to just three dimensions—commercial, legal, and political, ending with some specific action plans to further space enterprise. We will also revisit and amplify some of the themes discussed in our opening chapter.

9.1 CHALLENGES AND REALITIES

Globally, all of the existing national space programs are moving forward but with more economic constraint in the midst of varying degrees of chaos within their countries. Consider the contemporary realities for three major spacefaring partners ... . In this Wiley–Praxis Series, a companion volume on *The New Russian Space Programme* by Brian Harvey provides an excellent review of the challenges being faced by that country as it transitions into a market economy.[5] Yet, this magnificent space enterprise struggles to survive in the midst of national turmoil. As Dr Michael Fulda, president of the Institute for the Social Science Study of Space, reminds us, Russia is undergoing three simultaneous, traumatic changes—the breakdown of the Soviet empire, free market reforms, and democratization, all resulting in social, economic, and political instability.[6] Exhibit 81 on the next two pages provides an overview of current Russian policy for its civil space program (RSA), and does not include the program of the Military Space Forces (VKS).

---

† The author acknowledges the special contribution to this chapter of his esteemed colleague, Dr Nathan C. Goldman, space attorney and professor . The research reported here was initially undertaken when we were Faculty Fellows together in a NASA Summer Study on utilizing space resources.[4] See Notes on Contributors.

Russian Space Policy

On August 7, Decree #791 was signed by Prime Minister Viktor Chernomyrdin "On Carrying Out Space Activity in the Interests of the Russian Federation Economy, Science and Security." With the intention of improving conditions in the space industry, the main points of the decree are:

1) The Russian Space Agency (RSA), Russian Federation State Committee for Defense Sectors of Industry and Russian Federation Ministry of Defense are to put their main efforts to the following tasks:

   — Monitoring the environment and disaster areas
   — Global positioning systems
   — Telecommunications around the world and television broadcasts to Russia
   — Study of natural resources
   — Manned spaceflight and materials processing
   — Scientific research in astrophysics, geology, and solar studies
   — Verification of arms control treaties
   — International agreements, the International Space Station and planetary studies

2) The RSA, Academy of Sciences and Ministry of Defence will adjust progams in response to funding they receive.

3) The RSA will be the state contracting agency for scientific research and experimental programs for organizations it governs undergoing conversion from military projects.

4) The RSA will deliver a plan to the government by the end of 1995 describing how space industry developments can be exploited in other industries.

5) The Ministry of Defence and RSA are to provide a proposal to the government in one month describing the use of Russian territory as space booster impact zones.

6) A list of current funding accounts to industry for non-research and experimental work is to be compiled by the RSA.

7) An amortization rate of 80% is to be applied to fixed capital like test facilities at enterprises and research institutes.

8) A list of enterprises and organizations participating in space projects will be made by the RSA for examination by the Federal Energy Commission for which no reduction or restriction in fuel and energy will not be allowed.

Exhibit 81 continues on p. 245

Continued from p. 244

9) An intergovernmental agreement on customs and duty-free import/export will be developed by the RSA, Ministry of Foreign Affairs, and State Customs Committee with the governments of the ESA, Japan and China.

10) The Ministry of Finance is to make provisions in draft federal budgets for the maintenance of unique test facilities of organizations and enterprises participating in the space program.

NOTE: In the above points it should be noted that not all enterprises are contracted by the RSA for projects, making NASA's agreements with the RSA for International Space Station components an exception in the functioning of the RSA—it does partially grant the industry giants the power to contract directly to other customers bypassing the RSA.

Exhibit 81—Russian Space Agency Policy. Russian space undertakings undergo painful transition from the glory days of former U.S.S.R. to a market-oriented economy. For an update, refer back to both Chapter 1 and James Oberg's special report, "Russia's Space Program" in *IEEE Spectrum*, December 1995, p. 18–35. This exhibit summarizes current operational policy for the **Russian Space Agency** (RSA in English, RKA in the Russian language). *Source: Countdown*, September/October 1995, Vol. 13:5,, p. 27, bimonthly, published by CSPACE Press Inc (123 32nd St., SE, Grand Rapids, MI 48548, U.S.A.).

The European Space Agency is an international success story only three decades old, linking together 15 nations in a working model of scientific, technological, and political cooperation.[7] In October 1995, the Council of Ministers for the European Space Agency (ESA) met to confirm space policy and expenditures to the year 2000 for its fourteen states, including Canada, a cooperating state.[8] They approved ESA's long-term strategic plans, particularly with reference to the four-phase lunar program, described in Chapters 7 and 10. For the next four years, Council decisions financed the necessary level of resources for ongoing mandatory activities, including the Agency's general budget and scientific program. The bulk of this budget is spent on contracts awarded to industry and universities in member countries, ensuring that the contributors get a financial return on investment, as well as a share in technological spinoffs. Because of multinational currencies, this financial authorization is recorded in **MAU**—a million accounting unit paid by a member state into the Agency budget; it is roughly the equivalent of an **ECU,** the currency unit of the European Unit (1 AU=US$1.18). Exhibit 82 illustrates the ESA budget of 2679 **MAU** (approximately, $3 161 220 M) in terms of expenditures for the mid-decade on both mandatory and optional programs. (Macroprojects like space station/platforms and microgravity research are optional contributions.) Yet, this space investment occurs as plans for the European Union and economic integration slow down, states from Eastern Europe clamor for EU membership, and socioeconomic problems plague the member states.

France, Germany, and Italy are the largest contributors to this ESA budget. For example, in Germany, aerospace activities are supported by federal and regional government, but carried out by research institutes, industry and academia. Space makes up 17% of federal research and technology spending, in which the German Space Agency (DARA) is at the forefront.[9]

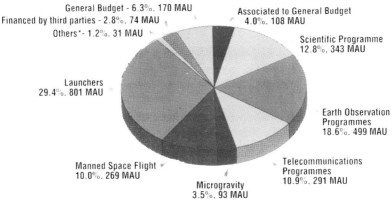

General Budget - 6.3%. 170 MAU
Financed by third parties - 2.8%. 74 MAU
Others*- 1.2%. 31 MAU

Associated to General Budget
4.0%. 108 MAU

Scientific Programme
12.8%, 343 MAU

Launchers
29.4%. 801 MAU

Earth Observation
Programmes
18.6%. 499 MAU

Manned Space Flight
10.0%. 269 MAU

Telecommunications
Programmes
10.9%. 291 MAU

Microgravity
3.5%. 93 MAU

*Others = TDP-1, TDP-2, GSTP, Prodex

Exhibit 82—ESA Budgeting Expenditure on Space. The ten categories of space investment by Europe are illustrated—the percentages represent **million accounting units (MAU)** contributed by the member states. Source: *ESA—All About the European Space Agency*, 1995, p. 2, and *ESA '94 Annual Report* published by the European Space Agency (Public Relations Division, 8–10, rue Mario Nikis, F-75738, Paris cedex 15, France).

As a **case in depth,** we again highlight the challenges being confronted by the United States space program, which in recent years has been driven more by **budget**, rather than policy. It is another argument for space scientists and engineers to study economic and financial modeling, so they can better estimate costs before submitting space proposals to legislators and investors.[10] In a technical symposium address at Air/Space America, Dr Elie Shneour (1988), president of Biosystems Institutes, Inc. of San Diego, astutely commented:

> The fact is that we do not talk much any more about the science and art of space missions. We argue instead about the relative merits of manned versus unmanned missions, about the advantages of a strictly civilian space adminis-tration versus one which combines resources with the military, and about how much all of this will cost. There is no denying that the costs of space develop-ment are huge, so large that it is not NASA which plans space missions today. It is largely the province of the Office of Management and Budget's decision-making. And while we defend ourselves for being impractical and unrealistic visionaries, the rest of the world is slowly but surely stealing a march on the United States in space as it has in other contested fields.

The Congressional Budget Office examined the projections of the National Aeronautics and Space Administration for the 1990s and beyond, offering these conclusions:[11]

(1) The Administration 1989 budget request for $6 billion (current dollars) for construction of a space station over three years or more will require increased investment beyond the $9 billion appropriated for all current NASA activities in 1988. To extend and exploit the technology and infrastructure created by NASA since the conclusion of the **Apollo** program would increase NASA's budget to $16.4 billion annually by the year 2000. The White House Administration requested $15.1 billion for the 1991 agency budget, and among the Congressional cuts was the $2.5 billion sought for the permanently inhabited station, **Freedom**. In a 1991 General Accounting Office (GAO) report on the history of NASA funding and plans for the International Space Station, the GAO points out the agency estimates of the latter's costs rose from $11 billion to $18.5, and recommended asking NASA to disclose annually the station's direct costs and revised estimate for first-phase assembly and first-year operations.

(2) The interdependence within NASA's core programs in developing the U.S. space infrastructure, especially space transportation and stations or spaceports, will continually cause costs to soar in the 1990s. This includes spending on space science, research and technology, and does not allow for funding of new space schemes, such as proposed in several recent analyses. The fiscal challenge is dramatically visualized in Exhibit 83, which compares the financing of NASA's core program with the recommendations of these three national studies.[12] If the U.S.A. is to build aerospace planes, settle the Moon, and explore Mars, these three reports would require public expenditures ranging from $15 to 30 billion in excess of NASA's planned funding for its core program. Thus, when President George Bush set forth his Space Exploration Initiative in 1989 for a return to the Moon and mission to Mars by 2019, he did not indicate *how* such macroprojects would be funded, so the proposals were largely ignored by Congress.

Contrast the two financial viewpoints presented in Exhibit 83. On the top, (a) are the recommendations for spending in three major U.S.A. space studies, and on the bottom (b) are the actual NASA budget authorizations. The latter came from Dr Buzz Aldrin, Chairman of the National Space Society, in a letter warning members that "The House Appropriations Committee responsible for NASA's budget is intent on dismantling our national space program" (July 24, 1995). If that Committee's scenario were adopted the NASA budget would drop to $11 billion by 1999 and seriously undermine most human space initiatives. Within the Agency itself, Alan Ladwig, Associate Administrator for Policy and Plans, forecasts a $13 billion budget for NASA in year 2000, still a downward trend in public financing.[13]

Clearly, the taxpayers **alone** from just one nation **cannot continue** to provide such huge investment funding without more international and private sector participation. The scope of desired spending *necessitates*:

- **a major restructuring in the annual budget process for NASA, and probably in the agency's organization, something that was studied in 1990 by the NASA Advisory Committee on the Future of the U.S. Space Program—although they did not resolve the problem, they recommended that OMB and Congress establish a reasonable share of NASA's budget for advance technology development ...** [14]

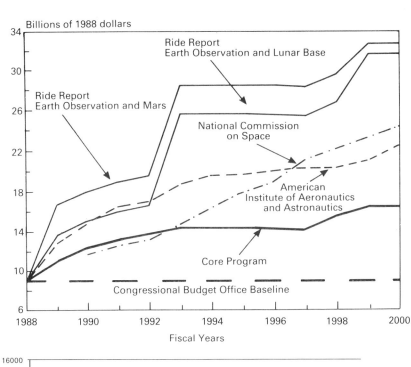

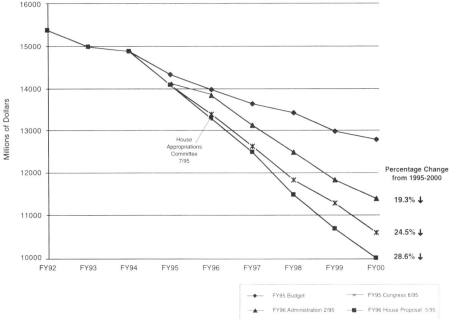

Exhibit 83—NASA Budget Options Until Year 2000. *Source*: (top diagram) U.S. Congressional Office estimates in *The NASA Program in the 1990s and Beyond* (May 1988) for comparison with *NASA Budget Authority in FY95 Dollars*, an analysis by the National Space Society (July 1995) for 1992 to year 2000.

- a larger partnership in the American space enterprise with the private sector and the public at large as a new means for sharing escalating costs ...
- a greater participation by the international investment community in funding U.S. space endeavors (possibly through stocks and bonds offerings, or other innovative financial investment mechanisms) ...
- an increase in joint space ventures with other nations and their nationals ...
- a willingness to reduce costs by purchase of space hardware and software in the global marketplace, such as from Russia or India.

Remarkably, many of the recommendations of that Committee chaired by Norman Augustine have been or are being implemented. By 1995, major organizational renewal is under way in NASA under the leadership of a new Administrator, Dan Goldin; a new Republican-dominated Congress is considering multiyear budgeting for space macroprojects like the ISS; partnership agreements have been increased by the Agency not only with the private commercial sector, but with other international space agencies. The most outstanding change since the first edition of this book has been the excellent working relationships developed by the American and Russian leadership in cooperative space ventures, particularly purchase of the latter's hardware and Russia's unique participation in the **International Space Station** program.

Now for a reality check—despite the technological advantage that space gives to the U.S.A., the NASA budget has decreased from $14.5 billion (1994) to $14.3 billion (1995) to $13.7 billion (1996); by FY 2000, the funding reduction will be 25% that of five years prior, possibly $13.2 billion or less. Exhibit 83 starkly compares these pragmatic figures with the projections of experts as to their **ambitions** for national space spending. All this is happening at a time when the new Republican Speaker of the U.S House of Representatives, Newt Gingrich, is calling for **rethinking** how we mobilize the American people:[15]

> We need a series of large projects. You don't hold together a free people of the planet by small things: "Let's get another 30 000 cars in this year." That's not exactly a battle cry ... . We ought to go back on the Moon. We ought to be on Mars. And we ought to do it with all the free nations of the planet participating, so that we build momentum of the human race ... .

The same Congressional Budget Office study cited above also observed that the civilian space program has been justified as a means for realizing human destiny, an investment in the international standing of the country, and a provider of economic benefits. The last-mentioned can be demonstrated in terms of public goods and services, stimulation of private sector R&D, and the creation of new industries, technologies, products, and services as a result of space activities. NASA has tried to make a case for space investment by highlighting the spinoff applications from its research, and underwriting commercial technology transfer.[16]

The country is still reaping technological returns from the **Apollo** legacy that ranges from laser heart surgery, magnetic resonance imaging, and voice-controlled wheelchairs for the handicapped, to cued speech devices for the deaf, reading machines for the blind, and "aquiculture" techniques for recycled water. Perhaps the president of the National Space Society, Ben Bova, said it best in a *Space World* editorial (Sept. 1987, p. 6):

Space is not a luxury. The space program is not intended merely for exploration and adventure. Space is an economic necessity ... and offers important economic paybacks ... . Since NASA's creation, Washington has
appropriated roughly $130 billion for the space agency—less than half of
one year's Defense budget ...Yet the money we taxpayers have invested in
space comes back to us magnified 20, 30, 50 times each year.
Space-derived technology is responsible for $500 billion per year in the
U.S. economy ... As space begins to develop, as industrial plants take advantage of low gravity, high vacuum, and endless solar energy, space
technologies will become the dominant force in 21st Century economies.

One U.S.A. estimate was that for every $1 invested in the space program, the ROI was $7.
In 1988, the European Space Agency did a similar analysis, concluding the return for that
continent to be $3.20 indirectly to economic activity for every dollar ESA spends (*Space
News*, April 8–14, 1991). The contractor interview study undertaken by the Bureau
d'Economie Theorique & Appliquée of Strasbourg, France, pointed up the diverse multi-

Exhibit 84—International Lunar Enterprises: **ISLEA**. Because of slowness by public sector
agencies in developing lunar resources, **private enterprise** has begun its own initiatives. Artistic
rendering of an **ISLEA** lunar landing vehicle for 1997 or beyond that results from a joint venture
agreement between International Space Enterprises of the U.S.A. and the Lavochkin Association
of the Russian Federation (Appendix C). *Source*: **International Space Enterprises** (4904
Murphy Canyon Road, Ste. 330, San Diego, CA 92123, U.S.A.).

plier effects of space contract—the technology spinoffs, improved productivity of companies that work for space industry, and the creation of new products with applications outside of space. To transform space dreams and plans into **realities** will require both innovative leadership, institutional change, and international technological ventures as depicted in Exhibit 84.[17]

Those within the **space community or movement** worldwide are convinced of the value of developing the high frontier. The Planetary Society, for example, has about 40 000 members internationally—4300 of them responded to a 1994 survey on their views.[19] About 95% of the respondents believed that space exploration is essential to the future of our society, but this selected sample have yet to appreciate the economic returns possible from the utilization of space resources, as Exhibit 85 confirms. Further, those surveyed generally expressed skepticism about the trustworthiness of government in informing the public about the risks and benefits of space exploration.

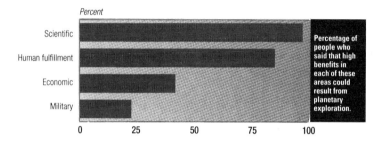

Exhibit 85—Benefits of Space Exploration. 1994 survey findings as to principal reasons to pursue space exploration as perceived by the 4300 members of **The Planetary Society** (65 North Catalina Ave., Pasadena, CA 91106, U.S.A.). The rationale of other groups might place **economic** as a higher priority than **scientific**. How would readers respond to this inquiry?

The first challenge, then, is to convince the masses of the value of space development, particularly the benefits to planet Earth and its inhabitants that will come from utilization of space resources!

9.2 CHALLENGES ON THE COMMERCIAL SPACE FRONTIER

As spacefarers take their culture and society into orbit, that process in a microgravity environment alters humanity, our social organization and institutions. One aspect of these historic changes is commercial development and the integration of outer space business activity into the Earth's global economy. A new era of free enterprise on the space frontier is being ushered in by private initiatives that go far beyond governmental efforts, thus creating an orbital economy. Already the outlines of **astrobusiness** are emerging and have been reported upon in a number of recent books.[20] A growing space-based economy presents unique financial, legal, and political problems because of the wealth potential and international implications of space commercialization. But Nathan C. Goldman maintains that the impact will dwarf the 15th and 16th centuries exploration and developments in the New World.[21] This Houston political scientist and space lawyer predicts that as we break our planetary bonds and migrate to the stars, **Made in Space** will be a label found increasingly on products sold down below.

Space enterprise will dominate global business in the 21st century. One indicator is a proliferation in publications on the subject from books to reports to journals and newsletters (see Appendix C for the publications). The American Institute of Astronautics and Aeronautics, for example, recently issued a three-volume series for innovators and entrepreneurs on space commercialization, and regularly cooperates with the Space Studies Institute in publishing proceedings of their space manufacturing conferences.[22] Whole annual directories are now published just on space companies and industries.[23]

9.2.1 Space Business Progress and Prospects

Forecasts on the future of space industrialization and commerce go back before the landmark three-year study for NASA on the subject by the Space Division of Rockwell International. In 1978, C. L. Gould, project director, reported the findings of experts who analyzed the subject in terms of humankind's needs and future trends.[24] The seminal document defined emerging nonterrestrial industry from the perspective of new products and services that would be immensely productive while providing a favorable return on investment. Their recommendations called for using the Space Shuttle not only for launch vehicle jobs, but to do science/applications experiments from materials processing to building large space structures for power systems, telecommunications, and remote sensing. The second proposal was to use space for profitable, information transmission and services. Finally, the Rockwell predictions supported using space to improve the management of Earth resources. Almost two decades later we now appreciate how prophetic this study was since so many of their prognostications are now being realized.

In 1982, Brian O'Leary, former astronaut, edited two volumes, previously cited, of expert opinion on space industrialization. The latter term was described as being of two types:

- near-term, small products which can be produced by machinery or by utilizing the uniqueness of weightlessness and zero gravity (i.e., crystal growth, ball bearings, and pharmaceuticals);
- long-term processing of nonterrestrial materials into large space structures (i.e., satellite power stations and space habitats).

Operations of the first kind require payloads on the order of a few tons and form an evolutionary foundation, while the latter involves millions of tons for enterprises in cis-lunar or asteroidal materials processing. O'Leary concluded that political, not technical or economic, considerations will probably determine the pace of such developments. That has proven to be true in both the American and the Russian space programs, where fiscal constraints on public monies have curbed both research and exploration.

The National Academy of Public Administration then did a significant study for NASA on encouraging business ventures in space technologies. Under the leadership of Erasmus Kloman, its project director, a distinguished panel recommended in 1984 that NASA:[25]

- declare and institutionalize a major commitment to business venturing in space technology;

- assist industry in pursuing opportunities for profitable investment in space;
- offer government space facilities and services for use by private companies under conditions that encourage commercial development;
- continue R&D studies for the full commercial exploitation of space;
- reduce risks and restrictions that impede commercial exploitation of space technologies.

To its credit, the National Aeronautics and Space Administration has implemented not only many of the above strategies, but acted upon other recommendations from the public or private sectors to advance space business. NASA has also funded numerous university and association studies, workshops, and publications on the industrial development of the high frontier. The topics have ranged from the exploitation of space resources to satellite services to space-based solar energy (these are usually published through the U.S. Government Printing Office). Frequently, such conferences and proceedings' volumes are deve-loped in conjunction with the American Institute of Aeronautics and Astronautics, as well as the American Astronautical Society. With Congressional prodding, NASA promoted further commercial applications of automation and robotics and has even contributed to the Foundation for Space Business Research to further space commerce studies. It also cooperates with the Space Business Roundtables of the World Space Foundation.

However, NASA's biggest contributions to astrobusiness may have been in the opening of an Office of Commercial Programs at its headquarters and appointing an Assistant Administrator for Commercial Programs. That official's primary effort was then devoted to promotion of NASA's sixteen Centers for the Commercial Development of Space. These innovative CCDS programs are still associated with industry or academia throughout the nation. Space-based, high-technology research was conducted in materials processing, remote sensing, life sciences, automation and robotics, space propulsion and power, space structures and materials. Some of the commercial center experiments flew as payloads on the Shuttle Orbiter. Although NASA provided the basic expertise and funding for these incubators of future commercial space ventures, financing was supplemented by industrial affiliates. In 1991, the General Accounting Office, the investigative arm of the U.S. Congress, recommended that NASA garner additional industry financing so these research centers become more self-sufficient or face severe termination deadlines for their government subsidies. Founded mostly with government grants, some of these centers failed to attract sufficient private sector investment. This may explain why NASA has reduced and transformed this program in 1995, placing it under the Office of Space Access and Technology and renaming it, Commercial Development and Technology Division (mail Code X at headquarters). Through its **National Commercial Technology Network** or **NCTN**, NASA fulfills an agency-wide goal of conducting 10% of its R&D through partnership with the private sector. The new attitude at NASA is reflected in this Commercial Technology Mission Statement:

> **The commercial technology mission will require that each NASA program be carried out in a way that proactively involves the private sector from the onset, through a new way of doing business, to ensure that the technol-**

ogy developed will have maximum commercial potential.
—Daniel Goldin, NASA Administrator, excerpt from *Agenda for Change*,
July 1994[26]

Although these imaginative efforts by government to "pump prime" businesses entry into space continues in a more limited way, the most promising is the Technology Commercialization Centers. The University of Texas—Austin through its Institute of Constructive Capitalism (IC2), for example, received a $5.4 million grant from NASA to develop two TCCs in conjunction with NASA field centers at the Johnson Space Center in Texas and the Ames Research Center in California. ICE has conducted market assessment of which NASA technologies are ready for licensing through these two "incubator" operations to encourage start-up enterprises and new jobs. By 1995, Johnson's TCC was supporting thirteen such companies engaged in applications of NASA technologies, while Ames' TCC was doing the same for eighteen corporations within Silicon Valley. In the next twenty-five years, such uses of the **lab-to-market model** will likely accelerate transfer of space technology into the world of commerce.

Other examples of the Federal government's endeavors in this regard are:

- *Small Business Innovation Research Program* to increase small business participation in development of NASA aeronautics and space technology by conferences and grants in two levels of funding—its publications include Pre-Solicitation Announcement, Small Business Innovation Research Solicitations, and the SBIR Reporter. This program to aid small business exists in every department of the Federal government.

- *Commercial Use of Space* encourages private investment in commercial ventures related to space manufacturing—in this endeavor, NASA supports research applications, provides access to its facilities, and promotes information exchanges through its 9 field and 16 commercial centers, as well as Industrial Applications Centers or Commercial Technology Centers throughout the nation. The varied activities of NASA with companies to promote commercial utilization of space include Joint Endeavor Agreements (JEA), Technical Exchange Agreements (TEA), and Industrial Guest Investigator (IGI), in addition to issuing publications for users. [27]

- *Privatization of Space Operations* is a policy being encouraged by the White House and Congress. However, the legislative provisions often set policy but not funding to achieve this goal. Currently, the NASA Authorization Bill in Congress restricts rather than giving the Agency the flexibility to achieve the mandate. At present, the hope is to transition the microgravity parabolic flight and other such operations to the private sector. In November 1995, NASA announced that the management of the Shuttle Transportation System was being contracted out to the United States Space Alliance, a joint venture of Rockwell International and Lockheed Martin corporations.

- *Space Offices in Other Government Agencies*—to use the U.S. government again as a case in point, offices of **space** technology, commercialization, manufacturing, or transportation have not only been established within NASA, but also in the U.S. Departments of Defense, Energy, Transportation, and Commerce.

The space age began with military applications of missile technology; the use of space technology in many countries originated through support of their armies, and later their air forces. **Military space** uses satellites to gather intelligence, as well as to control ground operations and to launch weapon systems, but many of these technology advances have commercial implications.[28] The aerospace industry worldwide has benefitted from military clients, and the case for defense from space has been made by Lt. Gen. Daniel Graham in numerous publications.[29] Military space activities in America are coordinated by the U.S. Space Command in Colorado Springs. One emerging market is satellite launching in which the Defense Advanced Research Projects Agency alone spent over $235 million during the last five years, while in the medium to small launch market, DARPA now expends about $300 million a year. The role of the U.S. Department of Defense in fostering commercial developments in space is a subject beyond the scope of this book. However, an excellent volume on *Commercializing Strategic Defense Technologies* has been edited by Dr Stewart Nozette of DOD's Ballistic Missile Agency.[30] Undoubtedly, DOD technology spinoffs, especially from the Army, Navy, and Air Force, have long benefitted the world of business. For astrobusiness, the driving force was the U.S. Space Defense Initiative Organization, whose Technology Applications Office, now the Ballistic Missile Agency, was a prime contractor for the recent **Clementine 1** mission to map the Moon. The follow-up **Clementine 2** mission is to be carried out through the U.S. Air Force Phillips Laboratory.

For those seeking information about space activities in various Federal agencies, readers should be aware that reorganization occurs regularly, so that names/functions of specific offices change frequently under new Presidential Administrations or through Congressional "reforms" (see directory listings in Appendix C under both "NASA" and "US"). The U.S. Office of Commercial Space Transportation is responsible for the promotion and regulation of commercial launch space activities. OCST also has a public affairs division to provide information on technical and program issues for private industry groups and the public at large. Their mission is summarized in Exhibit 86, and is primarily

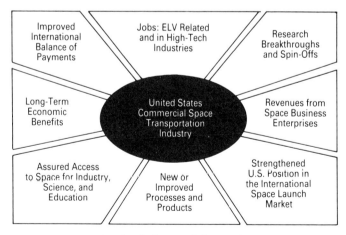

Benefits of a Strong U.S. Commercial
Space Transportation Industry

Exhibit 86—Benefits of a Space Transportation Industry. *Source*: Office of Commercial space Transportation, U.S. Department of Transportation (400 7th St. SW, S-50 (Rm. 5415), Washington, DC 20590, U.S.A.).

to implement the National Space Transportation Policy issued by the President (August 5, 1994). OCST's concerns are **space launch** vehicle technology, regulations, infrastructure, trade agreements, and applications. [31]

The U.S. Department of Energy has been slow to appreciate the possibilities of space-based energy; presently its Office of Space and Defense Power System is more concerned with the use of nuclear reactors in space ... . The U.S. Department of Commerce, itself under threat of abolishment by the Congress, sponsors the Office of Air & Space Commerce, which issues comprehensive space business reports. These not only underscore the importance of space to maintain U.S. global economic competitiveness, but envision the space frontier as a challenging place to do business, while creating new industries and jobs. Thus, DOC calls for more private investment in space-related goods and services, indicating why government should remove impediments to that end. Its publications view the prospects for astrobusiness within these major market areas: [32]

(1) *space transportation*—the foundation industry ranging from launch technology and services to multi-stage space vehicles, including future aerospace plane; currently U.S. policy opens up the private sector to the ELV market (expendable launch vehicles or rockets) ...

(2) *satellite communication*—the pioneer and proven business so vital to global telecommunications which today carries more than two-thirds of the international voice communication and most of the world's video programming; already a multi-billion-dollar-a-year industry, the operating, servicing, and replacement of orbiting satellites is already a flourishing business ...

(3) *satellite remote sensing*—commercial growth area capable of collecting data from space for weather forecasting, environmental monitoring and land-use, management of agriculture/fishing/forestry, mineral exploration, computer hardware/software manufacturers for analyzing data produced; more than l75 U.S. companies are so engaged in contrast to l25 elsewhere in 24 other countries ...

(4) *space-based facilities*—from platforms to orbital laboratories and habitats, such as Spacehab, a work/storage module which fits into the Shuttle cargo bay, and ISF, an industrial space facility to be used in conjunction with either the Shuttle or the **International Space Station** ...

(5) *space materials research and processing*—promising use of microgravity for improving ground-based methods, products, and services, as well as for manufacturing new ones, such as pharmaceuticals, crystals, and higher strength alloys ...

6) *space-based energy systems*—a 21st century business prospect to supply solar energy via satellites around or from the Moon (see Appendix B), as well as through nuclear fusion or helium-3 mining.

7) *space based commercial navigation and positioning receivers*—these will improve both sea and ground traffic and safety—already over $100 million business ...

8) *space entrepreneurialism*—will lead to the emergence of businesses both on the ground or aloft that are now only conceptualized in science fiction, as new technologies are invented and new needs arise from space settlement. The latest manifestation is **space entertainment**, either in the form of space simulations and virtual reality games on the ground within space exhibits and theme parks, or through remote control of video and rovers in space, such as on the Moon. [33]

The emerging private sector role in space amidst declining government funding requires innovative mechanisms for raising venture capital, from stocks to bonds. Addressing the 1994 International Lunar Exploration Conference in San Diego, California, John Harris reminded the audience that venture capitalists are looking for a return of five times their initial investment in five years. Space commerce usually is long-term payback at high risk—the space insurance industry incurred about $220 million in 1994 losses. At the 1995 ILEC, the emphasis was upon the use of robotics in space, whether for science, commerce, or entertainment. There was also some consensus about the need for a **Lunar Economic Development Authority** to advance human activities on the Moon, including the creation and management of business or industrial parks.[34]

B. Alex Howerton, space business writer, cautions space entrepreneurs to be: (1) bottom-line oriented; (2) not to exaggerate the developmental plans; and (3) build small, incremental products and services, the profits from which can underwrite further careful development. Howerton has written numerous case studies of space entrepreneurial companies who are providing such solutions.[35] Some of these start-up corporations leading the space growth areas which he describes are: International Space Enterprise, LunaCorp, Artemis Project, Kistler Aerospace, Hudson Engineering, and Forward Unlimited. Around the world, each year more ground-based space enterprises are founded, ranging from General Astronautics in Vancouver, Canada, which has a new low-cost **Urania** launcher for carrying medium-class communications spacecraft, to Lunar Enterprise Corporation's video which scans the Moon, to the Pacific Aloha Spaceport, to the non-profit state California Spaceport Authority at Vandenberg Air Force Base.[36] At the last-mentioned, a new partnership called Spaceport Systems International is attempting to renovate the launch facilities there for commercial use, namely putting medium-sized satellites into orbit.

Space commerce is growing throughout the global marketplace. Toward the end of the Soviet Union, its leaders began to push for the commercialization of its space program as a means of dealing with the country's economic difficulties. To gain foreign exchange, Glavkosmos offered attractive prices for low-cost launches, sale of rocket engines and equipment, and even rides for foreigners on their spacecraft to their orbiting **Mir** station. Russian marketing of their space expertise in the U.S.A. began early in the decade of the '90s as Exhibit 87 illustrates.

American government agencies did acquire Russian electric thrusters, Topaz 2 nuclear reactor, and today purchasing Russian space hardware and software, as well as technical expertise. U.S. companies are very interested not only to save launch costs, but to take advantage of Russian workmanship and innovation in such things as high-pressure pumps, engine cooling systems, and launch inspection/repair practices. Just as the Commonwealth of Independent States seeks to commercialize its space capabilities, so does the People's Republic of China and even India.

Other opportunities for increased astrobusiness may be found in the provision of insurance, legal, and tourism services aloft. International syndicates organized for this purpose will require new governmental legislations and provisions, possibly related to the UN Office of Outer Space Affairs, or a global space administration yet to be established. Today insurance premiums represent almost a quarter of launch costs; lowering them would require cooperation to delimit liability. The financial community has expanding opportunities for reasonable profit in space-related ventures. Having gained a return on investments

Exhibit 87—SOVTECH Advertising in the U.S.A. Just before the demise of the U.S.S.R., the Soviets attempted in the early '90s to derive commercial benefits for their space hardware/software and expertise. *Source*: Space Commerce Corporation (6900 Texas Commerce Tower, Houston, TX 77002, U.S.A.).

in satellites and launch vehicles, it must learn to raise more funding for long-term space business. Furthermore, financiers of all kinds need to exercise leadership in providing more creative ways for the public to invest in astrobusiness. Some of this investment opportunity will be ground-based, like the building of new spaceports or even space theme parks for entertainment purposes. For example, the State of Florida has already put out bonds to finance its own spaceport near Cape Kennedy. Some financing will be required for space-based facilities, like a lunar industrial park, mine, or solar power plant. The 21st century market for lunar products and services is long-term, and dependent upon the construction of a lunar base and possibly the formation of a **Lunar Economic Development Authority** to manage the undertaking (see Chapter 10 and Appendix B). In Exhibit 88, professor emeritus H. H. Koelle of Berlin's Technical University offers an interesting comparison with reference to the relations of lunar base outputs and its potential users.[37]

Relations of lunar base outputs and its potential users

Products	new knowledge	services	material goods and energy
lunar enterprises	new information relevant to lunar operations	maintenance and repair social needs of lunar crew	construction material, feedstock, consumables, propellants, el. & therm energy
in-space enterprises	research results relevant to space operations	communication, maintenance and repair, recreation	propellants, construction material, consumables, energy
Earth enterprises	research results relevant to life on earth	environmental observation adventure	HELIUM-3 electrical energy

Exhibit 88—The Lunar Market Prospects. The International Astronautical Federation has a Sub-committee on Lunar Development which is doing forecasting studies on the Moon market. The above table on **Relations of Lunar Base Outputs and Its Potential Users** is a typical model for commercial enterprise in space developed by Professor H. H. Koelle of the Berlin Technical University. *Source*: "A Frame of Reference for Extraterrestrial Enterprises," *Acta Astronautica*, 1993, Vol. 29: 10/11, pp. 735–741.

Increasingly, global space macroprojects will have to involve international partnerships and consortia, so packaged as to attract participation by venture capitalists, investors, and entrepreneurs. Instead of competition in a narrow space market, national agencies and corporations need to collaborate in joint technological ventures for "do-able" enterprises on the high frontier that will provide return on investment.

9.2.2 Model for Astrobusiness Analysis
One model for analyzing the future of space industrialization or commerce has been proposed by Dr Nathan Goldman and adapted by this author.[38] Exhibit 89 offers ten sectors for consideration that are somewhat comparable to the categories explained above in U.S. Department of Commerce publications.

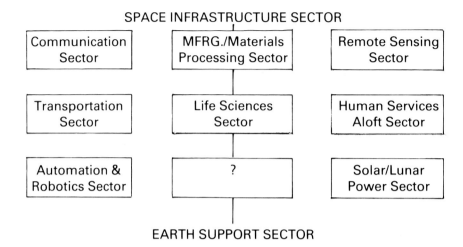

SPACE INFRASTRUCTURE SECTOR

Communication Sector	MFRG./Materials Processing Sector	Remote Sensing Sector
Transportation Sector	Life Sciences Sector	Human Services Aloft Sector
Automation & Robotics Sector	?	Solar/Lunar Power Sector

EARTH SUPPORT SECTOR

Exhibit 89—The Space Commerce Model. Created by Dr N. C. Goldman and adapted by Dr P. R. Harris. Refer to N. C. Goldman's "Space Law and Space Resources" in NASA SP-509, *Space Resources*. Washington, DC: U.S. Government Printing Office, 1992, Vol. 4, pp. 143–153 ... . *Space Commerce*. Cambridge, MA: Ballinger Publishing/Harper & Row, 1995.

Within that framework, Goldman originally described six sectors of space business as a major economic arena of the future:

- **Space infrastructure sector**—beginning in LEO (lower Earth orbit), the future requirements range from growing transportation systems and space stations or platforms to mechanical servicing of same, including orbiting spacecraft; and extending to GEO (geosynchronous orbit) for space-based nuclear or solar power stations, storage depots, assembly/fueling facilities leading to Moon and Mars bases and settlements beyond; movement toward a space-based economy that mines and processes indigenous resources. [The author would extend orbital power and services into separate classifications.] ...

- **Communications sector**—this $2–3 billion satellite business faces problems of launch failures and rising insurance costs, plus strong competition from ground-based use of fiber optical cables and networks; its competitive advantage may be in lowered STS costs, and mobile, multi-point communication linkages that will expand satellite tracking of people, vehicles, ships, facilities, and resources. [To this, the author would add new developments in small satellites, plus the computer hardware and software businesses for satellite and spacecraft operations, along with the space management information systems. This category might also include space publications of all type, whether on the ground or aloft.]

- **Transportation sector**—space transportation systems that extend from rockets to the Shuttle Orbiter, to single-stage-to-orbit vehicles, to an aerospace plane that climbs from Earth to orbit, aircraft like, becoming a spaceship in orbit. The next generation of spacecraft must lower the cost of space access, including advanced and cargo versions of the Orbiter, plus a variety of expendable launch vehicles, and orbiting spacecraft necessary for lunar or Mars operations.

- **Remote sensing sector**—the National Commission on Space (1986) predicted that advancing technology in observing systems and computers will revolutionize terrestrial science; enormous commercial applications range from weather forecasting and climate control, locating planetary resources/problems, to charting oceans and aiding archeology. Leadership in this emerging field is shared by many entities and countries—such as, NOAA (National Oceanic and Atmospheric Administration), private companies such as EOSAT (Earth Observation Satellites), and the French SPOT, which has high photographic resolution, as well as by ESA and Russian capabilities.[The author speculates enormous growth prospects generated by the Mission to the Planet Earth program.]
- **Manufacturing sector**—studies suggest this could become in the 21st century a multi-billion dollar industry, including the manufacturing of nonterrestrial materials. A new science of microgravity research, begun in industry and research institutes, has now extended to universities and NASA Technology Centers. [The author would expand this category to the whole realm of materials processing from advanced and composite materials to vacuum epitaxy technology.]
- **Earth support sector**—tomorrow this will include privately, as well as publically, owned launch facilities or spaceports; ground-based space vehicle maintenance centers; space services for propellants and energy sources; ground-based operations for satellite manufacturing, maintenance, and tracking; construction of habitat modules and equipment for space deployment; space educational centers or academies to prepare spacefarers. [In this category, the author would include such enterprises as the International Space University, as well as a whole ground support system to emerge for deployment and service of spacefarers and their families.]

To the original Goldman model, the author proposes four more classifications are emerging sectors:

- **Life sciences sector**—commercial application of research on protein crystal growth, macromolecular crystallography, bio-serve space technologies, drug processing, cell research, closed ecological systems, and life-support systems, etc. A NASA report suggests that the research will be in gravitational biology, biomedical, biospherics, environmental factors, operational medicine, physicochemical/bioregenerative life systems, exobiology, and flight programs.[39]
- **Human services aloft**—will range from supplying food, cleaning, and similar hotel services to those in orbit; to travel and tourist agencies; to in-space support services that are psychological, sociological, educational, political and social in scope; to entertainment and recreation, both in and from orbit. Any activities would be included here which ensure safety and survival in orbit, as well as to improve the quality of space life.
- **Automation and robotics sector**—will be more than supplying "tin collar workers"—many automated systems will provide information and assistance; others will be virtual reality systems. Some robots aloft may become friends and pets in isolated, confined environments. Primarily, A&R will be used for construction and maintenance, for transportation and monitoring, and a host of undreamed applications.
- **Solar/lunar power**—a 21st century business that can produce billions of dollars in income by sending solar power to Earth from the Moon or vicinity (refer to Appendix B).

Two methods advocated are solar power satellites (SPSs) from lunar orbit by Glaser; and a lunar power system (LPS) from the Moon by Criswell and Waldron.[40]

NOTE: A blank box has been left in the model above (EXHIBIT 89) for the reader to insert his or her own candidate for another sector of astrobusiness. Journals (such as *Space Commerce*), specialized newspapers (e.g., *Space News*), and newsletters (e.g., *Space Fax Daily*) often provide clues to emerging activities. One example is the Kentucky Center for Space Enterprise in Lexington which advances R&D technologies and processes related to inhabiting and using the resources of the Moon, Mars, and other extraterrestrial bodies.

Space offers unlimited economic potential for drug and large crystal production, cosmic photography and communication, varied commercial supply and repair services, medical and solar research, mining and construction, and environmental sciences applications. According to author David Gump, one firm is already on the road to winning—MRA (Microgravity Research Associates) with exceptional ground-based investigations for large-scale crystal manufacturing of gallium arsenide for superspeed computer chips.[41] Another is 3M corporation engaged in space crystal research to create smart adhesives and optical computers. The pace of commercial growth has led to the formation of CASE (Competitive Alliance for Space Enterprise) by Washington-based business people to influence government policy in creation of a single space office within the Federal agencies to help industry and entrepreneurs navigate the bureaucracy, and to streamline the regulatory process.

Furthermore, growth in space commercial activities on the ground or aloft will lead to new vocational fields and space careers, ranging from human factors and health services to architects and environmental engineers.[42] S. Norman Feingold, a psychologist and author of *Careers Today, Tomorrow and the 21st Century and Beyond*, is resolutely optimistic about future job openings in space. Within 30 years, he predicts the development of space colonies and orbital tourism; the need for space specialists in law, biology, botany, medicine, police, and other non-terrestrial vocational activities (*Los Angeles Times*, September 5, 1988, p. V/2).

Just as the early days of shipping, railroading, and flying were dependent upon government subsidy, so too is space commerce. Today governments in spacefaring nations support astrobusiness through research and operating grants, deferred payments of launch costs, free or cut-rate spacecraft rides, patent rights on government-sponsored research, and incubator programs to facilitate the transfer of space technology to private enterprise. The legislation from Congress that created the successful COMSAT entity through the Communication Act of 1962 may be a prototype for more space "public/private" corporations. In addition to reorganizing space agencies toward more commercial applications of their R&D, there may be need for another type of commercial space institution. For example, in France, there is CNES (Centre National d'Etudes Spatiales), a public establishment of industrial and commercial character that is responsible for formulating space programs, overseeing selected projects, and providing liaison with the country's foreign partners in joint space endeavors. CNES carries out its scientific activities through contracts with a dozen major research laboratories; it markets its products and services through companies it founded, such as, Arianespace and SpotImage. To commercialize space fully will require new organizational creations, even world corporations, possibly chartered by the United Nations or a global space agency.

The longer we operate in space and stay aloft, the more innovative and entrepreneurial will be the ways humans discover to capitalize upon space as a place and as a resource. At the Russian space station, cosmonauts have already learned to make more than a hundred different alloys and metal mixtures thought impossible to manufacture, plus very pure crystal and glass lenses. Space-business devotees are constantly coming up with unique ways for extraterrestrial exploitation, from burials, banks and Earth truck-tracking systems to lunar mass-drivers and large optical telescopes. Arthur Clarke, father of the communication satellite concept, proposes PEACESAT satellites to monitor military incursions, nuclear accidents, and other inappropriate activities which threaten world peace among the planet's 150 nations. He envisions two-dimensional communication networks from space contributing to the rise of the Global Family or Tribe in which members are linked electronically across national borders. The need is for more **non-aerospace** corporations and associations to get involved in astrobusiness. One example is the American Society of Civil Engineers, which has established an aerospace division which sponsors both conferences and publications on space engineering, construction, and operations.[43]

An aerospace consulting firm, C. S. P. Associates, forecasts the economic promise of commercial space activities to be $10 billion in revenues by the year 2000 (*Insight*, Dec.87/Jan. 88, p 48). Roy Gibson, former Director General of both the European Space Agency and the British National Space Center, offers an interesting observation to our discussion on commercial challenges aloft: [44]

> **The transition from developmental, government-funded activities to commercial exploitation is a gradual process. New methods of allowing for privatization, such as government solicitations based on performance capabilities rather than the purchase systems, have been suggested in ESA and NASA ... . In the future, the private sector may be more strongly encouraged to take a larger role in space activities. With incentives, industry will take advantage of opportunities offered.**

But it will be expanded human presence in space settlements and colonies that will truly open up the commercial potential aloft. The Lunar Power System planners, for instance, project a community on the Moon of up to 4000 "technauts" in the next century (Appendix B). Space traders will be among the leaders in humanity's trek through the Solar System.

9.2.3 *Lowering Space Transportation Access Costs*

Rockets have been called **dream machines** because they became the means for humankind to leave the home planet and go offworld.[45] They went from dream to reality in the mid century, but are still too expensive to build and operate if they are **really** to open the new frontier in the 21st century! As long as the cost of getting payloads into orbit remains excessive, space commerce and space settlement will be delimited. The Space Transportation System (STS) called the **Shuttle Orbiter** costs taxpayers some $3.2 billion a year to operate, and absorbs almost half NASA's budget. It is a very complex, technologically sophisticated system that is now costing about $4500 per kilogram into orbit and requires 35 000 workers to maintain the fleet of four reusable orbiting vehicles and their expendable external fuel tanks. Flying for more than a decade with only one launch failure, **Challenger**, the latter's replacement, **Endeavor** cost $2 billion to build. Averaging 5–16-day missions aloft, this orbiting laboratory can carry a crew of up to 8. In its next configuration, shuttle designers are to provide for a payload up to 11.5 tonnes to a 51° inclination orbit, so that it is capable of servicing the **International Space Station**.[46] According to Dr John Mansfield, Associate Administrator for Office of Space Access and Technology, NASA is not only expecting to reduce costs by contracting STS operations to the private sector, but is actively searching for a shuttle replacement system with reusable engines that could turn around between flights in two weeks with only 50 maintenance crew workers. With the aim of reducing spaceflight costs to $450 per kilogram to LEO, future space transportation will feature: (a) new composite materials in their construction, such as in engines, and with cryogenic tanks; (b) changed architecture by miniaturization of instruments within the small spacecraft initiative (costing $50 million per vehicle versus $400 million for the larger spacecraft)—possible because of microcircuitry or applications-specific integrated chips.[47]

In August 1994, the U.S.A. issued a new **National Space Transportation Policy** to ensure reliable and affordable access to space. In addition to **expendable launch vehicles or ELVs**, NASA was directed to lead in the technology and development of next-generation **reusable launch vehicles** or **RLVs**, particularly **single-stage-to-orbit** or **SSTO** vehicles.[48] The latter R&D program is called within NASA, the **X-33** and **X-34**, and provides the initial underwriting for industry to develop advanced launch systems for the Agency at reduced technical and business risk. In the contest for a Shuttle replacement by year 2012, two corporate giants—Boeing and McDonnell-Douglas, now considering merging some or all of their operations—teamed up to build a vertical take-off and landing vehicle. Exhibit 90 is an artist's concept of what this future craft might look like—a strong resemblance to the latter company's flight tested **DC-X**. In the same exhibit is also a rendering of the **Scorpius**, an innovative low cost launch concept of Microcosm, Inc., which uses 49 engines in 7 clusters or pods for 4 or more stages into orbit, carrying varying payloads (100 kg to 27 000 kg).[49] To reduce weight, a **DC-XA** is underway containing an all-composite cryogenic tank, one that is not only light-weight, but also reusable.

The **X-33** innovative strategy will hopefully produce by 1999 an advanced, flight-tested, fully operational RLV, so that both NASA and its contractors can go to financiers with an STS that is not a technical but only a market risk. Invest $20 billion in this venture, and a new spaceplane could be ready for passengers before 2010! This approach has been endorsed by the Space Transportation Association, which advocates that government use its annual purchases of space transportation services to **leverage** private sector STS development at lower cost.[50] The **Scorpius** concept, on the other hand, is a low cost approach both in its developmental and in its operational stages; it has proven 99% reliable under contracts with the USAF Phillips Laboratory, and offers the advantage of launch on demand in multiple sites. Made of composite parts, it features fewer parts than conventional rockets, ground-level servicing, and unique engine and propulsion systems. What the Ford "Tin Lizzie" did for automobile travel, this may do for space transportation.

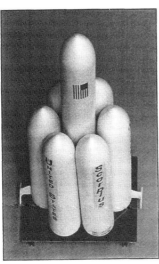

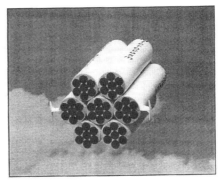

* 49 engines arranged in 7 clusters, or pods

* 4 stages plus optional upper stage

* No moving parts other than on/off valves

Exhibit 90—Single-Stage-to-orbit Reusable Vehicles. To cut transportation costs into space, innovative spacecraft and rockets will be essential, such as illustrated above. The **DC-X** is an artistic concept of the joint venture proposal from corporate giants, McDonnell-Douglas/Boeing—a vertical take-off/vertical landing launch vehicle that is reusable. *Source: Ad Astra*, September/October 1995, p. 14 ... . Smaller entrepreneurial companies have other less expensive lunar vehicle concepts, such as **The Scorpius,** 49 engines arranged in 7 clusters or pods, that take four stages to reach upper orbit. *Source: Space Mission Engineering*, Microcosm Inc. (2601 Airport Dr., Ste. 230, Torrance, CA 90505, U.S.A.).

James Sloan of Information Universe envisions great **synergism** between the low cost launch vehicle (LCLV) and the SSTO, offering mission models for a wide range of payloads to orbit. The LCLV would carry payloads in the 0.5- to 4.5-tonne class.[51] In Europe, planning and research are also under way for a variety of advance space vehicles, such as: **Hotol** (British), **Hermes** (French), **Sanger** (German). In the arena of expendable rockets, the European Space Agency will be launching the new **Ariane 5** in 1996, pictured in Exhibit 91, and has research under way on a **Crew Transport Vehicle**, a capsule for crew or cargo transport to use on the former. ESA is working on **FESTIP** (Future European Space Transportation System Programme), which includes both reusable and winged launchers using air-breathing engines. The only heavy-lift vehicle is Russia's **Energia**, and that country ... . The Japanese also have under way an orbiting plane called **Hope** ... Meanwhile, among the really far-sighted, like the Interstellar Propulsion Society, plans are being considered for manned transportation to other **star systems**.[52]

The second challenge is to drastically reduce the cost of accessing space so that space industrialization and settlement may progress through mass migration aloft!

> **There are countless planets, like many island Earths ... Man occupies one of them. But why could he not avail himself of others, and of the numberless suns? ... . When the sun has exhausted its energy, it would be logical to leave it and look for another, newly kindled star in its prime.**
>
> —Konstantin Tsiolkovsky

9.3 LEGAL SPACE FRONTIER CHALLENGES

To ensure the industrialization and habitation of space requires legal institutions, laws, and experts. However, our planetary legal system and enactments need to be transformed for meaningful nonterrestrial applications. Just as in past centuries, the New World built upon the legal heritage of Europe, but created something unique to the American continent, so too the **space law** now being created, as well as the **astro law** that will someday emerge in orbit, must manifest space culture (refer to Chapters 2 and 4). Those seeking to engage in space commerce currently face, both domestically and globally, a complex, convoluted, and often contradictory regulatory scheme. In the U.S.A. centralization of laws, regulations, and agencies affecting space business practice is only in a primitive stage. Attorney Nathan Goldman has provided a review of both national and international practice, as well as future needs regarding space law.[53] Relative to international space treaties under the auspices of the United Nations, Declan J. O'Donnell, president of the World Bar Association, is among those who believe that some of these, such as the Moon Agreement, have proven to be inadequate if not anti-space development in wording and need to be replaced. [54]

Nearly thirty years ago, a colonel in the U.S. Air Force legal division published an essay on "astronautical law;" it is now recognized as the first comprehensive examination of the legal issues related to human movement beyond this planet. Its author, Martin Menter, today is honorary director of the International Institute of Space Law, an ongoing forum for lawyers which sponsors colloquiums and publications on the subject of Space

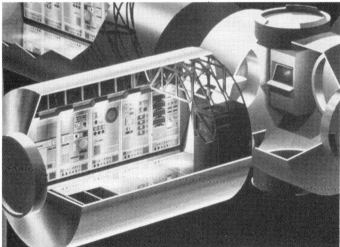

Exhibit 91—ESA's New **Ariane 5** and its **Columbus** Module. ESA's 1996 generation of rocket is **Ariane 5,** launching from Kourou, French Guiana, with a lift capacity to lower Earth orbit of 23 tonnes. An imposing cryogenic stage over 5 m in diameter, it contains a 100-tonne thrust engine that carries 155 tonnes of liquid hydrogen/oxygen as fuel, flanked by two solid boosters each 3 m in diameter and more than 10 m high. It can carry and then orbit three satellites at a time, or launch automatic platforms, such as the **Columbus** Attached Pressurized Module (8.5 m long by 4.5 m diameter), which is also pictured above. This **ISS** space station laboratory will enable astronauts to work without spacesuits while engaged in microgravity and life sciences experiments. *Source: Europe in Space,* European Space Agency, 1995.

Laws and Regulations (see Appendix C). The organization is part of the International Aeronautical Federation headquartered in Paris. Many IISL members have published and continue to publish scholarly books on the legal dimensions of human actions in outer space.[55]

9.3.1 Space Laws and Regulations

> Space law is actually a complex mixture of international and domestic laws that govern a wide spectrum of activities. They can range from the exotic, like creating the institutional framework for an international mining consortium, to the more routine, like drafting telecommunications agreements. The fields of law these activities can involve include administrative law, intellectual property, arms control, insurance law, environmental, criminal, and commercial laws, as well as the treaties and legislation written specifically for space.[56]

So explains attorney Joanne Irene Gabrynowicz, who teaches graduate courses in space law and policy at the University of North Dakota. She believes that space law provisions provide insights into the forces that come to bear on space programs and mission ... . Presently, a private company or a nation needs to be concerned with both the international regimes and the regulations regarding space activities. These include, for instance:

• **UNCOPUOS**, the United Nations Committee on the Peaceful Use of Outer Space and its key treaties—"The Outer Space Treaty on Principles Governing Activities of States in the Exploration and Use of Outer Space, Including the Moon and Other Celestial Bodies" (October 10, 1967); the "Liability Convention on International Liability for Damage Caused by Space Objects" (October 9, 1973); "Convention on Registration of Objects Launched into Outer Space" (October 15, 1976); and the Moon Treaty or "Agreement Governing the Activities of States on the Moon and other Celestial Bodies" (December 5, 1979).

Of the five UN Treaties put forth by UNCOPUOS, the U.S.A. and former U.S.S.R. have signed four, excluding the last-mentioned Moon Treaty, in addition to bilateral agreements and the International Telecommunications Union. The agreements reached underscore the theme of international cooperation and that space is to be used for the benefit of mankind, not for national annexation, and reserved for peaceful purposes; that nations are responsible for the space activities of their nationals, but liability on fault pertains not only to damages in space. In the case of joint space ventures among nations, the Liability Convention holds countries jointly and individually liable for damages caused by such collaborative efforts. Even the environmental protection of celestial bodies, as well as back-to-Earth-contamination, is considered within the UN treaties. When the Sierra Club did sponsor a conference and proceedings on environmental ethics and the solar system, space law and its applications were examined, but not in depth.[57]

Unfortunately, national interpretations of these international space agreements often differ. The American position is that nations are permitted to use, not own, space, and can mine and claim resources; nations may not have the right to own a site, only ownership to ore extracted. An open question is whether "peaceful purposes" in space means non-aggressive and permits defensive uses, like the Strategic Defense Initiative of President Ronald Reagan's administration.[58] There are other problems with these UN agreements. First, how representative of the global space and legal communities is the make-up of the **COPUOS**? Second, the times and situations have radically changed since these treaties were written—do they represent today the best international thinking and consensus on their subject matter? Some of the international space legislation was a product of the "Cold War" between East/West ideologies ... . A case in point is the so-called "Moon Treaty", which was not agreed to by any of the spacefaring nations; in the United Nations only eight, mostly developing, countries actually signed it. Its content has met hardy opposition and controversy, often along North/South lines—scholars in many countries found its language deficient because of inadequate wording, vague concepts, and anti-developmental approaches to space resources.[59] In the U.S.A., the National Space Society actively argues against its provisions. Yet, when the "treaty" came up for revision in 1994, the thirty-seventh session of **COPUOS** met for ten days in June, and then decided to keep the status quo. The Committee's chairman, Ambassador Peter Hohenfellner, wrote to this author that **COPUOS** recommended to the forty-ninth UN General Assembly that no further action should be taken. In fact, his letter states (May 18,1995), "This was endorsed by the General Assembly and, therefore, no developments are anticipated on this matter, in the near future." How informed were the Committee's members of the alternative viewpoints and recommendations for change in that 1979 document? (Refer to Appendix D for a possible replacement treaty.)

- **International Agreements**, especially at a regional level, can occur among parties apart from the United Nations involvement. For example, these are examples of such multilateral accords that have evolved: The European Space Agency Convention, the Intelsat Agreement, the Eumetsat Convention, and the International Space Station Intergovernmental Agreement. The ESA and its multinational agreements have played an important role in the evolution of space law, and may even provide a prototype for future government entities in orbit.[60] Such understandings set legal foundations among the signatories for a variety of space activities from administration to satellites, from design to maintenance. Sometimes the pact is bilateral in scope, as with the Treaty Banning Nuclear Weapons Tests in the Atmosphere and Underwater or the Liability Convention between the U.S.A. and former U.S.S.R. The last-mentioned holds a spacefaring nation liable for injuries caused by its space-related activities, whether on the ground or in orbit.

At the national level, space business and entrepreneurs face additional legal guidelines and constraints. Again using the U. S. as a case in point, generally, a lack of coordination is inherent to government policy and practice, complicated by traditional bureaucracy. Finch and Moore have summarized the policy and regulatory situation regarding astrobusiness:[61]

- **Administration Policy**—Since Eisenhower, U.S. Presidents have issued national space policy directives. On July 4, 1982, for example, NCS-42 of President Ronald Reagan authorized private sector participation in civil space activities with appropriate government supervision and regulation. On May 16, 1983, that same President issued another directive for the commercialization of expendable launch vehicles. Again on November 18, l983, President Reagan designated the Department of Transportation as the lead regulatory agency for this purpose and promoted a Strategic Defense Initiative for military space. But President George Bush failed to get the support of Congress behind his Strategic Exploration Initiative for civilian space, while President Bill Clinton has yet to articulate a real national space policy.

- **Regulatory Agencies** set their own organizational policies and regulations within national policy and law which impact space commerce and industrialization. Currently, these are the major agencies influencing space business—NASA, Federal Communication Commission, and the United States Department of State, Defense, Transportation, Commerce, and Energy. Within DOD, business may have to deal with varied positions of the military services, and within a department, such as Commerce, there are subentities that can legally affect astrobusiness (e.g. , National Oceanic and Atmospheric Administration, National Telecommunications and Information Agency). Instead of coordinating space regulations for business, the opening of space offices in various Federal agencies has only complicated bureaucratic actions.

- **Congressional Actions** mainly influence astrobusiness in two ways. First, by monitoring and influencing governmental agencies that relate to space business activities (e.g., various statutes requiring congressional approval before Landsat was commercialized or a study of the space commercialization process by the Commerce Department). Secondly, by legislation directly regulating space activities, such as the streamlining of the approval process for new space ventures, or bills to encourage joint R&D activities to increase industrial innovations, such as in the use by NASA of automation and robotics. The President of the United States signs such law into being, as with the case of the National Aeronautics and Space Act of 1958 that separates military and civilian space activities; it was amended in 1988. Congress also passed the Comsat Launch Act of 1962 which authorized U.S.A. participation in Intelsat, an international communication satellite system; the Commercial Space Launch Act of 1984 to promote economic growth, peaceful entrepreneurial activity in space, and to authorize the Department of Transportation to regulate commercial launches within the country (amended in 1988 to require launch providers to obtain insurance to coverage damage due to accidents and failures, such as the **Challenger** shuttle system); the Remote Sensing Policy Act of 1992 which provides for the issuances of licenses for private operation and analysis of sensing data, requiring that only unenhanced data be made available to the governments of countries which have been imaged by this technology.

- **Court Precedent** has always been a factor in legality. As people or institutions worldwide sue in court concerning space-related activities on the ground or in orbit, the judgments rendered contribute to the body of space law. For example, the case of *Martin Marietta vs Intelsat* in a U.S. Federal court resulted in a finding that broadened interpretation of the Launch Act to the effect that it was the intent of Congress to provide protection for U.S. launch providers; therefore waivers based on that Act and used by Martin, the provider, with Intelsat, the owner, were enforceable.

Two examples from the past illustrate how space business can be hamstrung by obsolete national policy and regulations:

- Art Dula, the Houston space lawyer who headed up the Space Commerce Corporation, had been trying before 1992 to get the State Department to lift "Cold War" constraints on U. S.A. companies launching payloads on the then Soviet rockets. Similarly, if previous negotiations with Brazil had been completed to utilize Soviet **Proton** launch vehicles at Alcantara near the equator, Dula would then have needed policy changes for American corporations to take advantage of these lower launch costs ...

- The Cape York Space Agency in Australia not only had a contract with Glavkosmos to use Soviet rockets, but wanted an American firm, United Technologies' USBI Division, to manage its facilities. Again, U.S.A. policy and regulations in the '80s mitigated against American companies wishing to take advantage of such opportunities. It had been hoped that President Bush's new Commercial Launch Policy would ameliorate both situations, while protecting somewhat the infant commercial launch industry in the U.S.A. by assuring that the government's payloads will be launched only on American vehicles.

In this post-Cold War era the above are moot points, for both the U.S. government and industry are seeking to cooperate with the Russian Federation and CIS countries, so as to foster democracy and market economies in these former totalitarian societies. Furthermore, entrepreneurs have been frustrated in their attempts to promote spaceports both in Australia and in Hawaii for lack of both government and public support. But these also illustrate that conditions, policies, and legislation must change to be relevant to contemporary human needs aloft.

Presently, there is a growing body of space law and expertise worldwide which sometimes facilitates space enterprise, but more often constrains it. Professor Gabrynowicz, previously cited, observes that what much of space law requires is unknown, and will change as technology matures, human experience grows, and new situations are encountered. She envisions space law as dynamic, with the next generation of such law agreeing on specific norms and addressing such vital questions as:

- **Is sovereignty necessary to establish property rights?**

- **Are space resources, as well as space itself, the province of all humankind?**

- **If so, how are they to be allocated?**

- **How can non-spacefaring nations be assured the use of space and its resource?**

- **How will the investments of spacefaring nations be respected?**

- **How will private space activities be allowed to operate?**

9.3.2 Space Settlements and Governance

At the moment, there is no planetary policy for offworld colonization and administration. Yet by year 2002, an **International Space Station** with a multinational crew may be operational—a prototype of orbiting colonies envisioned by some.[62] Further, by year 2010, government space agencies are tentatively planning a lunar outpost, and private enterprise has grandiose schemes for a lunar base and industrial park before 2025. Many space planners worldwide expect an outpost on Mars or its moons by 2050. But even at the United Nations, there has been no serious discussion about how international partnership could be formed and managed to ensure permanent settlements on the high frontier.

One of the most significant legislative proposals in this regard was presented to the U.S. Congress, as the Space Settlements Act of 1988. Introduced by Rep. George E. Brown, Jr.(D-CA), of the House Science, Space, and Technology Committee, this failed attempt was discussed in our opening chapters as a forerunner of what yet may come to be. The proposed H.R. 4218 would have amended the National Aeronautics and Space Act of 1958 to set the establishment of space settlements as a long-term mission for the space agency. Further, NASA would have to conduct a steady, low-level effort to explore all the technical and sociological issues relating to the achievement of settlements in space (*Space World*, June 1988, p. 3). The proposal came in response to various national recommendations (National Space Commission, 1986; Ride Report, 1987) and President Bush's new policy that a long-range goal for the United States is "to expand human presence and activity beyond Earth orbit into the Solar System." Although not enacted as originally proposed, Congressman Brown did manage to get the substance of his bill passed by means of an amendment to a NASA funding authorization during the 100th Congress session. S.2209 is now law and provides national legislative commitment to space exploration and settlement, requiring NASA to report every two years on its progress in that regard. Unfortunately, NASA has been rather lax in purposing this new mandate because of the defeat of the Space Exploration Initiative. Many space experts forecast that a whole new type of legislation will have to be enacted regarding space settlement and industrialization, administered outside the space agency itself. Conceivably, this undertaking may require an international agreement resulting in a global administration for this purpose.

Apart from the hundreds of man-made satellites, as well as unmanned spacecraft, orbiting the Earth, space habitation raises the most intriguing non-terrestrial legal issues. In conjunction with the celebration of the 200th anniversary for the U.S. Constitution, the Smithsonian and its National Air and Space Museum organized two conferences of thirty law specialists regarding basic rights and freedoms which apply to space. The outcome was a "Declaration of First Principles for the Governance of Outer Space Societies" (refer to Appendix A). William J. Brennan, former chief justice of the U.S. Supreme Court, called for UN approval and insistence of these initial eleven principles to "ensure that the fundamental needs for life, individual freedom, liberty, justice, dignity and responsibilities inherent in self-determination are integral parts of humanity's exploration and settlement of space." (*Los Angeles Times*, May 22, 1987, I:27). In that address to the American Law Institute, Brennan noted that space law principles have to be developed in anticipation of future space settlements.

Smithsonian's legal counsel, Dr George S. Robinson, co-authored *Envoys of Mankind*, which offers a conceptual foundation for a new body of space law and constitution, as well as a review of the evolving body of national and international law concerning space[63] In the opening address to the Legal Forum on Space Law of Air/Space America (May 17, 1988), organized by the author, Robinson warned that establishing space habitats for lengthy human duration demands that planners and designers of space stations and future space settlements focus seriously on the social organization of space societies, including the values underlying whatever organizational principles are ultimately adopted. He contends that space law should clearly recognize the unique physical and psychological environment aloft where human biology and technology are totally integrated for survival. He also quoted Harrison Schmitt, former lunar astronaut and senator, "I have been governed in space by earthbound authorities, and within the constraints of survival, even rebelled against that governance." Future spacekind progeny, Robinson advocates the need for the unparalleled principles of law to safeguard their lives as the truly bio-cultural envoys of earthkind.[64]

A summary is provided below of other speakers at that 1988 forum co-sponsored with San Diego's National University Law School and under the chairmanship of Chuck Stovitz, a Beverly Hills attorney. Unfortunately, the proceedings were not published, but the author was present to take these notes which in retrospect provide unique insights:

- Neil Hosenball, former NASA general counsel—Among the foreign competition to be faced in commercial space, consider that the People's Republic of China has a new Ministry of Aerospace, which now offers launch services at less than U.S.A. suppliers. The Soviets not only offer similar low cost launches, but provide quality remote sensing photography services. The Japanese are planning to invest in an Australian international spaceport ... . Currently the U.S. Congress is "hamstringing" private investment in space activities ...We need to get NASA out of operations which could be taken over by the private sector, thus freeing up the Agency to concentrate on science, technology, and exploration.

- Art Dula, partner in Dula, Shields, Ecknert, Houston attorneys—We have recently set up Space Commerce Corporation to market Soviet rocket services. Our foreign technical team was the first to be permitted to check out the Baikonur Cosmodrome near Tyuratam in Central Asia. We are not doing this for the Russians, but for American business who have a need to buy less expensive transportation services. Although this may only be a temporary service, we have to overcome opposition from U.S. governmental officials and existing policies on export licenses.

- H. Cushman Dow, vice president and general counsel of General Dynamics, San Diego—In attempting to market the services of our Atlas-Centaur, we face competition not only from other American launch craft, the Shuttle itself, but from foreign competitors, principally the French Ariane and possibly the Japanese. The latter are backed by their governments, which means they need not make a profit and may be state subsidized. It is only in the United States that "commercial" really means "commercial" or private sector. The expendable launch business in the U.S.A. is dependent upon cooperation with NASA and the military, who control the facilities to be used, and upon Congressional action regarding liability insurance.

- Theodore Harper, Graham & James, attorneys, Los Angeles—In building their aerospace plane, the Japanese seek international partners, but not the United States. The trend in space business is toward many new players and more international partnership, but not with Americans, who insist on applying U.S. contract law and export control regulations.

- Dan Byrnes, adjunct professor, Pepperdine University Law School—Presently, only $500 million is available in liability insurance for space accidents. We need the establishment of a Space Industrial Advisory Council to counsel the U.S. President on space commercial policy.

[NOTE from author: the current U.S. rate on liability insurance is 18%, which is enforced by the U.S. Department of Transportation.]

These excerpts from the past, only eight years ago, underscore accelerating changes with planetary societies and alterations in perspectives which will impact emerging space law. Conditions now are more favorable to international cooperation in space, and global joint technological venture that may really open the high frontier (see Exhibit 92).

Exhibit 92—Interagency Agreements: **SOHO** Observatory. This is an artist's depiction of **SOHO**, a solar observatory that resulted from a legal agreement between ESA and NASA. It was launched from Cape Canaveral on December 2, 1995, on an **Atlas IIAS** rocket. Located 1.5 million kilometers from Earth, **SOHO** will enable investigators from the U.S.A. and Europe to study the structure of the solar interior, as well as the physical processes that form and heat the solar corona. *Source*: ESA Solar-Terrestrial Science Programme **(STEP)**.

One such change in the matter of **space governance** has been promoted by both the World Bar Association and the United Societies in Space Inc. of Englewood, Colorado. This strategy, discussed in Chapter 7, would create a **space metanation** to protect the *common heritage of humankind* (CHOM). Such a concept was not foreseen in the 1967 Outer Space Treaty signed by 100 nations, including the U.S.A., but was later incorporated into the Moon Agreement of 1979. CHOM as a "term of art" was defined in the UN agreement on the Law of the Seas in 1982, signed by the U.S.A. in 1995; under the rules of treaty law, it would also be applicable to astro law. The difficulty comes in the application of this concept which requires an active sharing of space resources and properties for the "benefit of mankind." Some interpret such provisions as curbing space development, for there would be no incentives to investors and entrepreneurs to develop the high frontier with a hope of getting a return on their efforts aloft.

In contrast, the proposal of United Societies in Space, Inc. would place the new space nation under the UN Trusteeship Council provisions, with trustees appointed from the spacefaring nations, for a hundred years. [65] Both the principles of international law and the concept of CHOM would be protected under this new entity under a constitution acceptable to all nations, including those in developing economies. This innovative plan suggests a new Treaty on Jurisdiction in Outer Space to be passed under the auspices of the United Nations (Appendix D). It also envisions the establishment of **space authorities** that may or may not be placed under metanation for the purpose of issuing bonds, processing leases on land and facilities, as well as promoting new enterprises on various planets and asteroids. The first such would be a **Lunar Economic Development Authority (LEDA)**, its legislation and scope being comparable to that of the Tennessee Valley Authority which came into being with an Act of the U.S. Congress. The **space metanation** approach would provide answers to some of the critical questions raised above by Professor Gabrynowicz, and be more pro-development.

> *The challenge for members of the legal profession worldwide is whether the law as currently applied will be used to impede or facilitate space commerce, whether its practitioners will build upon terrestrial legal foundations to fashion systems of laws relevant to space travelers and dwellers!*

9.4 POLITICAL CHANGES ON THE SPACE FRONTIER

The Space Age literally came into being and turned into a space race because of political decisions connected with the "Cold War." It was the Soviet launching of **Sputnik 1**, the world's first artificial satellite on October 4, 1957, followed by the first unmanned probes of the Moon with the Russian **Lunas**, and the first man in space, cosmonaut Yuri Gagarin, that prompted political responses within the American administration. John Logsdon, the premier space policy analyst, has documented the political influences leading to national decisions to send men to the Moon through project **Apollo** to the current Space Shuttle program. [66] As director of the Space Policy Institute at George Washington University, he led the political debate for internationalization of the U.S. space program, namely bring global partners into the costly Shuttle and Space Station programs. In an ABC network television discussion of the matter , Logsdon is quoted as saying that the country does not have to be "first" in every aspect of space exploration (a position seemingly adopted by later Reagan and Bush administrations):

> Why shouldn't we go with the Japanese and the Germans and so forth? This is a human problem, not a United States problem. I think the days of racing to a place are over ... . We can't cooperate [on a space station project] with others until we have a strong program ourselves. And that—a strong articulation of purpose and priorities—can only come from the President.
> —*Los Angeles Times,* September 12, 1988, p. 4.

As the two superpowers struggled for preeminence as spacefaring nations, the political debate over the issue of international participation increased, but is seen today as quaint history.[67] Americans are no longer afraid that joint space ventures with foreigners, especially the former Soviets, may weaken the nation, militarily or technologically, by giving away secrets. On the other hand, the sheer complexity and cost of new space technological ventures are forcing many countries to form consortia, as we discussed in Chapters 1 and 2. The competition became less between the space programs of various countries than with other parts of an economy seeking their share of shrinking national resources. Thus, in the same media program cited above, Senator William Proxmire (D-Wis.) talked about the exorbitant costs to build and run the shuttle/space station programs, suggesting in the '80s that NASA seek alternatives by sharing the projects with other nations, as well as the private sector.

In the last decades, related political issues exacerbated the situation between the superpowers and their endeavors on the high frontier. For example in the U.S.A., leaders disputed whether to invest public monies primarily in "Star Wars" or commercial space activities, or both. A 1988 Congressional Budget Office report rightly observed that the U.S. space program was at a crossroads, noting that "civilian space is becoming an all or nothing proposition" because of the large price tags associated with proposed space projects. [68] It was a combination of such factors that made it impossible for President George Bush to garner sufficient support for his policy on the Space Exploration Initiative. Geopolitical changes in the present decade have altered the myopic national focus of space planners—now Russian cosmonauts and ESA/Japanese astronauts use the Shuttle, and the Shuttle docks at the **Mir** station with both astronauts and cosmonauts. The new era of collaboration is symbolized in the picture of Exhibit 93. The political reality of the mid-90s is that governments and space agencies seek **space synergy**, primarily to share expertise and save costs. Henceforth space missions will be multinational in scope, and the crews, consisting of both genders, will be multicultural and multidisciplinary in backgrounds! Furthermore, there are long-duration spaceflight policy issues to be resolved that range from cross-cultural training of crews to matters of privacy.[69]

9.4.1 Influencing Space Policy

Policy has been defined as a definite course of action taken to achieve a goal, to facilitate, to build consensus, or for expediency. In adopting policy that gains public support, the art of compromise is often practiced, Politics, on the other hand, is viewed as the art or science of governing, of managing political affairs, power and decisions. Whether policy or

Exhibit 93—Historic Link-up: Crews of STS71 and **Mir** 18/19. In June 1995, the American and Russian crews of these missions gathered aboard the Shuttle in the space science lab module for this inflight portrait. To identify those in this NASA photograph, hold the picture vertically, with the socked feet of Anatoly Solovyev at the bottom. Then go clockwise to view Gregory Harbaugh, Robert Gibson, Charles Precourt, Nikolai Budarin, Ellen Baker, Bonnie Dunbar, Norman Thagard, Gennady Strekalov, and Vladimir Dezhurov. Try to distinguish the astronauts from the cosmonauts!

politics, the key is developing and implementing strategies for moving human systems ahead, for accomplishing tasks. Political scientists and politicians are supposed to be the experts in such matters, but increasingly those in leadership positions are turning to those with expertise in strategic issues management.

Dr Nathan Goldman believes there is a political dimension to all human enterprise whether on this planet or in space.[70] In the U.S.A., for instance, this University of Houston and Rice University professor maintains that policy decisions are involved in everything from getting Presidential or Congressional support for a particular mission scenario to setting a national priority, such as to build a lunar base or explore Mars.[71] In Europe, ESA used to set policy through its Council of Ministers. However in 1995, the European Union notified the Agency that it would take charge of Europe's space Policy. The story in *Space News* stated: [72]

> **Frustrated that billions of dollars of government investment has not brought a secure commercial position for Europe in space-based telecommunications—either in satellites or the lucrative ground equipment market—the EU Commission and the European Parliament propose to assume control of Europe's long-term space direction.**

Edith Cresson, the EU's research commissioner and a former French Prime Minster, ...stated, "The strategic, economic and industrial implications of space are too great to be left outside the European Union."

Apparently, the EU is pressuring ESA, primarily a research agency, to pay more attention to space commerce and less to space technology. The proposal is expected to be enacted by mid 1996, and will likely result in the EU Commission licensing global satellite telephone systems.

Goldman identifies three models of policy science which apply to the space field:

(1) **incremental decision-making** or business-as-usual—present arrangements are the starting point for future alterations, changes come in increments through debate and compromise, as well as forming coalitions of support to realign matters, as happened with the long-standing efforts to build a second American Space Station;[73]

(2) **non-incremental decision-making** or Schulman model caused by big projects that need a "critical mass" of support before receiving full public funding, such as the major leap forward with the decision to invest in the building of a Space Shuttle where the very size and scope of the program makes it non-divisible once started;[74]

(3) **strategic decision-making** or Huntington model when a crisis causes a major political choice, as when President Kennedy, reacting to Soviet space advances, made the decision to go to the Moon after stating, "We are in a strategic space race with the Russians, and we have been losing"[75]

This last model advocated by John Logsdon tends to place the emphasis upon external factors, while the other two focus on internal factors. For this latter approach to succeed, Logsdon, North American editor for *Space Policy*, the prestigious journal published in Oxford, believes that well-developed preliminary plans and consensus on feasibility have to exist before the decision is made.

Perhaps, this is the case today as budget reductions and escalating costs stimulate all spacefaring government officials to rethink the "go-it-alone" policy in space. The current position forms alliances with other nations to share in space undertakings, encouraging the private sector to join with the public sector in space development.[76] However, analysis of a recent American space policy initiative demonstrates the difficulties: "Project Pathfinder" was announced by President Ronald Reagan in 1988 as a policy to develop "pathfinder" space technologies which would enable the country to return to the Moon and eventually fly to Mars successfully. The President proposed to spend £1 billion on this program; the Congress in FY 1989 appropriated only $40 million, a 60% cut in the original request. This initiative was not a pathfinder. Obviously, the Federal government will no longer monopolize U.S. space business because it will not fund such programs alone.

A Resources of the Future symposium held in June 1986 concluded that space policy makers need to understand better the interrelationship of economics and technology (Macaulay, previously cited). There are more political and economic benefits to be gained—for example in Earth observation, space transportation, and lunar industrialization—by adopting a policy of space collaboration over competition.

9.4.2 *Influencing Lunar Policy*

Returning to the Moon for a permanent settlement and development of its resources is a policy issue to which one or a combination of the above theoretical models may be applied. Many organizations have taken a policy stance on this, such as the National Space Society; major meetings, such as the International Lunar Exploration Conferences, bring together a cross-section of supporters for this strategy. In 1993, the NSS's San Diego chapter promoted a "Back to the Moon" bill in the U.S. Congress to authorize government purchase of lunar science data from private vendors. In 1995, the NSS got on the information highway by establishing a **Return to the Moon Homepage** on the World Wide Web (**http:/www.ari.net/back2moon.html**). As part of their mission to advance people living and working in space, the Society's officers and members regularly testify on behalf of space legislation in the Congress. For example, on May 25, 1995, the NSS Executive Director testified that the "ultimate purpose of NASA should be to empower individuals and private organizations to go into space for their own reason." At the Second Annual International Exploration Conference sponsored by NSS in 1995, the participants agreed to promote a **Lunar Consortia**, made up of organizations represented at this meeting. They will promote a variety of lunar return strategies, including the **Lunar Economic Development Authority**.

Your author with many others believes a key rationale for lunar industrialization will come from the need for clean, non-polluting solar energy to replace organic fuels, such as provided by a Lunar Power System (Appendix B) or Solar Power Satellites.[77] Building a lunar base is the kind of 21st century macroproject that would require a broader and more integrated space constituency than now exists. Since this massive undertaking involves spending over £100 billion for twenty years or more, a case **for investment** has to be communicated that will attract political and public backing, especially on the part of the media and the young generation who will pay for an implement the policy. The only group with a viable plan in this regard is the European Space Agency, but its initial scientific focus is too narrow for gaining mass enthusiasm, no less the necessary expenditures by financially pressed member states.[78] However, ESA's useful lunar reports do go beyond previous American and Russian studies, and recommend participation by other international partners, as well as the need for eventual industrial development on the lunar surface.

Establishing a Moon base by a spacefaring consortium would necessitate development of a **global space ethos** among the public in the nations involved, so as to achieve the following·

(a) a convincing **rationale** for this endeavor that includes a combination of "pay backs" to the investors and taxpayers, whether in terms of science, commerce, resource utilization, habitation, or whatever arguments make sense ...

(b) a legal, financial and administrative mechanism for a variety of nations and their institutions to work effectively together in such a venture, such as **Lunar Economic Development Authority** ...

(c) a fully operational **Space Transportation System** would have to be in place, able to move people and cargo at more reasonable cost than today's shuttles and rockets ...

(d) a **site plan** that takes advantage of the best location in terms of geology, concerns for

biological and sociocultural problems, infrastructure provisions to ensure human survival and quality of life for an expanding population in a gravity-free environment.
(e) the **technology** for innovative robotic applications, habitat construction, mining , oxygen production, and mass-driver activities, but particularly for lunar telecommunica-
tion on the Moon and with Earth ...
(f) a **lunar personnel deployment system** that provides for recruitment/assessment, orientation, on-site support, and re-entry guidance.

The montage of NASA photographs in Exhibit 94 illustrates some of the "infrastructure"
needs if both humans and robots are to function effectively on the Moon. Building first a
lunar outpost, then a base with an industrial park, infers a commitment by **earthkind** toward space colonization—lunar foundation should become the "launch pad" for human
expansion into the Solar System.

For that type of commitment to long-range developments on the Moon by an international consortium of both public and private sectors would require strategic policy decisions within the countries and organizations participating. To accomplish the kinds of
lunar enterprize described above, as well as in Chapter 10 and Appendix B, policy decisions for financing and governing the operations must come first. At present in the U.S.A.,
again as a case in point, there is no such **national policy** toward space exploration and
exploitation. My colleague, Dr Goldman, predicts new players from private enterprise will
arise and positively influence such decisions—they are to found in the communication
satellite industry, commercial launching companies, space service firms and entrepreneurs, solar and nuclear energy scientists, and even in the hotel, tourism, and entertainment businesses. They will be found in downsized aerospace and defense complex
seeking a new outlet for their technologies and well-trained technicians. This space policy
expert believes such an economic network will translate into a political network, forming
coalitions with space advocates and scientists, and possibly defense supporters.

Using again the case of the U.S.A., for a critical mass to coalesce in support of such
decisions regarding farsighted, mega-space programs, more is needed than Presidential or
Congressional statements and studies with visionary goals and initiatives. The basis for
national consensus may already by found in the findings and recommendations of those
same studies paid for by tax-payers—such actions as proposed by the Synthesis Group on
America's Space Exploration Initiative (1991), the Report to the President of the National
Space Council (1990), the Advisory Committee on the Future of the U.S. Space Program
(1990), the Congressional Budget Office (1987), the Ride Report (1987), the National
Space Commission (1986).[79] What the U.S.A. needs is not more studies on where our
space program should go, but decisions on implementing some of the recommendations in
these previous six reports. Leadership and action by the President on **national space policy** also call for majority support of that policy in the Congress and among the citizenry.
To ensure "national will" toward the creation of a **spacefaring civilization**, supporters
must expand to be recruited beyond members of space agencies and defense departments,
beyond aerospace contractors and space activists. Globally, those favoring development
of the space frontier now number only several million when one counts only patrons in
government and industry employees, pertinent trade and professional associations or citizen societies, plus fans of space movies or television programs (such as *Star Trek*). By

banding together on occasion, minor political space battles have been won, like changing the name of the first shuttle, to save the **Galileo** mission, to kill the Moon Treaty, or to establish **Challenger** centers of science education. One researcher discovered that the level of public support for NASA programs varies owing to economic conditions, perceived benefits, and success of on-going space events.

To get national and worldwide public investment in the enormous expenditures and risk behind present space plans demands mobilization of civic champions on a scale yet to be experienced. It requires assertive, concertive leadership by the pro-space movement to reach beyond their own membership, so as to enlist pubic backing at a level beyond that given by the peoples in their World War II efforts. Mass communication and education is essential to get cultural commitment by the citizenry as a whole to the peaceful exploration and exploitation of the next frontier! Further, utilizing space resources will demand new institutional arrangements by which people may **actually** invest money in space enterprise, be it through purchase of stocks, bonds, tax incentives, or whatever it takes, even lottery tickets!

In 1990, the Denver firm of Strattor, Reiter, Dupree & Durante made a proposal to the American Aerospace Industry on "Building a Broader Space Constituency." I urged industry to engage in more direct, aggressive advocacy of space undertakings, such as McDonnell-Douglas did with "Project Inform" when it sponsored a multifaceted campaign to enlighten the public about the value of space resources through network television, brochures targeted to specific audiences, and displays for use at conferences and conventions. These public relations experts recommended that the aerospace industry-wide back a national marketing endeavor to:

- re-establish strong Congressional support for the space program;
- undertake a review of all previous public opinion surveys on attitudes toward space projects, and promote an additional study of the pubic, media, and Congress to ascertain possible linkages, particularly with reference to space exploration and settlement;
- chart, based on a full evaluation of the current attitudes, a clear course to enlist great public space support, possibly through targeted direct mail, phone banks, road shows, paid and free media advertising.

Unfortunately, that counsel was ignored and the American aerospace industry continues to lose market share and jobs within the space field.

In a lecture at the California Space Institute (May 11, 1988), Dr Brenda Forman of Lockheed Corporation confirmed that the civil space program needs a well organized lobbying effort to move the President, Congress, and voters. Such a strategy to enlist the silent majority is under way with **Spacecause**, a non-profit organization founded in 1987 to build a lobbying network among public, academic, and commercial space industries so that government will give policy priority to civilian space activities; it complements SPACEPAC, formed in 1982 to support political candidates with outstanding civilian space voting records or to endorse public servants who merit recognition. In 1992, **Spacecause** joined with the National Space Society in obtaining the passage in the U.S. Congress of the **Commercial Space Competitiveness Act** provisions to:

Exhibit 94—Providing Lunar Infrastructure. In the upper depiction SAIC artist, Pat Rawlings, shows a possible reusable **Oberth Lunar Lander** firing its advanced cryogenic fuel engines to resupply the needs of a lunar outpost. In the lower rendition a robotic factory is operated by **Selenian technauts**. *Source*: NASA/Johnson Space Center, Houston, Texas, U.S.A.

Exhibit 94 (continued)—In the upper depiction Rawlings illustrates a future **space telescope** which has an automated walking mobility platform. In the lower picture we view commercial activities in a 21st century lunar industrial park. *Source*: NASA/Johnson Space Center, Houston, Texas, U.S.A.

- extend the limitations on launch liability ...
- launch voucher demonstration program to improve competition among companies ...
- space transportation infrastructure matching grants to develop private spaceports, payload integration facilities, etc. ...
- anchor tenancy and termination liability for private companies when government defaults on legal contracts, allowing for more predictability in space venture ...
- opening of surplus or otherwise available federal launch and payload support facilities to use by private companies on a cost-reimbursable basis ...
- protection of intellectual property of companies doing business with NASA ...
- creation of a **Commercial Space Achievement Award** to be administered by the U.S. Secretary of Commerce.

Such endeavors are attune to Forman's contention that politics and technology should be mutually supportive. For an adequate and coherent national space policy, this international marketing expert advocates building bridges of information and understanding between legislators and space scientists. For her, political action on behalf of space development must come from the bottom up in society, focusing upon space-related government committees or agencies at all levels. Forman believes that the American public loves space, but has yet to perceive the need for political activism if space resources are going to be effectively utilized to resolve Earth-side problems, related to poverty, hunger, and health. In a democracy, the political system is a heritage that must be skillfully worked if space industrialization and settlement is to become a mainstream reality.

From his research, Alan Marshall of Massey University in New Zealand has concluded that once extraterrestrial materials are perceived as economically valuable, Solar System development will expand rapidly. But he cautions that politico-legal regimes should not proceed with it in an imperialistic manner, and advocates creation of **space environmental ethics** to protect offterrestrial bodies.[80]

> *The challenge is to awaken the world's political leaders to the potential of space resources and human destiny on the high frontier, so they will enable legislation and mechanisms for creating a spacefaring civilization!*

9.5   CONCLUSIONS ON SPACE INDUSTRIALIZATION AND SETTLEMENT

Safely establishing human presence on any frontier involves coping with challenges and risks. Moving humanity beyond its planetary cradle may represent the boldest venture in the evolution of our species. Ingenuity and innovation will be stretched in the transformation of Earth-bound mindsets and culture, policies and procedures, laws and regulations, education and business. The task requires multidimensional actions, demanding multidisciplinary thinking and methodologies. This chapter reviewed only three dimension of the challenges aloft, and sought to show their interrelatedness. Space commerce, law, and politics cannot be viewed separately, but require a more integrated approach by business persons, lawyers, and politicians in conjunction with scientists and academicians. This analysis, though, demonstrates some of the complexities and opportunities inherent in space industrialization and settlement, particularly for large-scale technology venturing and entrepreneurialism.

Space Education

In this book, I have referred many times to **space education** as it is being pursued in colleges and universities, through the Universities Research Association and the International Space University. Congratulations to the many youth space camps and schools which are now in existence, prototypes of programs, like the **Challenger Centers for Space Science Education**, that should be expanded globally. However, there is a critical need for **mass education** at the primary and secondary levels worldwide to the younger generation to consider the opportunities of *living and working in space*! After all, these children may be the future Martians, so education department within space agencies are to be encouraged in their outreach programs. As part of its **EuroMir 95** mission, the ESA arranged for young people in its member states to come to Disneyland Paris for a space education/entertainment event that culminated with a live satellite link-up with the Russian space station, **Mir**. Dr Sally Ride, first American woman astronaut in space and now

Exhibit 95—Today's Youth: Tomorrow's Spacefarers. Space education of youth on the ground or in orbit, such as is being done through **ISU**, creates the next generation of **spacefarers**. Original artwork by Howard Cook/Air Works (719/472-076). *Source*: International Space University (See Appendix C).

director of the California Space Institute, has a NASA project going to install cameras on the Shuttle **Orbiter**, which will exclusively be operated by students; these cameras will instantaneously relay high-resolution data to classrooms in high and middle schools throughout the country. Dr Ride hopes by this method to "give students their own piece of the space program." The Young Astronauts Council is now operating "Space School," a distance learning method that uses space-age technology to teach math and science, as well as to beam quality educational adventures. The educational programming comes from Seattle, Washington, and through satellite television is transmitted into classrooms nationally. Entrepreneurial companies are already designing games for kids to learn about space living by simulations and virtual reality. International Space Enterprises of San Diego has such a space education/entertainment program under way—to "keep the dream alive," as its president, Mike Simon, would say.[81] Expansion of such creative efforts, including youth space publications, to introduce the new generation to the challenges of space will ensure humanity's expansion aloft for Millennia III (see Exhibit 95)!

The high frontier prospects in the 21st century are only dimly perceived, as humankind struggles like infants to leave our cradle, Earth. For human enterprise in space to succeed and flourish, synergy or cooperation becomes the key ingredient between the generations, between the public and private sector, between planners and policy makers, between professionals and technicians, as well as among organizations and nations.

WE CLIMB TREES STILL*
They climbed trees once
To get closer
To the stars
—Jumped and fell short.

So, they dreamed of
Winged chariots and horses and men,
But wax wings melt
And paper wings burn
—Again they fell short.

So, they returned to the trees
To gaze at the stars and to think.
We built balloons then, and
Planes and rockets.

Although this time we did not fall,
The stars remain beyond our reach;
—And still we climb trees
To get closer

*In honor of Robert Goddard's Anniversary Day.

By Nathan C. Goldman, Ph.D., Attorney/Author, Houston, Texas, U.S.A.

10

Strategies for Lunar Economic Development

Now that we have celebrated the 25th anniversary of humanity's lunar landing, are we any closer to returning to the Moon permanently? No **united** vision, plan or strategy for exploring, settling, and industrializing the lunar surface has come forth from the world's space agencies, despite numerous conferences, studies and reports on the subject. To build a lunar base within two decades, harsh economic realities and political conditions indicate that the only way **earthkind** can afford that undertaking is by means of international cooperation with major participation by private enterprise. If the limited resources of the U.S.A., Russia, Europe, Japan and other spacefaring nations were combined into a joint technological venture on the Moon, then synergistic development there would encourage **competing** space constituencies to collaborate.[1] Whether an international space station is constructed or not in lower Earth orbit by the U.S.A., Russia, and other partners, the Moon can serve such purposes and more. Space science proponents would then be able to use the lunar platform as a site for their experiments, both in astronomy and in life sciences, as well as to launch from there additional planetary missions. Simultaneously, those favoring human exploration and "manned" missions might use the Moon for study of extreme environments and habitats, comparable to what is happening today among cooperating countries in Antarctica.[2] In addition, champions of lunar industrialization could undertake macroprojects relating to manufacturing, helium-3, or energy transmission by power beaming, and other uses of *in situ* resources, such as lunar oxygen produced from ilmenite.[3] The uniqueness of the Earth–Moon system provides an advantage for humanity to use this lunar laboratory to create a spacefaring civilization for the New Millennium! The Moon's characteristics as a resource to humanity are illustrated in Exhibit 96.

The Moon is our "beachhead" for exploring and settling our Solar System! The author presents here a near-term strategy for utilization of its resources for the benefit of our planet's inhabitants by establishment of a **Lunar Economic Development Authority. ***

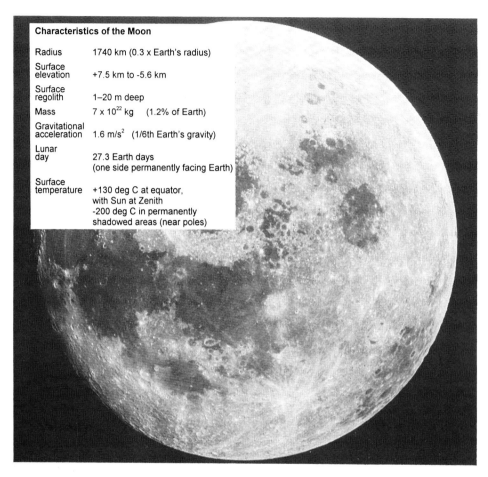

Characteristics of the Moon

Radius	1740 km (0.3 x Earth's radius)
Surface elevation	+7.5 km to -5.6 km
Surface regolith	1–20 m deep
Mass	7×10^{22} kg (1.2% of Earth)
Gravitational acceleration	1.6 m/s^2 (1/6th Earth's gravity)
Lunar day	27.3 Earth days (one side permanently facing Earth)
Surface temperature	+130 deg C at equator, with Sun at Zenith -200 deg C in permanently shadowed areas (near poles)

Exhibit 96—Characteristics of the Moon. Utilizing lunar resources is feasible, but the problem has been well summarized by Werner von Braun: **We can lick gravity, but sometimes the paperwork is overwhelming!** For further information, contact the International Lunar Exploration Working Group through Dr David McKay at NASA/JSC (**dmckay@snmail.jsc.nasa.gov**).

10.1   CONTEMPORARY GLOBAL SITUATION

Consider these converging forces which favor global action in support of a lunar-oriented space policy:

10.1.1   *Economic conversion*
The end of the Cold War and downsizing of defense and aerospace technologies favor converting this expertise toward peaceful lunar development that builds upon and expands current technological capabilities, while turning humanity outward and upward ...

10.1.2   International lunar research and initiatives

These are already under way within both Asia and Europe. In its bold 5-year space plans, the Republic of China is calling for a lunar satellite launch about year 2000, while Japan will fly Lunar-A, a lunar orbiter and penetrator probe scheduled for 1997. Furthermore, Japan's Lunar and Planetary Society proposes to institute evolutionary lunar programs, involving orbiters and landers, roving robotics and telepresence, astronomical projects, habitat studies, and even tourism. On May 31, 1994, Japanese business leaders and scientists suggested that their government's Space Activities Commission invest in a 30-year, 3 trillion yen undertaking to build a Moon station by 2024, entirely constructed by robotics! Japan's Science and Technology Agency welcomed the proposal which included solar power generation. In July 1994, a task force report of that Commission accepted those recommendations for Japan's space policy to include the building of a manned station and observatory on the Moon, preferably with its international spacefaring partners. In March 1996, Mitsubishi Corporation made a major investment in the ambitious plan of LunaCorp (Arlington, Virginia) to provide interactive robotic exploration of the Moon, as well as high-definition lunar video by 1999 for theme parks, television networks, and scientific research. Exhibit 97 indicates the scope of Japanese macroplanning.

Similarly, a European Space Agency's International Lunar Workshop concluded on June 6, 1994 that both logic and timing make it apparent that the Moon is "a natural, long-term space station" and the "testbed for any plans of human expansion into the solar system," as long as such endeavors protect the lunar environment. As described in section 7.3A, the ESA study concluded that "current international space treaties provide a con-

Exhibit 97—Japanese Lunar Macroplanning. Japan is calling for an international lunar initiative to include projects such as pictured here related to astronomy, habitat studies, and solar power. *Source*: Boeing Corporation's artistic rendering by Jack Olsen.

structive legal regime within which to conduct scientific exploration and economic utilization of the Moon, including establishment of scientific bases and observatories." The
"Beatenberg Declaration" that emerged endorsed the four-phase lunar initiative which we
explained in that chapter on macrothinking. Exhibit 98 illustrates two such science
prospects.

   Several outcomes from the above workshop point to a growing global consensus
emerging on lunar development, particularly relative to exobiology on the Moon. The
Japanese delegates agreed to host a l996 International Lunar Symposium, so an International Lunar Exploration Working Group (ILEWG) was formed under the chairmanship
of Prof. Hitoshi Mizuatani. Through the Planetary Missions and Material Branch of Houston's Johnson Space Center, Dr Michael Duke is developing there an information network
to support ILEWG, so that the world's scientific community will have access to the latest
lunar data. The American delegation's Dr Wes Huntress of NASA passed a proposal to
hold in 1995 a Workshop on the Moon as an International Scientific Resource sponsored
by the Interagency Consultative Group for Space Science. (For further information on
these developments, request the free newsletter *Beyond Lower Earth Orbit* from editor
John Connolly, Planetary Office, NASA/JSC, Houston, Texas 77058, U.S.A. or jcon-
nolly@snmail.jsc.nasa.gov (also review Appendix B of this book).

   Concurrently, energy scientists worldwide, including those within the Russian
Academy of Science and Japan, are showing exceptional interest in lunar solar energy,
whether beamed from the Moon or its orbiting satellites. Leadership is coming from a
strategic and international partnership being formed by proponents of wireless power
transmission. Speaking at the International Astronautical Federation's 45th Congress, Dr
David R. Criswell, co-inventor of the Lunar Solar Power System (see Appendix B), observed: [4]

> **By mid century 2050, lunar power industries can be sufficiently experi-
> enced and profitable to diversify into a wide range of other products and
> locations, other than solar power beaming. Specialized industries on as-
> teroids and other moons will arise. Mankind can begin the transition to
> living independently off Earth. People can afford to move to space and
> return the womb of biosphere Earth to the evolution of other life.**

(For further information, contact the Space Power Network (SPNET) and inquire about
past and future proceedings of the WIRELESS POWER TRANSMISSION CONFER-
ENCES—either: Dr Nobuyuki Kaya, Kobe University, Kobe 657, Japan [e-mail
kaya@kobe-u.ac.jp] or Dr Gay Canough, editor of *Space Power,* ETM Solar Works
Inc., PO Box 67, Endicott, NY 13760, U.S.A.
[e-mail canough@bingvaxa.cc.binghamton.edu].)

10.1.3 Private enterprise
Multinational lunar ventures, such as those envisioned by ISELA (the U.S.A.'s International Space Enterprise and Russia's Lavochkin Association), are already being planned.
An agreement signed in May 1993 calls for joint missions before the end of the decade to
launch data satellites and commercial payloads by lunar landers (see Exhibit 84).[5] Further,
LunaCorp proposes to send two teleoperated rovers to the Moon to visit historic landing
sites and artifacts. Its president, David Gump, in conjunction with Carnegie Mellon Uni-

Exhibit 98—Lunar Science Opportunities. European Space Agency studies identify specific as-tronomic/astrophysics prospects on the Moon's surface, such as illustrated above in the lunar telescope and the solar interferometer. *Source*: ESA reports, *Mission to the Moon* (SP-1150, 1992) and *International Lunar Workshop* ((SP-1170, 1994). Also refer to N. A. Budden's *Tools of Tomorrow: Catalog of Lunar and Mars Science Payloads* (RP-1345) available from NASA/JSC (Mail Code SN2, Houston, TX 77058, U.S.A. or e-mail **good-ing@snmail.jsc.nasa.gov**).

versity's Dr William L. Whittaker, is planning a terrestrial lunar theme park to serve as mission control and host for the televised programs from the Moon which may be produced in CD-ROM, and broadcast on cable. Gump and Whittaker envision customers utilizing a computer simulator to qualify in actually driving the lunar rovers via telepresence. This is the project Mitsubishi is underwriting.

10.1.4   United States lunar initiatives

Despite two decades of national reports relative to U.S.A. space goals and policy, there appears to be no *national consensus* in that regard, especially relative to lunar development. After expenditures of billions on planning an orbiting space station, the country is divided in its support for that program, as detailed in Chapter 8. At the moment, it is still questionable whether Congress will actually fund the construction and operation of a redesigned orbiting platform, for a time called *Alpha*, even with the commitment of international partners. As of this writing, the *Los Angeles Times* (December 27, 1995) reports that the Russians are further jeopardizing the **International Space Station** by their latest proposal to redesign it once again, but assembling the new station around **Mir**. Not only is this politically unwise for the U.S.A., but NASA complains that the approach would delay construction by a year, while boosting project costs by as much as $2 billion. And so the station drama goes on, with the Agency developing contingency plans should the Russians drop out, and Congress crying that such redesign would undermine American support for the whole program. Such station dithering and debate is why your author published an article in the *Earth Space Review* (October–December 1995, Vol. 4:4, pp. 9–12) on **Why Not Use the Moon as a Space Station?**

On the occasion of the **Apollo 11** milestone (July 20, 1994), public enthusiasm could not be galvanized, and NASA continues to cope with huge cuts in its budget. As one columnist, Charles Krauthammer, lamented: [6]

> **Ours was the generation that first escaped gravity, walked on the Moon, visited Saturn—and then, overtaken by an inexplicable lassitude and narrowness of vision, turned cathedrals of flight into wind tunnels.**

Yet, there **are a few hopeful signs** that the U.S. government may soon capitalize on the original investment of the American taxpayers in the **Apollo** missions:[7]

(a)   Since going back to the Moon permanently requires a robust, less expensive *space transportation system* with capability in both LEO and GEO, R&D funding is being spent on new rocket technology which utilizes composite materials that will lower the cost to orbital access. Most promising is a single-stage-to-orbit vehicle, such as the Delta Clipper program which McDonnell-Douglas Aerospace project engineer Jeff Laube assures us can be adapted to fly to the Moon (see Exhibit 90) ... .

(b)   Further, the low cost **Clementine 1** mission sponsored by the U.S. Department of Defense's Ballistic Missile Agency, in conjunction with the Navy and NASA, took 1500 pictures in February/March 1994, producing the first global digital map of the Moon—multispectral imaging data of 34 million square kilometers. Among the findings was a mountain top that might sometime be valuable for the first human settlement aloft—in this plateau the Sun never sets and a future colony would have full-time solar power by building a high collecting tower there ... .Further, three government agencies (NASA, U.S. Department of Energy, and the Air Force) have proposed another "small, low cost" **Clementine 2** mission for 1996 or beyond. Under the leadership of the Jet Propulsion Laboratory in Pasadena, California, the astronomical challenge is to land a spacecraft at the site of **Apollo 15** site in the Apennine Mountains, then release three robotic landers for a picture-taking expedition. Two moving rovers would carry a Cassegrain telescope conducting astronomy experiments; the orbiter itself would have in-struments (neutron and gamma-ray spectrometers) to detect the signature of water on the lunar surface, as well as a laser altimeter for more detailed studies of the Moon's topogra-phy. The Air Force also hopes to test a new propulsion system for the small spacecraft, which uses a high-energy liquid fuel (chlorine parafluoride). The synergistic mission would be launched on a Titan 2 rocket, and it remains to be seen if the U.S. Congress will fund the undertaking ... .

(c)   After more than 25 years from the **Apollo** missions, NASA expects to end the hiatus and launch **Lunar Prospector** into polar lunar orbit in June 1997 (refer to Exhibit 100). It is part of the **Discovery Program**, a new way to do smaller, faster, cheaper innovative missions via competition among private contractors—in this case the award went to Lock-heed Corporation, whose Dr Alan Binder conceived, designed, and developed the pro-posal for using a new LLV2 spacecraft. Supplementing previous automated lunar data collection missions, this one will provide a global, low-altitude mapping of the lunar sur-face composition, gravity fields, and gas release events. The **Prospector** payload will collect information that significantly improves our understanding of the evolution of the lunar highland crust, basaltic volcanism there, and lunar resource mapping. Mission ad-viser Dr James Arnold, University of California—San Diego, notes that the new, simpler approach contrasts with previous NASA solar system exploratory missions, which often took too long and were too costly. Exhibit 99 provides a mission profile ... .

(d) *National space legislation*—in 1995, a "Back to the Moon Bill," part of an Omnibus Commercial Space Act, apparently is near passage in the U.S. Congress—while limited in scope, it would create a legal regime for NASA to purchase lunar data from private enter-prise, allowing commercial companies to conduct innovative lunar probes on their own designs.

10.1.5 International Lunar Agreements

In the summer of l994, the Moon Treaty or Agreement came up for review in the United Nations General Assembly, and within the U.S. State Department. Because of its anti-developmental provisions, many of the leading spacefaring nations, including the U.S.A. and the former U.S.S.R., did not sign this document prepared by the Committee on the Peaceful Uses of Outer Space (COPUOS). In a private letter to the author (May 18, 1995), its chairman, Peter Hohenfellner, informs us that at the thirty-seventh session of COPUOS (June 6–16, 1994) this Moon Agreement revision was considered and it was recom-mended that no further action be take, which the 49th General Assembly of the United Nations endorsed. While he says no new developments are expected on the issue in the near future, Hohenfellner did find interesting materials on the subject which **United Soci-eties in Space, Inc. (USIS)** submitted and commended their efforts. That Englewood,

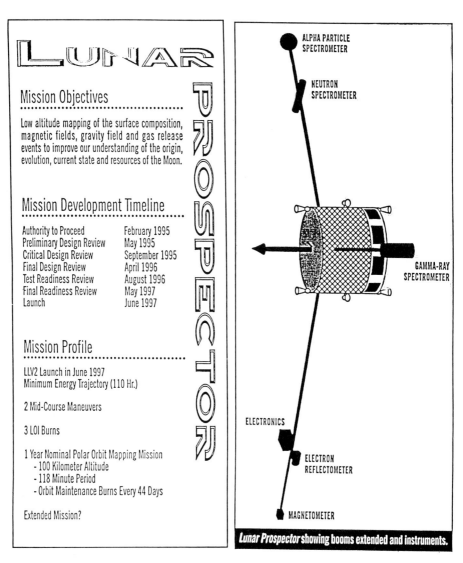

Exhibit 99—Lunar **Prospector** Mission Profile. The next unmanned lunar mission by NASA is summarized above. *Source*: P. Dasch, "Lunar Prospector," *Ad Astra*, May/June 1995, pp. 32/33—the magazine of the National Space Society (see Appendix C).

Colorado, organization has a **creative proposal** to amend or replace the 1979 limited "agreement" with a new Treaty on Jurisdiction in Outer Space (refer to Appendix D). This pro-space development approach, discussed previously in section 7.3B as an example of "macrothinking," would preserve the concept of "Common Heritage of Mankind" by establishing a space **Metanation** under the UN's Trusteeship Council for New Territories, with spacefaring nations acting as trustees for one hundred years (see next section).[8]

10.2   FOUNDING A LUNAR ECONOMIC DEVELOPMENT AUTHORITY

The above plan would also create a **Lunar Economic Development Authority (LEDA)** to:

(a) issue bonds to underwrite lunar enterprises, including a transportation system, a base and an industrial park ...

(b) lease surface and mining rights for private or public sector lunar macroprojects and collect fees therefrom ...

(c) coordinate and facilitate international endeavors by space agencies, scientific organizations, and private corporations on the Moon or its vicinity ...

(d) provide a lunar administrative structure for oversight supervision of terrestrially sponsored undertakings and communities on the Moon, so as to protect the environment and interests of its owners, humanity ...

(e) contract and supervise necessary infrastructure provisions, such as a **lunar transportation system**, a **lunar power system,** and a **lunar personnel deployment system** ...

(f) act as a clearinghouse of data about lunar information, such as conditions, resources, sites, and programs for future investors, project sponsors, and settlers ...

(g) conduct public information, outreach, and development programs on this planet to encourage investments in lunar resource utilization and lunar resettlement.

Rather than depending on Earth's taxpayers to finance construction of lunar infrastructure, **LEDA** would underwrite such development through income-producing activities such as outlined above. On Earth, ships and airplanes pay for the privilege of sailing or flying into urban ports; it would seem reasonable, then, that spacecraft from both public and private sectors worldwide might also be charged a fee someday by the **LEDA** for the privilege of landing on the Moon. When tourism reaches the lunar surface, such fees could contribute significantly to lunar development.

Right now for development purposes, there is no legal or financial mechanism, no less technological infrastructure, for such interplanetary undertakings that provide transnational, global participation. Even those who drafted the 1979 Moon "Treaty" envisioned some type of outer space regime or **authority** to oversee and regulate the "orderly development and exploitation" of extraterrestrial resources. Writing in 1994 on "Lunar Industrialization," Prof. Haym Benaroya of Rutgers University forecasts that such commercialism could employ 3–12% of our population in new jobs! But he too foresees the need for some type of Space or Lunar Industrialization Board (but within the U.S. government) to set policy, as well as coordinate and oversee economic, legal, and technical aspects of resource development on the Moon, and later Mars. In that same year, Lubos Perek, a former chief of the UN Outer Space Affairs Division, made a significant case for improving the management of outer space activities. He argued that to manage exterrestrial resources, a UN International Space Centre (UNISC) needed to be formulated. [9] More recently, Dr George S. Robinson, former associate counsel of the Smithsonian Institution, called for a **Declaration of Interdependence** between Earthkind and Spacekind, perhaps establishing an International Organization for Spacekind Cultures (IOSC—refer to Appendix A). [10] This space law scholar sees as its purposes to:

(1) provide an interdisciplinary, international, and transnational body of recognized experts to continuously review interactive relationships between Earth dwellers and spacefarers ...
(2) grant international agreements of recognitions and capacity (IARCS) to those space communities that satisfy the requisites for home rule as set by IOSC or something comparable (such as, the proposal of **United Societies in Space** to found a space **Metanation**, possibly under UN auspices) ...
(3) refer case situations of conflict to the International Court of Justice, or a transnational court yet to be founded for this purpose.

The above citations underscore the growing consensus that utilization and development of space resources require creation of a new entity to coordinate global space enterprise and governance, whether within or outside the existing United Nations. [11]

Certainly, some institution has to be devised to foster lunar development, preferably one which in scope goes beyond governments sponsorship or even a combination of national space agencies. A **Lunar Economic Development Authority** should be intersectoral, representing the interests of public and private sectors on a planetary scale. It should encourage participation of transnational consortia, whether from universities, corporations, space associations, or agencies. The macrothinking here should go beyond the European Space Agency proposal for an International Lunar Quinquennium meeting every five years to discuss lunar projects (refer to section 7.4). As a case in point, consider the matter of constructing a lunar base, such as the one depicted in Exhibit 100. This is a proposal of Lockheed Missiles & Space Company, Inc., in cooperation with Bechtel and Science Applications International Corporation. The artist's conception is based on assumptions that it will serve as a permanent center for scientific, industrial, and mining operations. Capable of expansion, the first stage begins as a lunar outpost with the following features:

• Living, working, and recreational facilities to support a crew ranging from 20 to 50 people.
• Greenhouses designed to recycle life-support air and water supplies while supplementing food requirements.
• Shielded plant growth facilities to assure an adequate seedling population for the greenhouses if a crop fails.
• Utility workshops that enable technicians to repair and maintain equipment in a shirtsleeves environment.
• Solar array power augmented by advanced fuel cells for lunar night cycles.
• Burial of the outpost in lunar soil for added protection against hazardous solar-flare radiation.

This is an innovative, macroengineering plan for early 21st century lunar development by three private corporations. Now consider answers to these practical questions:

(a) Who is the customer/s and how is this macroproject going to be paid for?
(b) Under whose authority is this base to be built? That is, assuming the proponents could raise the money for this endeavor, how do these private companies get international permission to use the Moon for this purpose? Under existing space treaties, the U.S.A. and NASA would not seem to have the power to authorize or even to contract for such

Exhibit 100—Lunar World. The 21st century lunar base, beginnings of a major industrial park and settlement on the Moon. *Source*: Lockheed Missiles & Space Company, Inc., Sunnyvale, California, U.S.A.

development. Right now, Article II of the UN's Outer Space Treaty (1967) would seem to preclude it:

Outer space, including the Moon and other Celestial Bodies, is not subject to national appropriation by claim of sovereignty, by means of use or occupation, or by any other means.

(c) If and when such a base were functional, whether it is by private or public or combined initiative, who or what supervises this operation?

Although the writer believes that the founding of a new **Metanation** to further human enterprise in space is the ultimate solution to this challenge, the establishment now of **space authorities** for the Moon, Mars, and other planets, can be immediately pursued. Then, we would put institutions in place to empower scientists, engineers, entrepreneurs, or settlers to go aloft and utilize space resources. [12] To facilitate living and working in an isolated, confined, sensitive environment like the Moon, the legal and governance prototypes already exist in the Antarctic Treaty (1958–1961), with its protocols and organizations, as well as in the Tennessee Valley Authority. [13] The Antarctic Treaty provides the legal framework for the area south of 60 degrees south latitude on this planet, reserving the region for peaceful purposes and encouraging international cooperation in scientific research there. The other model is the T.V.A., authorized by the U.S. Congress in 1947 with a Board of Governors appointed by the President and confirmed by the U.S. Senate.

When it was founded, the United States Government not only donated land and facilities for the new entity, but vested enough sovereignty in the T.V.A. so that it might obtain more land, including by eminent domain. The Authority's objectives were to conserve assets for the benefit of the American public in general, and specifically to provide electric power for the benefit of the people in the region served. To achieve its objectives, in 1948, T.V.A. issued $50 000 000 (U.S.) worth of bonds @ 3.2% interest rate, secured by a blanket debenture of its assets. Thirty years later these debentures were retired with no defaults, rollovers, or commissions having been paid. Today, the T.V.A. is one of the largest, most successful power producers in the world, a strategy worth emulating if resources on the space frontier are to be transformed for the betterment of the people of "Spaceship Earth."

This type of quasi-governmental **service authority** is a proven, respected, and traditional venue for underwriting and managing both public and private undertakings across jurisdictions and borders. It has been gainfully used to construct terrestrial infrastructure from airports and bridges to toll roads and convention centers. Why not adopt a comparable mechanism to finance and promote a **space infrastructure** which might supplement or replace direct taxation for space exploration and commerce? Port authorities have been successfully constituted across the nation from New York to San Diego. The New York Port Authority, for instance, crosses state lines to serve a metropolitan area's transportation needs. NYPA has its internal police force, which can arrest those who fail to comply with state, local, and Authority regulations. The approach is justified because of the size, value, and complexity of port facilities relating to transportation, safety, docking, food spoilage, longshore personnel traditions, and union contracts ... . The new Denver International Airport Authority was also financed by Muncipal Airport Revenue Bonds totaling $275 000 000, but these were government guaranteed ... . Spaceport Authorities are another example of the same strategy in use from Florida to Australia ... . An interesting historical point is that the Federal Statutes authorizing the inauguration of the U.S. space program began with a section on Police Authority (42 U.S.C. 2456) with the power for personnel to arrest citizens and bear arms for that purpose; that statute also created the agency that eventually became the National Aeronautics and Space Administration.

Thus, legal precedent exists and might be tested for application in space by the immediate incorporation of a **Lunar Economic Development Authority**, whether this be accomplished by private enterprise, government, or a combination thereof; whether it be within or without the United Nations, whether it be under national or international law. If instituted by year 2000, this would be a demonstration model that may eventually become part of a space **Metanation** were it to be formulated under the United Nations. Should **LEDA** prove to be a successful prototype, then it might be replicated next by the establishment of a **Mars Economic Development Authority (MEDA)**, and the model eventually repeated for the development of other planets or asteroids in our solar system. There are various scenarios as to how such **space authorities** might come into being within this decade:

(1) The U.S. Congress provides legislation constituting a **Lunar Economic Development Authority**, essentially to conserve national interests and promote development of the Moon and its resources for the benefit of our citizenry and to cooperate with other nations in this goal. The charter might be similar to that of T.V.A., and possibly some existing

space assets might be transferred from NASA or DOD to the new Authority to provide security for the lunar bonds sold for investments on the Moon. **LEDA**, in turn, might legally contract for services from NASA or other federal and state agencies, or from universities and corporations in the private sector at home or abroad. The Communication Satellite Act of 1962 is another precedent for such action, for it established COMSAT to cooperate with other countries to develop an operational satellite system, as well as to provide services on a global scale to others. Given the trend toward "privatization" of public properties, imagine if the assets turned over by the Congress to the new lunar Authority were to be two spaceports now functioning at Cape Canaveral in Florida and Vandenburg AFB in California; both built and paid for by taxpayers could then produce bond revenue and other income flows if operated by **LEDA**!

(2)   Another scenario would form a consortium by spacefaring nations committed to lunar development who sign an international agreement to establish a **Lunar Economic Development Authority**. In this approach, **LEDA** acts on behalf of the participating countries in financing and macromanaging resources on the Moon. The precedent for this already also exists in such agreements as INTELSAT, which established a global satellite communication system signed by governments or their designated public or private telecommunications entities. [14]

(3)   Still another possibility is for private transnational enterprise to act synergistically by legally incorporating a **Lunar Economic Development Authority** in one or more states or nations. Thus, profit and/or non-profit organizations might combine their strengths to undertake macroprojects on the Moon. There is ample precedent for this among world corporations and foundations seeking to protect the global commons. [15]

(4)   Although any of the above solutions would precipitate desired **action** toward near-term lunar development, many prefer a strategy whereby spacefaring nations work through the United Nations to found the **LEDA**, and/or a space **Metanation** under the Trusteeship Council for New Territories. Although getting an international consensus through the UN Committee on the Peaceful Use of Outer Space may be a slower process, the concept of space port authorities under such auspices might be to the advantage of all humanity seeking to use our "interplanetary commons" for the 21st century and beyond. [16]

10.3 CONCLUSIONS ON LUNAR ECONOMIC DEVELOPMENT

Flush with excitement of initial Space Age successes, Patrick Moore, British broadcaster and author, wrote a book in 1975 about *The Next Fifty Years in Space*. One chapter was devoted to **The Lunar Base**, which he forecast would become operational between the years 1995 and 2000. With hindsight, one might ask why the U.S.A. did not follow up with plans for such an undertaking after the **Apollo** mission series ended. There were many political, economic, and social factors that prevented the nation from achieving such a vision by the end of this century. Space experts from NASA and the aerospace industry still estimate that a lunar outpost might be functional by year 2010. From my viewpoint both as a management consultant and as a space psychologist, it would seem that **mindset** may prove to be the stumbling block delaying and hindering the exploitation of space resources in general, and lunar resources in particular. Positive mindsets in both the world's political and space communities which need to be cultivated would:

- *place individual space missions in a larger context*—to think in terms of just an automated lunar prospector mission is inadequate unless it is positioned within a broader strategic plan for lunar science and industrialization; the public expects near-term return on investment ...
- *act synergistically in global cooperation*—today no one country or space agency can effectively undertake space macroprojects without forming international partnerships; the sharing of talent and resources ensure long-term "paybacks" ...
- *facilitate intersectoral and interdisciplinary planning*—space macroprojects require going beyond traditional interfaces, such as public and private, science and business; investment will be forthcoming, for example, if goals include both science and industrialization, involving industry and universities with government agencies.

The above review of current trends underscores why the next major **space** investment and undertaking by the U.S.A. and its spacefaring partners should include a return to the Moon to stay and commitment to lunar development. Many "Mission to Planet Earth" goals can be achieved best by using the Moon as a platform for scientific, environmental, and economic benefit. The Moon can become the laboratory for international cooperation, the launching pad of humanity into the universe. Expert consensus for this is emerging as confirmed in this statement from an International Academy of Astronautics report:

> **We believe the time has come that these global trends should induce responsible governments to take action deciding to continue the development of lunar resources and consequently to assign an existing multinational space organization (or establish a new one) the responsibility of returning people to the Moon permanently and developing its resources for the benefit of mankind.**[17]

The author concurs in these recommendations, but is convinced that a "new multinational space organization" must be formed, one that goes beyond but collaborates with existing national space agencies. To this end, we have proposed here the creation of a **Lunar Economic Development Authority**, and have explored several alternatives for accomplishing this goal. **LEDA** is considered a prototype for future space authorities that could be constituted to develop other planets and asteroids in our solar system, as well as for constructing orbiting colonies as proposed by visionaries, such as the late Wernher von Braun and Gerard K. O'Neill, or today's Marshall T. Savage. [18] This authority concept offers a bridge over troubled waters in contemporary space policy and law, and a mechanism for constructing and really financing lunar infrastructure through public participation. This legal entity with its own Board of Directors enables national sovereignties to act synergistically in the exploration and development of the high frontier. It could not only issue revenue bonds for this purpose, but **LEDA** could literally build the "bridge between the two worlds of the Earth/Moon system" by:

- leasing land, facility, and equipment rights;
- fund raising and fee collection from investors and developers;

- site planning and permits for habitats and industrial parks;
- zoning and inspection to protect lunar environmental and ecological concerns;
- long-term management and policing of lunar settlements;
- administering a lunar personnel deployment system and regulating tourism.

Such practical matters have already been researched by Charles Lauer at the University of Michigan College of Architecture with reference to real estate aloft. He has written extensively on the financial, legal, regulatory, and design aspects of business parks in orbit that have implications for a lunar industrial park.[19]

Yet another strategy for moving toward industrialization of the Moon would be for the U.S. Administration and Congress to convene a **White House Conference on Lunar**

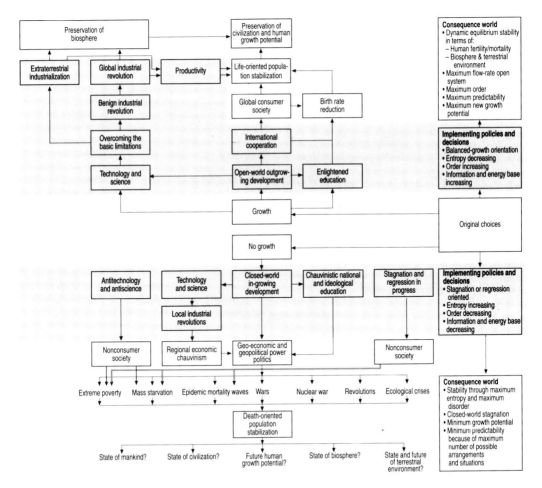

Exhibit 101—The Consequences of Growth and No Growth.: **Krafft Ehricke**. *Source*: Reprinted with permission of Marsha Freeman from her "A Memoir of Krafft Ehricke's Extraterrestrial Imperative," *21st Century Science and Technology*, Vol. 7:4, Winter 1994–1995, p. 34 (PO Box 16285, Washington, DC 20041, U.S.A.).

Enterprise that would build appropriate consensus, as well as lunar policy and strategic planning (see Epilogue). One outcome might be a legislative draft for creation of a **Lunar Economic Development Authority**[20] .... . Another action would be for the United Nations to sponsor a global convocation for the world community to consider a similar agenda, possibly under the aegis of the UN Office for Outer Space Affairs. Like the ISY '92 convention, this might be called **World Space Congress 2000** ... . If such official leadership is not forthcoming, then transnational enterprise will act to fill the vacuum and promote space resource development and commercialization.[21] Krafft Ehricke envisioned lunar industrialization as our **extraterrestrial imperative**, warning of the consequences of growth or no-growth policies (described in Exhibit 101).

AUTHOR ACKNOWLEDGEMENT

In the preparation of this chapter, the author acknowledges the exceptional contribution of **Declan J. O'Donnell,** co-founder with him of **United Societies in Space Inc.** (See Notes on Contributors).

The exploration and use of outer space, including the Moon and other celestial bodies, shall be carried out for the benefit and interest of all countries, irrespective of their degree of economic or scientific development, and shall be the province of all mankind.

Outer space, including the Moon and other celestial bodies, shall be free for exploration and use by all States without discrimination of any kind, on the basis of equality and in accordance with international law, and there shall be free access to all areas of celestial bodies. There shall be freedom of scientific investigation in outer space, including the Moon and other celestial bodies, and the States shall facilitate and encourage international cooperation in such investigation.

—Article I, the United Nations 1967 Outer Space Treaty.

The purpose of this Corporation (Lunar Economic Development Authority) shall be to promote the Moon as a place to live and work as a society of peoples and to help create and maintain a consensus governance authority at the venue of the Moon, including its useable orbits, and to educate people on the benefits, burdens, and responsibilities of living and working in space. The Corporation will serve as the agent of humankind in space and at the Moon, as well as the agent for all of the sponsor nations, to develop the Moon for humankind. It is the intended business of the Authority to administer each nation's rights under the *1967 Outer Space Treaty* ... .

—Article III, Articles of Incorporation for the Lunar Economic Development Authority, Inc.

Exhibit 102—Space Light*. **Spatium Lux** is an original painting by Dennis M. Davidson—the Moon as seen from Earth beckons us.

The opening of a new, high frontier will challenge the best that is in us. The new lands waiting to be built in space will give us new freedom to search for better governments, social systems, and ways of life, so that by our efforts during the decades ahead, our children may thereby find a world richer in opportunity.

Dr Gerard K. O'Neil, Author *The High Frontier—Human Colonies in Space*

Space Studies Institute, Princeton, NJ (1989)

Epilogue

TRANSFORMING SPACE VISIONS INTO REALITIES

Human dreams and ideas span time, often taking centuries before being transformed into worthwhile activities. Some of our forebears dimly perceived the spectacular achievements which this generation has witnessed since the dawn of the Space Age. Hopefully, *Living and Working in Space* has provided readers with insights into the challenges which lie ahead in exploring and settling the high frontier.

In the future, our descendants may remember the 20th century primarily for proving that our species was not Earth-bound and could live in the microgravity environment. The decades of the 1980s and 90s may be viewed as a watershed period for commercial space and living aloft. It was the period when humanity shifted from space competition to cooperation, from a space race to forming joint ventures for international macroprojects. The satellite industry not only turned our world into a **global village** by its communication capabilities, but it demonstrated that it could be a profitable process. Furthermore, orbital imaging and sensing has shown myriad practical applications on Earth, even in protecting our planet. The Russian space station **Mir**, in orbit for ten years, became a platform for true international cooperation by agreements which brought abroad Europeans, Japanese, and even Americans via their Shuttle Orbiter dockings. Spacefaring nations have much to gain in forming partnerships with Russia, the country with the most expertise in launches and extended spaceflight (in March 1995, cosmonaut Valeri Polyakov had spent a record 438 days aloft). Within the context of the U.S. space program, for instance, 1986 may be seen eventually as a time of both set-back and vision. In that year the high risk of space exploration was demonstrated by the **Challenger** Orbiter tragedy and loss of lives, hence this book's dedication. But the resulting investigations and report of a Presidential Commission under Chairman William Rogers contributed to reforms which reinvigorated the program—a turning point had been reached—NASA, the public sector, will no longer dominate and totally control access to space. Instead, international business, the private sector, was challenged to assume a larger role in global space developments, particularly space transportation. The biggest drawback to development of the space frontier has been the excessive launch costs, getting up the first hundred miles to the "high ground." With the above policy changes, transportation expenses to orbit began to be reduced, and the

Shuttle was to be supplemented by a variety of expendable and reusable launch vehicles made by private enterprise in the U.S.A., Europe, Russia, Japan, China, and even India. Simultaneously, a worldwide **space ethos** began to emerge, and in America it was sharpened by the vision enunciated by the National Commission on Space. As previously indicated, their report articulated peaceful, civilian goals and strategies in outer space for the next fifty years:

> A Pioneering Mission for 21st Century America—To lead the exploration and development of the space frontier, advancing science, technology, and enterprise, by building institutions and systems that make accessible vast new resources and support human settlements beyond Earth orbit, from the highlands of the Moon to the plains of Mars.
>
> (*Pioneering the Space Frontier,* 1986 p. 2.)

To actually implement such lofty goals requires **transformational leadership** in both the public and the private sectors globally now and during the 21st century. The business community at large, not just aerospace and communication satellite companies, must lead in the creation of a space ethos that supports an enlarged and well-funded space program. As this is written, the European Union is demanding more control over space policy and expenditures, and advising the European Space Agency to pay more attention to space commerce. Yes, space is a place for fulfilling dreams, as well as for promoting free enterprise and realizing profit!

But how? Specifically, as our case in point, how can America further capitalize upon its $20 billion investment in the Apollo lunar landings? How can the nation get payback on its total space expenditures from the *utilization of space-based resources*? Some innovative answers may be gleaned from the reports and recommendations of various space studies previously cited. For example, an assemblage of space scholars and specialists gathered for a NASA summer study at the California Space Institute in 1984 on the La Jolla campus of the Scripps Institution of Oceanography, University of California—San Diego. The proceedings were revised and updated into five highly readable and concise volumes for International Space Year in 1992 under the title *Space Resources* (NASA Sp-509), and are available through the U.S. Government Printing Office. Apart from the technical and economic insights, especially for the establishment of a lunar base, that report includes proposals for:

- **building public consensus and financial support for the space program;**
- **initiatives within the private sector to foster the peaceful use of space by its exploration and industrialization;**
- **legislation that would transform the nation's space agency, as well as its policies and procedures.**

The scholars at the "Cal Space" sessions made specific proposals for the mobilization of the scientific, business, and political constituencies. At this juncture, they found that the justification for peaceful and commercial development of space resources is more human and scientific than economic or political. Their rationale for moving forward on the space frontier had to do with discoveries which maintain the American technological excellence, security, and leadership. Their recommendations were aimed at benefitting the Earth's peoples, especially in the Third World, by technology transfer from space. Their aspirations were that human potential could be better realized by extending human presence permanently into our Solar System.

Envisioning the next century in space, the Cal Space study group (in which this writer participated as a faculty fellow) made *a case for investment in space*. It went from the success stories of commercial satellites and sensors to the identification of space-growth industries in materials processing, manufacturing and mining, solar energy and power satellite networks, and tourism. Much of their prognosis was later to be confirmed by the National Commission on Space report (1986). But to take advantage of these and other undreamed opportunities aloft demands more than vision and boldness. It means pragmatic steps to raise capital, especially in the private sector, for high-risk technological venturing that will build the necessary space infrastructure. Therefore, three strategies proposed in the forementioned analysis will be summarized here in the hope that readers may decide to participate in the implementation of one or more approaches that will advance human habitation aloft:

1. White House Conference on Space Enterprises

With so many domestic, as well as international, needs and concerns, it is difficult to rouse the average citizen to financial support of the space program. Understandably, members of Congress cut recommended NASA appropriations to solve more pressing public problems in the homeland. In a *Commercial Space* editorial (Summer 1986, p. 15), Donald E. Fink described how other nations are awakening to the commercial prospects of space and are pursuing ambitious programs to realize this potential for their countries. He bewailed the lack of such domestic awareness, calling upon U.S. business, industry, and government leaders to revitalize what should be the world's outstanding space program. One way of focusing public attention and raising consciousness on national needs is through the calling of a White House Conference, followed by publication of its proceedings, as well as issuance of new policy and legislation. To facilitate civilian space involvement and leadership, while broadening participation in the national space effort, readers may wish to promote the holding of a White House Conference on Space Enterprises. One recommendation made at the 1995 International Lunar Exploration Conference was that the event be devoted to the theme of **lunar development**.

For such an undertaking to occur within a few years, citizens would have to communicate with their Congressional representatives and request the introduction of a bill (a) supporting the event and (b) requesting the Administration to convene such a meeting. The White House staff would then be mandated to establish a planning/operations entity who would then set dates and agenda, as well as issue invitations. The primary conference objective could be to examine ways to implement the recommendations of recent national space studies relative to opening up the final frontier as a way to improve the quality of life here for all the planet's inhabitants. The secondary purpose could be to foster a national consensus on peaceful, commercial exploration and utilization of space-based resources. This proposed White House Conference might be structured around the theme set forth by the National Space Commission in its 1986 Presidential Report, namely: "Stimulating space enterprise for the direct benefits of the people on Earth." The sessions might be organized on the four parts of that report - implementing civilian space goals; strategies to provide low-cost access to the Solar System; action plans for the next twenty years for opening the space frontier; American leadership on the space frontier.

A decision by the President to sponsor such a conference would boost public morale, turn national energies constructively outward, and ensure the country's reliability in international space agreements. Through this meeting in the nation's capitol, both intellectuals and implementers would be challenged to contribute to strategic planning which advances the human migration into space. To communicate the message for such a White House Conference and enlist public involvement, a planning task force would have to be constituted, perhaps chaired by an outstanding space media professional, like Walter Cronkite or Hugh Downs. To recharge national enthusiasm for space, distinguished citizens and other international participants would be invited to discuss immediate and pragmatic means for achieving large-scale programs in orbit. The invitees might include not only space professionals, writers, and advocates, but people of competence and distinction in a position to influence the citizenry in their support of space activities. Furthermore, corporations and organizations already in the space business might be invited to join with the government as co-sponsors of the event. The year chosen might also have special significance in terms of a space achievement anniversary or a new undertaking like the launching of the **International Space Station**. The main point is to hold such a White House Conference before the end of this decade, so that deliberations might also be televised and a national teleconference held in conjunction with the Washington, DC meeting.

Other nations should consider a comparable convocation in their homelands. For example, the European Union and the Commonwealth of Independent States might convene such a space conference for the countries within their alliance. Ideally, a second **World Space Congress** should be organized in year 2000 to announce policies and commitments for Millennia III!

2. Alternative Funding of Space Enterprises

New options must be pursued for financing space ventures, other than through the taxes and the annual governmental budget allocations. That traditional public sector approach will not obtain the $700 billion which the National Commission on Space estimates is required over the next five decades to open up the space frontier. Where are funds of that magnitude to come from, especially with a huge national deficit and legislative spending restrictions? The history of both the Shuttle and the Space Station to date has been that of government cut-backs which undermine NASA designs and safety in mission planning.

Creating a space ethos implies getting the masses of citizens involved, in some manner or other in space ventures, as indicated in chapter 1. In a democratic, free enterprise society, what better way to accomplish this than as a "financial investor." The scholars and experts that NASA assembled at the California Space Institute considered innovative ways for space financing, and urged that a joint task force be established by the President and Congress to analyze the options and recommend legislation. Citizens need financial incentives, like tax rebates, to invest in opening the space frontier, just as was done in the opening of the Western Frontier. To capitalize upon the enormous public interest and good will generated by the American space program in the past forty years, *alternative* or supplementary funding possibilities should be explored, including the issuance of public bonds, and the authorization of stock sales in limited R&D technological space partnerships or trading companies. Recall that back in the **Sputnik** days, the COMSAT offering on the stock exchange was oversubscribed by the public! Specifically, the Cal Space group also urged consideration of a national space lottery.

Public lotteries to support scientific exploration and civilizing ventures in newly opened frontiers are part of the nation's tradition. Since the 15th century, European countries have used the lottery device to raise capital for public works. In 1612, the English used this means to support the Jamestown settlement. In the New World, the colonists and first citizens of this republic employed the mechanism to fund the establishment of higher education, including Harvard, Kings College (Columbia), Dartmouth, Yale, and other universities. In the 19th century, Americans again used lotteries to open up the Western frontier. During the present decade in the U.S.A., for instance, lotteries have become popular again within states to fund public services, particularly education. Today, many foreign countries, such as Australia and Mexico, successfully utilize lotteries as a means of raising money to accomplish social goals.

If income produced from new funding sources is to alleviate the Federal tax burden relative to space expenditures, the investment scope must be vastly broadened beyond the government budgeting process. That is what underlies the proposal of United Societies in Space, Inc., to establish space authorities, such as a **Lunar Economic Development Authority**, discussed in chapter 10. More creative methods of external financing of space enterprise are necessary. With the proper *space ethos* in a country, extraterrestrial endeavors would be perceived as a primary national interest and given the required financial priority. According to author, Ben Bova, the public does not fully appreciate the handsome pay- backs to the nation on previous space investments which warrant long-term funding of future space developments. Thus, the national task force examining space financing should recommend ways to ensure citizen involvement in underwriting civilian space ventures.

Were more private capitalization encouraged, then public policy makers and space leaders would be challenged to cooperate in setting disbursement objectives for the money so raised. The public is more likely to contribute enthusiastically by purchasing space bonds, stocks, or lottery tickets if the initial funds raised were devoted exclusively or primarily to economic, international, and scientific use, in preference to "star wars." For example, the initial target might be in the area of *space transportation systems*. That is, to build the space "highway" for the first few hundred kilometers up into Lower Earth Orbit, which is the most difficult part of interplanetary travel. Global participation in financing joint space ventures could provide the means for replacing the present shuttle fleet with advanced aerospace planes and reusable launch vehicles capable of operating in geosynchronous orbit or beyond ... . Just as the Conestoga wagons and railroad opened up Western resources to the nation, so will these less expensive spacefaring vehicles bring resources from orbit back to benefit the home planet. In time, innovative tax incentive provisions on behalf of space macroprojects may stimulate the construction of Moon/Martian bases, as well as start the whole process of colonization in the next century!

There already exist basic constituencies to enhance the success for alternative forms of space financing:

- hundreds of individuals in fifteen public forums who contributed input to the National Commission on Space report and whose names are actually listed therein;
- 300 000 members of fifty space advocacy groups with an estimated aggregate budget today of more than $30 million;
- millions of space media fans from *Star Trek* television viewers and other numerous motion pictures like *2001* and *Apollo 13*, to the worldwide audience who witness the satellite televising of NASA space feats or watch new productions, such as Disney's *Plymouth* series about the first lunar community.

Before his death, Gerard O'Neill, the visionary scientist, predicted that it will be private capital that will eventually finance space industrialization and colonization. Although this may begin with U.S. citizens, the internationalization of space will attract global investment ... . Readers from other countries may possibly adapt the above two strategies to promote their space enterprise within their own homelands, or to further global space developments, possibly in conjunction with the United Nations Office of Outer Space in Vienna.

The final proposal of the momentous California Space Institute summer study centers around restructuring of the nation's space entity.

3. Reorganization of the National Aeronautics and Space Agency

The emergence of a new work culture requires metaindustrial management (Harris, 1996). If the goals and recommendations set forth in numerous high-level space studies are to be pursued, then the U.S. Space Agency needs to be re-renewed and re-directed, as in the days of the **Apollo** missions. Under the past and present agency administrators, internal reorganization of NASA's culture and management is already under way, thanks in part to the recommendations of several external expert studies. To free NASA somewhat from undue political/military pressure and federal bureaucracy, as well as the tyranny of meeting annual budget demands and deficit restrictions, changes in the agency's charter and structure might enable the institution to regain its prominence in space. Initially readers might begin by lobbying the U.S. Congress to adopt one of the several reform proposals to provide NASA with multi-year authorization for funding long-term space objectives. Ensuring space achievements during the next century ultimately requires a reconstituted American space **authority.**

The assembled scholars and space experts at the California Space Institute recommended consideration of such legislation to strengthen NASA by making it more autonomous. Models already exist for this possibility in the U.S. Postal Service, the Tennessee Valley Authority, and the New York Port Authority. By creating a *National Space and Aeronautics Authority* as a semi-autonomous and quasi-governmental corporation, the revitalized system would be more capable of space macromanagement. Its budget could be set for long-term, large-scale project development over decades; the funding of research and development could be separated from operations. The outcome might be not only to liberate the redesigned NASA for pursuit of necessary planetary science explorations, but for cooperating internationally in expanded *civilian* space enterprises. Many scientists argue that NASA can function best when its role is primarily R&D, leaving to other agencies or the private sector the business of transportation, shelter and service in space. A modernized charter might empower NASA to seek more creative financing of space activities, particularly with reference to space technology transfer, as well as attracting more venture capital and to licensing **space trading corporations**. In past centuries, great trading corporations were formed by private investors to facilitate exploration and commerce in unknown or foreign lands. The 21st century may replicate this approach by international space trading entities, comparable to existing multinational communication satellite corporations.

Such legislative innovation might facilitate joint ventures between NASA and the private organizations in the U.S.A. and abroad, as well as with other national space entities. Michael Simon (1987), president of International Space Enterprises, has a thesis that government and industry should do more real joint venturing, so that corporations assume true risks to share in returns. Citing the industrial age model and history of aerospace and defense contractors with defused responsibility and bloated costs, Simon makes a case for a new rationale for space commercialization. In this free enterprise mode, government would encourage the private sector to greater responsibility and risk by:

- incentives for tax payers who invest in space enterprise;
- policies promoting innovative space entrepreneurialism;
- mechanisms for improving space market responsiveness;
- opportunities for achieving large-scale commercial benefits;
- initiatives which encourage synergy among companies, universities, and government entities engaged in working together to apply space research and transfer technology.

Perhaps Simon best states the case for investment in space development at the closing of his volume, *Keeping the Dream Alive*:

> The era in which we live presents humanity with three great challenges: to live in peace, to bring economic prosperity to all people, and to offer tomorrow's generations an exciting future of physical and spiritual growth. During its relatively brief existence, the Space Program has emerged as a central force in our quest to meet all of these challenges. By breaching the bonds of our home planet, we have taken the tentative early steps to become an advanced interplanetary civilization. The impact of the embryonic space age on our lives, already great, will expand and intensify in the years to come, as our horizons become as limitless as the Universe itself.

Re-examining one's space policy and program is a dynamic process that applies to all space agencies and organizations, regardless of country of origin. The European Union, for example, is now doing this with reference to the rapidly growing, European Space Agency. Ultimately, national space endeavors need a coordinating entity, such as a world or global space administration or authority. Steinbronn and Cordell (1990) have described an innovative concept for international management of the space undertakings. They propose an umbrella organization, not unlike INTELSAT and modeled upon ESA, which would be called INTERSPACE. Through this entity, any interested and qualified country could participate in the multinational activities of its consortia, as long as the mandatory space science requirements were adhered to by member nations. In Chapter 7, we reported on another legal and financial strategy recommended by O'Donnell (1994), namely to found a **space metanation**, possibly under the United Nation's charter provisions for new territories, which would represent the common heritage of humankind. The UN has already designated those who go aloft as humanity's **envoys**. In creating a spacefaring civilization, ponder these words of Robinson and White (1986) which highlight the paradigm shift under way:

> **Our embryonic envoys have been essential intelligence agents for greater understanding of this survival vision—a total view. Through our efforts to propagate our envoys into the cosmos, through their own personal preparation and adjustments, and also through our remote biotechnological reception of their new transglobal outlook, our envoys have helped us begin to understand the systematic, dynamic, multidimension, and continuous nature of the cosmos.**

EXERCISING TRANSFORMATIONAL LEADERSHIP

Since our species is in transition to space-based living, Robinson and White argue that this necessitates profound changes in sociology, biology, philosophy, government, and law. Space technological advances are drivers of a wholly new human environment and the creation of a space culture. Thoughtful citizens concerned about a people's future want to participate in the process, beginning with the formulation of a space ethos. But it also requires a new type of leadership which has been characterized as *transformational*.

The transformational leaders, according to Tichy and DeVanna (1986), recognize the need for changes, such as has been examined here in *Living and Working in Space*. Furthermore, such leaders create and communicate the vision of these desired changes so that a critical mass of people find them acceptable. Then, this leadership personally mobilizes this commitment into foresighted strategies which are converted into actualities. So, too, can transformational leadership renew the space program worldwide, restructure space agencies, refinance space undertakings. Transformational leadership can promote synergy among spacefaring nations, as in joint transnational missions to Mars and Venus.

When humans are engaged in such missions of long duration, Bormanis and Logsdon (1992) and their colleagues remind us that a whole range of space policy issues need to be addressed, such as:

(1) the uniqueness of the space environment;
(2) the selection, composition, and interactions of space crews;
(3) the space inhabitants as microsocieties with standards, laws, ethics, and values;
(4) the medical and scientific experimentation under way;
(5) the spacefarer's survival and quality of life, including communication and privacy rights, health care, pregnancy, deviant acts, death and risk management;
(6) the space explorer's environmental responsibilities relative to contamination, management of waste and debris, and other such ecology issues.

The exercise of authentic global leadership within all segments of both the public and private sectors could transform citizen goodwill into a space ethos that permeates our lives toward opening up the high frontier. When the majority of the world's population perceive the economic and human advantage to be gained there, then energies will be directed into its development and settlement As astrophysicist and author David Brin reminds us, *science* and its child, *technology*, are cooperative endeavors requiring knowledge to be shared, especially when applied aloft (Hargrove, 1986). The message of this book is simply that space is the place where **human emergence** can truly occur!

> For travelers, it is not enough to see the horizon alone. We must make sure of what is beyond the horizon.
>
> **(Kemal Ataturk)**

A SPACEFARER'S CREDO

When we fly around the Earth at eight kilometers a second, 400 kilometers up, we see our Earth as a whole planet. We observe the oceans, forests, mountains, cities, and roads—we absolutely do not see the borders between nations. The time has come for all people of the Earth to work together to build a bright future. Let's start!

(Yuri Romaneko, October I0, I989—cosmonaut with the then record aloft of 420 days, including 326 consecutive days in an orbital environment!)

APPENDICES

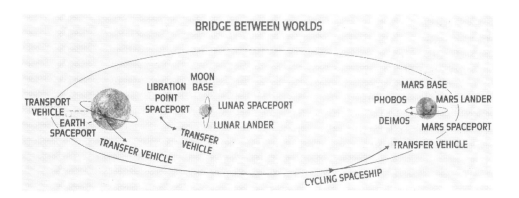

Exhibit 103—Bridge between Two Worlds: Earth/Moon System. Interplanetary exploration for the New Millennium depends upon first developing a space infrastructure in our own Solar System. *Source*: National Commission on space, *Pioneering the Space Frontier*. New York, NY: Bantam Books, 1986.

Appendix A[†]
Declaration of first principles for the governance of outer space societies

PREAMBLE

On the occasion of the Bicentennial of the Constitution of the United States of America and in commemoration and furtherance of its values we, the undersigned petitioners,

- Bearing witness to the exploration and inevitable settlement of outer space;
- Recognizing the universal longing for life, liberty, equality, peace and security;
- Expressing our unshakable belief in the dignity of the individual;
- placing our trust in societies that guarantee their members full protection of law, due process and equal protection under the law;
- Reaffirming our faith in fundamental freedoms;
- Mindful, as were our nation's founders, of the self-evident truth that we are endowed by our Creator with certain inalienable rights;
- Recognizing the responsibility of a government to protect the rights of the governed to exist and to evolve;

Do assert and declare in this petition the intrinsic value of a set of First Principles for the Governance of Outer Space Societies and, at the beginning of this third century of nationhood under our Constitution, resolutely urge all people of the United States of America to acknowledge, accept and apply such First Principles as hereinafter set forth.

ARTICLE 1

A. The rule of law and the fundamental values embodied in the United States Constitution shall apply to all individuals living in outer space societies under United States jurisdiction.

† Reprinted with permission by the Smithsonian Institution Press from 'Declaration of First Principles for the Governance of Outer Space Societies', in *Envoys of Mankind* by G. S. Robinson and H. M. White, ©Smithsonian Institution, Washington, DC, 1986, pp. 266–270.

B. Appropriate constraints upon and limitations of authority shall be defined so as to protect the personal freedom of each individual, such as the right to reasonable privacy, freedom from self-incrimination, freedom from unreasonable intrusion, search and seizure, and freedom from cruel and unusual punishment.

C. Toward this end, the imperatives of community safety and individual survival within the unique environment of outer space shall be guaranteed in harmony with the exercise of such fundamental individual rights as freedom of speech, religion, association, assembly, contract, travel to, in and from outer space, media and communications, as well as the rights of petition, informed consent and private ownership of property.

D. The principles set forth here should not be construed to exclude any other such rights possessed by individuals.

ARTICLE II

A. Authority in outer space societies, exercised under principles of representative government appropriate to the circumstances and degree of community development, shall reflect the will of the people of those societies.

B. All petitions to the United States Government from outer space societies under its jurisdiction shall be accepted and receive prompt consideration.

C. The United States shall provide for an orderly and peaceful transition to self-governance by outer space societies under its jurisdiction at such time as their inhabitants shall manifest clearly a belief that such a transition is both necessary and appropriate.

D. In response to aggression, threats of aggression, or hostile actions, outer space societies may provide for their common defense and for the maintenance of essential public order.

E. Outer space societies shall assume all rights and obligations set forth in treaties and international agreements, relevant to the activities of such societies, to which the United States is a party and which further freedom, peace and security.

F. the advancement of science and technology shall be encouraged in outer space societies for the benefit of all humanity.

G. Outer space societies shall protect from abuse the environment and natural resources of Earth and space.

The Declaration was based on the United States Constitution, and was drafted by scholars assembled for two conferences at the National Air and Space Museum, Washington, DC, in December 1986 and November 1987. It is reprinted here as a basis for discussion on governance of space communities. The above book by attorneys Robinson and White also contains a proposed 'Treaty Governing Social Order of Long-Duration or Permanent Inhabitants of Near and Deep space,' as well as a proposed 'Spacekind Declaration of Independence'.

The Declaration also was discussed in detail within an article by Dr George S. Robinson, when Associate General Counsel of the Smithsonian Institution, which we recommend to readers: 'Essay—Re-examining Our Constitutional Heritage: A Declaration of First Principles for the Governance of Outer Space Societies', in **High Technology Law** Journal, 1989, pp. 81–97. The journal is published by the School of Law, University of California (Boalt Hall, Berkely, CA. 90024, U.S.A.)

For 1996, Dr Robinson is chairing the panel of judges for the Global Space Essay Contest. It has as its theme composing a *Declaration of Space Interdependence*. To assist entrants in the contest, he prepared a two-part article which we also recommend to readers. It was entitled "Natural Law and a Declaration of Humankind Interdependence," and was published in the June and December 1995 issues of *Space Governance*. This is the journal of the United Societies In Space, Inc., which also sponsored the contest in conjunction with the World Bar Association. (Copies are available from USIS, 6841 South Yosemite, Englewood, CO 80112, U.S.A.; tel: 1-800/895-META; fax: 1-303/721-1182.)

PIONEERS

Columbus and Daniel Boone
   Would go to the Moon
If they were alive today
   —And the rest of us
wouldstay behind
   Until it was time
   To plan suburbs
And film movies about
   The *pioneers*!
but Columbus and Daniel Boone
   Would be on Mars
   By then.

—Nathan C. Goldman, Attorney/Author, Houston, Texas, U.S.A.

Exhibit 104—A Demonstration Lunar Power Base: Artist Concept. *Source*: Dr David R. Criswell and Dr Robert D. Waldron © 1991. From "Results of Analysis of a Lunar-Based Power System to Supply Earth with 20 000 GW of Electric Power," paper presented at the Solar Power Satellite Congress, Paris, France (August 26–30, 1991). [Refer to another rendition in Chapter 7, Exhibit 70.]

Appendix B
Case Study—A Lunar Power System

David R. Criswell and Philip R. Harris*

B.1 INTRODUCTION

Since the l960s, extensive lunar studies have been under way by both Russian and American scientists in conjunction with the Lunik and Apollo missions. Adjoining NASA's Johnson Space Center, the Lunar and Planetary Institute has continued further research, conferences and publications on the Moon and its resources (Mendell, 1985). NASA has also provided both universities and contractors with funding to further such investigations (McKay, 1992). For example, over three decades ago, General Dynamics Space Systems Division completed a NASA report in l965, "Operations and Logistics Study for Lunar Exploration Systems for Apollo (LESA)," and in l979, "Lunar Resources Utilization for Space Construction."

In this decade the Stafford Synthesis Group report prediction has proven accurate that investigations will examine life-support and propulsion technologies needed for Moon and Mars manned missions. Yet, unmanned missions in the 90s, such as **Clementine** and **Prospector,** are furthering the mapping of lunar resources and prospects. The Europeans have also laid out their long-term program for returning permanently to the Moon as a means to stimulate global economic development (Ockels and Battrick, 1994). In two International Lunar Exploration Conferences of 1994/5, the emphasis has been on commercial enterprises on or about the Moon featuring both human and robotic activities. On these occasions, the assembled delegates in San Diego, California, concluded that for technologies to reduce access costs to the Moon and its resources, space agencies should commit to acquiring commercial products and services (Simon, 1994/5). Indeed, the principal rationalization for a return to the Moon seems to be economic by means of lunar industrialization for the benefit of Earth's peoples (Angelo & Easterwood, 1989). One such justification has been solar power, which was highlighted in the July l989 *Report of*

---

* For biographies, see *Notes on Contributors* section upfront.

NASA Lunar Energy Enterprise Task Force. The rekindling of such interest has even resulted in a new book, *Solar Power Satellites—The Emerging Energy Option* (Glaser, Davidson, and Csigi, 1992). Chapter 10 has amply discussed the possibilities for lunar economic development, but this case will focus on one viable promise—using the Moon as a platform for energy production of several types.

The development of a lunar solar power (LSP) system presents manifold challenges as a large-scale, space-based enterprise. First, is the need for global communications about the rationale and case for investment to supply solar power to the Earth from the Moon. Second, the technical community must cooperate in *integrated planning* of an operational lunar power base which provides quick growth rate, lower power costs, and higher energy capacity (Criswell, 1995). To collect solar energy on the Moon and beam it to Earth is a macroengineering project calling for research demonstrations, as well as formation of joint ventures for major funding of innovative space infrastructures.

The concept of lunar solar energy transmission is a 21st century global power system which will not introduce new pollution into the biosphere, nor deplete existing organic resources. A window of opportunity now exists to provide this clean-energy solution to improve the human condition and add new net worth to Earth. The LSP system could offer a viable electric power supply to enhance the standard of living on the home planet, especially in economically developing countries. Engineers and technicians worldwide should be encouraged to initiate R&D into space energy resources. The challenge goes beyond engineering and extends into many other realms, such as finance, communication, sociology, law, and politics (O'Donnell and Harris, 1994). For instance, any beaming of solar energy to the home planet will have considerable legal ramifications (IAF/IISL, 1989, 1988, 1987). By implication, the undertaking means creation of a new type of both **macrothinking** and **macromanagement** as described previously in Chapters 7 and 8.

B.2 RATIONALE FOR A LUNAR POWER SYSTEM

The advantages of this 21st century energy system over conventional power have been summarized in Exhibit 105. Looking along its last row, the case for lunar power is highlighted as follows:

- offers maximum useful power level of >100 000 GW;
- the time to develop this resource is 10 years for an almost endless supply; its only limiting factor appears to be stray microwaves, while producing no pollution;
- after a high upfront investment, the long-term cost trends are downward with robust ROI forecasted;
- the risks in achieving project goals are relatively low.

(Criswell and Waldron, 1990b, 1991)

A decade and a half ago, 4.43 billion inhabitants used 10 300 GW of power (average 2.33 kW) in 1980. The developed nations used most of that energy (6.3 kW/person), while other countries used less (1.0 kW). Today only 10% of the world's population consumes most of the energy, but dramatic population growth in the Third World is producing escalating demands for energy, and so an increase in polluting fossil fuels. For a projected planetary population of 10 billion by the year 2050, the global electric power capacity must

POWER SYSTEM	MAXIMUM OUTPUT (GW*YRS)	MAXIMUM USEFUL POWER LEVEL (GW)	TIME TO EXHAUST (YRS)	LIMITING FACTORS	POLLUTING PRODUCTS	LONG TERM COST TREND	RISKS TO REACH GOAL

BIO-RESOURCES	100*Yrs	1,000	<10	• Mass Handling
				• Nutrients
				• Water
				• Land Use
					• CO_2
					• Biohazards
					  – Methane
					  – Disease
						Up	Not Possible

COAL	<1,500,000	20,000	100	• Supply • Pollution	• CO_2 • Ash Acids • Waste Heat	Up	Not Possible
OIL	<100,000	5,000	<30	• Supply	• CO_2, Acids • Waste Heat	Up	Not Possible

NUCLEAR FUSION	>500*Yrs	500
		
		Base load
			100s	• Accidents and Terrorists
				• Social Acceptance
					• Radioactives
					  – Spent Fuels
					  – Components
					• Waste Heat
						Up	High

NUCLEAR FUSION	>20,000*Yrs	>20,000 Base load	1000s (D-T) <100s (D-He$_3$)	• Engineering Demo • First Wall Life (D-T) • Power Balancing*	• Radioactive Components • Waste Heat	Un-known	High
HYDRO-ELECTRIC	2,000*Yrs	2,000	<1000s	• Sites & Fill-Up • Rainfall • Dam Failure	• Sediment • Flue Water	Up	Not Possible
OCEAN THERMAL	<1,000,000	<20,000	<100 •	Deep Cold Waters	• Power Balancing	Up	Not Possible
TERRESTRIAL SOLAR POWER	>20,000*Yrs	>20,000 Not tied to peak or base load	>10^8	• Clouds • Power Storage • Power Distribution • Power Balancing*	• Waste Heat • Production Wastes	Down	Moderate
SOLAR POWER SATELLITES	2,000*Yrs	2,000 Base load	>10^9	• Orbital Debris • Shadowing	• Sky Lighting • Orbital Debris	Down	Not Possible
LUNAR POWER SYSTEM	>>20,000*Yrs	>100,000 Base & Load following	>10^9	Stray Microwaves	None	Down	Low

* Balancing Local and InterHemispheric Heat Loads

Exhibit 105—Comparison of 21st Century Power Systems. This analysis shows the advantage of setting a goal by 2040 to obtain 20 000 GW of electricity from lunar solar power. Source: David R. Criswell.

© David R. Criswell 6/10/90

be increased tenfold at a time when coal and oil as a source of power may be eliminated. Exhibit 106 shows the power needs of the next century in terms of population by the billions, kilowatts of electrical power per person (at 1000 watts per kW), and total power necessary in gigawatts.

Earthlings now operate within a thermal economy in which 15% of the gross world product is invested in energy sources that are continually in crisis because of declining resources, increasing prices and pollution, terrorism and war. Sometimes, international trade and economic development are hostage to Middle East oil supplies and regional problems. In a July 1990 presentation to the first Lunar Power System coalition workshop,

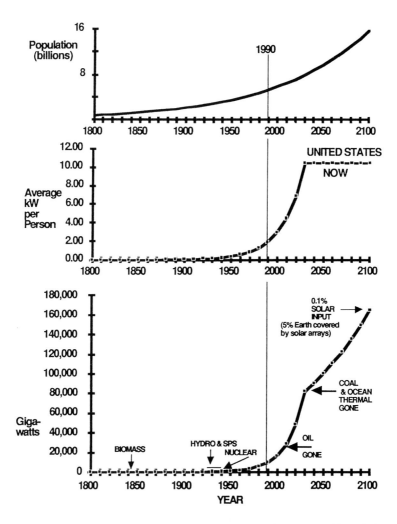

Exhibit 106—21st Century Power Crisis. Further comparison of growing world population and the energy needs of individuals and how these could best be met by lunar solar power. *Source*: David R. Criswell and Robert D. Waldron.

Dr M. N. A. Peterson, professor emeritus of the Scripps Institution of Oceanography, emphasized that most concerns underlying Earth observation and modeled predictions are *directly related to energy needs*, such as:

- atmospheric CO_2 and other greenhouse gases;
- air quality, acid precipitation, and oxidants;
- fossil fuel reserves and associated tensions;
- petrochemical needs and other energy options;
- wastes and disposal;
- stratospheric ozone depletion;
- water resources and river use;
- transportation and food production;
- levels of economic well being and incentives for environmental degradation and population growth

(Bodie, 1990)

Peterson, a former chief scientist for the National Oceanic and Atmospheric Administration, maintained that all these factors can be related to economics and the availability of energy. Then in a paper presented to a national conference on Earth Observation and Global Decision Making (October 24, 1990), this distinguished geologist and oceanographer expanded on this theme that every environmental concern addressed there had a connection to energy, its conversion to useful forms, its distribution, or the efficiency and costs of its use. The late Dr Peterson realized that the proposed LSP program can solve the problems posed in the above situations by providing a major new source of energy that is dependable, sustainable, economical, environmentally enhancing, available worldwide, and capable of growth (Peterson, Criswell, and Greenwood, 1990). When director of Pacific Ocean Policy Institute in Honolulu in 1993, Peterson began examining the connection between the oceans and alternative energy sources, including possible offshore rectennas that might receive lunar power beams.

Preliminary analysis by the system inventors, Criswell and Waldron (1993a, 1990b), indicate a projected annual net revenue from lunar power could be a billion dollars a year, with an internal rate of return on investment exceeding 30% if the power is sold at 0.1$/kWh (Criswell & Waldron, 1990b). Exhibit 107 shows the power build-up to 20 000 GW (GWe) and spread out beyond the year 2050. Only a lunar solar power system can meet those requirements at reasonable cost! The LSP strategy is timely and synergistic because:

(1) it fits within the NASA "Mission to the Planet Earth" program, while offering the space agency reasons for returning to the Moon that will gain public support, and provide some economic justification for further space exploration ...

(2) it coincides with national needs to become less dependent upon foreign petroleum and gas, especially from OPEC, as well as to halt the environmental damage of fossil fuels, especially depletion of the world's forest and ecological reserves ...

(3) it provides for a transfer of defense technology, scientists, engineers, resources, and investment to a "defense of the planet" that will foster leading-edge technology, research, and development ...

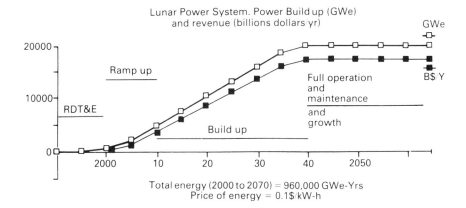

Exhibit 107—Lunar Power System. Power build-upon in GW for the next 50 years with forecasts of anticipated revenues fro an LPS system. *Source*: David R. Criswell and Robert D. Waldron.

(4) similarly, it offers the hard-pressed aerospace industry and high-technology companies a new market for investment, income, and jobs by transfer of technology to lunar power and related technology ...

(5) it presents a new leadership opportunity to the U.S. Department of Energy and the Environmental Protection Agency for solving energy shortage problems with an environmentally safe alternative ...

(6) it builds upon existing research and technology, such as in microwaves and beaming, so as to encourage new applications, even of DOD's Strategic Defense Initiative studies ...

(7) it will not only renew existing industry, such as the corporations in aerospace and satellite communications, but will create wholly new "astrobusinesses" from lunar construction to lunar law ...

(8) it contributes to the development of a **two-planet economy.**

Lunar industries lay the foundation for migration of humanity to space habitats. The needs, time, resources, knowledge, educated people, and growth path now exist to move mankind to more interesting engineered venues off Earth. The next fifty years can see the continued stifling poverty of billions of people, or we can create a wealthy, healthy future aloft.
David R. Criswell, "Net Growth in a Two-Planet Economy," 45th Congress of International Astronautical Federation, Jerusalem, Israel, October 9–14, 1994 (IAA-94-IAA.8.1.704).

Lunar solar power challenges many existing institutions to consider its implications for furthering their objectives—such as:

- the Electric Power Research Institute seeing how LSP could benefit the utility industry;
- the Off-Shore Industry Association considering how ocean platforms could be utilized as rectennas to receive microwave power beams from the Moon;
- the Sandia National Laboratory applying its expertise in terrestrial solar energy to studies of solar power from the Moon or its vicinity.

No wonder that in a letter to one of the authors (Harris—August 10, 1990), Dr Michael B. Duke, then Deputy for Science, NASA Lunar and Mars Exploration Program Office, stated:

> I have a feeling that it is a concept like the Lunar Power System that will get us back to the Moon and into space in a big way.

In an article on "Tomorrow's Energy" (*Parade Magazine*, Nov. 25, 1990, pp. 10–15), Dr Carl Sagan, the noted astronomer, observed that solar energy is a safe, promising solution to the world energy dilemma. In the long term, there are severe limitations on the current means of producing electrical power. Offterrestrial energy options are promising and inexhaustible, whether Solar Power Satellites (SPS) as advocated by Peter Glaser (1977, 1989, 1992) or the Lunar Solar Power (LSP) system described by Criswell and Waldron (1993, 1991, 1990). In the 21st century, helium-3 mined from lunar soil in fusion reactors may also offer a potential power source for use on Earth or in distant parts of the Solar System. With major adverse impacts on the Earth's environment at a minimum, the technological feasibility for exploiting solar energy at lunar facilities is worth pursuing, particularly for its economic potential.

Such observations have recently been confirmed by eminent scientists and engineers in the 1989 task force report on the NASA Lunar Energy Enterprise Case Study, the 1990 proceedings of the Lunar-Based Solar Power Planning Workshop, as well as in the 1991 report of The Synthesis Group.

B.3 LUNAR SOLAR POWER SYSTEM PROPOSAL

The Criswell/Waldron concept as reviewed at numerous scientific gatherings worldwide utilizes solar power for the benefit of humanity and is illustrated in Exhibit 108. A mature system would channel tens of terrawatts of free and pure energy, using proven basic technology with passive and low mass equipment. Exhibit 109 elaborates further on the lunar power details in terms of the major elements involved, their function and advantages, as well as the specific challenges related to each. The transmissions would be at the speed of light over great distances for use where and when needed without physical connections. Essentially, the system involves the use of solar reflectors in lunar orbit to collect sunlight by thin-film photovoltaics which are then converted into thousands of low-intensity microwave beams at power stations on the Moon. From large diameter, shared synthetic apertures on the lunar surface, these microwaves are beamed to receivers located anywhere on the Earth, including economically developing nations. Finally, ground electric power distribution systems complete the process begun aloft in the Solar System.

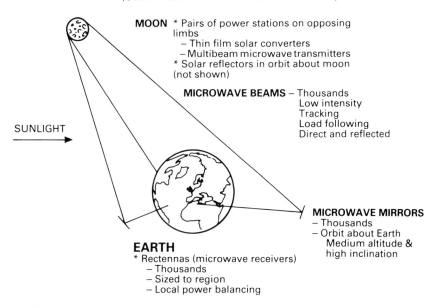

Exhibit 108—Lunar Solar Power Concept. The LSP system is projected in this illustration to a capacity of producing 100 000 GW. *Source*: David R. Criswell and Robert D. Waldron.

Exhibit 110 offers another analysis of an LSP development stages in terms of costs over a 20-year period, starting with a 1990 base. It should be noted that most of the investment is ground based for rectenna construction. A national effort would be necessary which integrates private and public sector contributions to this macroproject, beginning with an adequate government-funded space transportation system to the Moon and back.

Johnson-Freese (1990) examined the changing patterns of international cooperation in space. For a technological venture of this scope to succeed requires the formation of global partnerships in **space-based energy macroprojects**, not only among nations and public sector agencies, but with transnational organizations in the private sector. An energy consortia might be developed along the model that already exists to construct the International Space Station—namely, the U.S.A., Canada, Japan, European Space Agency, and expecially Russia. The Russians have unique capabilities in lunar studies, heavy-lift launch systems and long-duration space living experience. There has been some previous Russian interest in lunar solar power, particularly among the fusion engineering community (Sarkisyan, 1985). However, recent interviews by one of the authors (Harris) with key members of their Academy of Sciences confirmed that although Russian space scientists are fixated on Mars missions, their energy scientists are interested in pursuing studies on lunar solar power. Given the current socioeconomic crises throughout the Commonwealth of Independent States, the Russians would have to be convinced: (a) that their research would be underwritten by external sources; (b) of lunar profits in the future before they would undertake technological venturing to the Moon with other nations.

Major Element	Function	Advantage	Challenges
SUN	Power Source	Functioning and no ash Immense energy supply	Day/Night Cycle
MOON	Platform Resources	Natural resource – Place – Components – Logistics – Good electronic environment Minimal transportation Non-Intrusive	Construction & Logistics – Remoteness – Exotic environment Utilization of resources
* Thin film photovoltaics	Solar to electric power ($\geq$ 5%)	Little material for much power Fast production	High production rate
* Power plot	Collect electric power	Local small area (100 m^2)	Iron gathering & refining Wire making
* Microwave transmitters	Electric to microwave power	Highly redundant Many beams Beam pointing & bore sight control Limit maximum beam intensity	Precision beam forming
* Segmented lunar reflectors	Redirect sunlight	Focused beams Minimal stray energy Low cost per beam Immense beam range Limit maximum beam intensity	Surface shaping Efficient production

Exhibit 109—LSP System: Details and Challenges. The major elements for a lunar solar power system are compared as to function, advantage, and challenges. *Source*: David R. Criswell and Robert D. Waldron (continued next page).

Major Element	Function	Advantage	Challenges
* Industrial base (option)	Make emplacement machinery	Option to minimize transport and start-up cost	Design and low volume diverse production
EARTH RECTENNAS	Microwave to electric power	Output high grade electricity Base and load following Minimal waste heat Multi-use of land	Minimize production costs (LPS cost driver) Power balance
MICROWAVE REFLECTORS	Illuminate rectennas hidden from moon	Very low mass Simple Wide range of orbits	Pointing and control
LUNAR ORBITAL MIRRORS (option)	Illuminate lunar power bases – New moon – Eclipse	Eliminate power storage – on Moon – on Earth Increase output	Materials transport to orbit Manufacturing in lunar orbit Pointing and control
SPACE RECTENNAS (option)	Power facilities Power fast ships	Service inner solar system Very low mass, simple, & rugged	Minor
ORGANIZATION	ANALYSES COMMITMENTS FUND IMPLEMENT	NEW NET WEALTH TO EARTH ELIMINATE POLLUTION EXPANSION BEYOND EARTH ADVANCE SCIENCE	INTERNATIONAL COOPERATION UP-FRONT INVESTMENT

Source: © David R. Criswell, 1990

Exhibit 109 (continued)—LSP System: Details and Challenges. The major elements for a lunar solar power system are compared as to function, advantage, and challenges. *Source*: David R. Criswell and Robert D. Waldron.

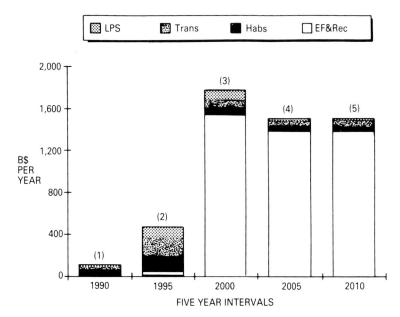

(1) Engineering research, analyses, design and development
        LPS — production processes/options (LPS)
        Transportation elements (Trans)
        Lunar base & space facilities (Habs)
        LPS Earth facilities and rectennas (EF&Rec)
(2) Build and deploy
        Rockets (governments)
        Scientific demonstration lunar base (governments)
        Demonstration production facilities (governments & private)
        Build up rectenna production capacities (private)
(3) Start up and build-up of implacement rate of lunar power capacity
        LPS production expansion (private consortium)
        Rectenna production (private — many terrestrial groups)
        * Start power sales on Earth
(4) Continue build-up of rate installation of lunar power capacity
(5) Steady production of new power capacity (>500 GWe/Year)

Exhibit 110—Lunar Power System: Development Stages. The LSP system co-inventors project its development in five initial stages until the year 2010. Delays in research fundings have already caused slippage in these estimates. *Source*: David R. Criswell and Robert D. Waldron.

To transform this innovative concept into a realistic plan for implementation, readers are encouraged to consider what they might do to further development of a Lunar Solar Power System. Perhaps, such future LSP activities may involve undertaking a research study, or writing a professional paper, or networking with others similarly interested in this macroproject. Co-inventors Criswell and Waldron have outlined some proposed near-term actions:

(1) convening through an appropriate sponsor, a national and/or international action workshop on lunar solar energy ...
(2) additional technology and economic investigations by other investigators of LSP macroeconomic models, and demonstrations of technological possibilities (e.g., mi-

crowave power beaming, orbital reflectors, recetenna options, lunar construction of solar cells and reflectors) ...

3) action planning for large boosters for transportation into orbit, for lunar construction materials, for large-scale habitats and facilities on the Moon.

Exhibit 111 illustrates some of the lunar possibilities.

The first step was taken in July 1990 when a significant planning workshop was held in LaJolla, California. Some twenty-five distinguished scientists, engineers, and other professionals established a Lunar Power Systems Coalition (LPSC) for interested individuals and institutions under the leadership of Dan Greenwood, CEO of NETROLOGIC INC. in San Diego (see Appendix C). Research papers and proposals have been prepared and distributed, while sponsors and investors are sought. For space-based energy to come into being in the next century, support must be obtained beyond government space agencies and energy departments. A **Lunar Economic Development Authority** would facilitate this process.

Prototype lunar power research would seem appropriate for integrated doctoral studies. The need is for multilevel technical analysis, interdisciplinary research that produces SPS and LSP pilot or demonstration projects. For example, the Aerospace Division of the American Society of Civil Engineers conducts annual conferences and publishes proceedings relative to **Engineering, Construction, and Operations in Space**. These multidisciplinary gatherings consider lunar challenges in construction, industrialization, and colonization, as well as solar power to Earth. There are also **Wireless Power and Transmission Conferences** which have been held periodically to develop a strategic international partnership in that regard. In 1995, these sessions were held at Kobe University and the papers were published (e-mail: **kaya@kobe-u.ac.jp**). The SUNSAT Energy Council and its journal, *Space Power*, have organized the **Space Power Network (SPNet)** to promote Internet discussion of future power generation and supply systems (canough@bingvaxa.cc.binghamton, edu). [See Appendix C for addresses of professional development activities cited above.]

B.4 LUNAR POWER BASE TECHNOLOGIES

When Peter Glaser (1977) originated the concept of solar power from space (SPS), it was to come from many satellites in geosynchronous orbit. The NASA Lunar Energy Task Force (1989) confirmed that this is technically feasible if satellites are placed in lunar orbit. While SPS does not require a return to the Moon, economic feasibility may necessitate the use of lunar materials and support services. LSP, on the other hand, envisions the Moon as a natural platform that is unique, suitable and available as a power station. Lunar power will not only site large arrays of solar collectors on its surface, but will utilize its resources as well. It is predicated on establishing a lunar base through a combination of public and private funding. The visionary General Dynamics (1979) study assumed the production of 10 GW of solar power per year requiring 400 humans on the Moon. The Criswell and Waldron (1991) LSP model is based on 560 GW annually or 56 times more lunar power than that earlier analysis. Assuming the same level of automation to human workers, that would project up to 4000 people living and working on the Moon for Millennia III! That

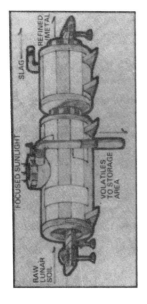

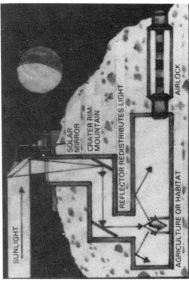

Exhibit 111—Lunar Industrialization Possibilities. In addition to the forecasts described in Chapter 10 and this case, other technical prospects are illustrated here while operating a lunar base. In addition to advanced lunar transportation and habitat systems, liquid propellants tank farms, storage areas for volatiles could be developed and lunar materials processed while solar energy is concentrated by a mirror system used to heat, fuse, and partially vaporize lunar material. Source: Space Resources (NASA SP-509). Washington, DC: U.S. Government Printing Office, 1992 (5 volumes).

implies space colonization on a grand scale, far beyond the outpost and base now conceived by NASA. To build facilities sufficient for that number to live and work safely on the Moon will tax the ingenuity of more than building contractors. Biological and behavioral scientists will have to be involved in providing for life-support systems and human factors planning much beyond traditional industrial engineering (such as was discussed previously in Chapters 2, 7, 8, and 9). Obviously, experts in space communications and high frontier defense technology can contribute much insight on antennas, rectennas, satellites, linking systems, transponders, and other applications for the peaceful and commercial applications of microwave and beaming technologies to lunar power.

In a previous study for a preliminary supply base on the Moon, the senior author (Criswell, 1979) projected a communications and control center within the lunar habitat area. Initially, the needs and deployment of the lunar crews would not be unlike those at scientific outposts in Antarctica or other polar regions (Harrison, Clearwater & McKay, 1991).

B.5 CONCLUSIONS

We have sampled the Moon and possess sufficient understanding to make use of many lunar resources. At the end of the inaugural LPS workshop, our colleague from Rockwell International, Dr Robert Waldron, provided a statement, *A Lunar Power System Coalition Charter,* worth repeating here:

> Global population trends coupled with legitimate aspirations for improved standard of living for current and future inhabitants of Earth will require substantial increase in energy consumption well into the 21st Century.
>
> (Bodie, 1990)

Waldron, now a retired engineer from Rockwell International and president of LPSC, concluded that only a lunar power system can provide that amount of energy in an affordable manner, while avoiding the environmental and ecological damage of fossil fuels. For this conversion to exterrestrial energy sources to occur will take generations of commitment and investment. The time has come for the technical community *to lead* in this change toward greater utilization of lunar resources. No other global investment on the horizon could provide such enormous returns over many decades as LPS. If *space is the place for synergy,* then why not prove it by joint, interdisciplinary studies of space-based energy utilization. If humans are to return to the Moon permanently before year 2010, the 1961 words of President John F. Kennedy bear repetition now:

> We choose to go to the Moon in this decade to organize and measure the best of our energies and skills.

CASE STUDY DISCUSSION QUESTIONS

(1) Why does lunar industrialization, such as a Lunar Solar Power System, provide a rationalization for return to the Moon missions, and help justify further space exploration?

(2)   Why do present terrestrial environmental concerns make study of LSP system feasible as an alternative form of 21st century energy?

(3)   Why do current research developments in automation/robotics, microbeaming, thin-film solar arrays, solid-state microwave electronics, and microcomputers have implications for the successful installation of the LSP system and facilities on the Moon?

(4)   Why is so much of the upfront funding for LSP required on the ground?

(5)   Why would such LSP terrestrial infrastructure investment on Earth be of benefit to the developing economies and their peoples?

(6)   Why are national and international consortia essential to develop, procure, and implement elements for LSP production and rectenna research?

(7)   Why are graduate students and industrial investigators well advised to undertake research in some aspect of developing space-based energy?

(8)   What can be done to convince the executive and legislative branches of the government, the public utilities and energy associations, the general public and media, that support now for an LSP system makes sense and is good public policy?

(9)   Why does the Moon offer a good platform for: (a) scientific observation; (b) energy production; (c) stepping stone into the Universe?

(10)   Would you like to live and work on the Moon? If "yes," why?

Exhibit 112—Spacecraft—In going offworld, there will be an evolution in spacecraft. *Source*: Rockwell International Corporation, Los Angeles, California, U.S.A.

Appendix C
Directory of Space Resources and Publications

INTRODUCTION

Among numerous space-related organizations and periodicals, this is a *select* directory of those which the author hopes may prove useful to scholars, professionals, and activists interested in the industrialization and settlement of the high frontier. The listing is alphabetical by either organizations or publication, and does not normally include commercial space companies.

[Reader feedback for the next edition is welcome on information to be added or corrected in this directory. The author may be contacted by fax: 619/454-4712 or e-mail: **pegb@connectnet.com**—in both cases, mark "Attn. P. R. Harris."]

© 1996 by P. R. Harris

1. ORGANIZATIONS AND PUBLICATIONS

Aerospace Industries Association, 1250 "I" St. NW, Washington, DC 20005, USA (202/371-8400). Industry, trade and public interest association for aerospace manufacturers in the U.S.A. Internal publications, plus a research center which publishes two studies a year (e.g., *Aerospace Facts and Figures, 1995–6*).

Aerospace Industries Association of Canada, 60 Queen St., Ste. 1200, Ottawa, ON K1P 5Y7, Canada (613/232-4297; fax: 232-1142). Comparable to above trade association, but for Canadian aerospace manufacturers. Publishes *Aerospace News*.

Agenzia Spaziale Italiana (ASI), Via de Villa Patrizi. 13, Roma 00161, Italy. Italian space agency within ESA which manages satellite, launch and tether systems, as well as projects like the **Cassini** probe and the **Columbus** space module.

American Astronautical Society, 6352 Rolling Mill Pl., Ste, 102, Springfield, VI 22152, U.S.A. (703/866-0020; fax: 866-3526). Premier scientific and technical society devoted to space activities. Publishes directory, newsletter *Space Times* and *Journal of Astronauti-*

cal Sciences. Sponsors professional conferences with proceedings published as books by UNIVELT INC. (PO Box 28130, San Diego, CA 92128—619/746-4005).

American Defense Preparedness Association, Rosslyn Center # 800, 1700 North Moore St., Arlington, VA 22209, U.S.A. (703/522-1820). Seeks to promote peace through industrial preparedness. Sponsors annual conferences on themes, such as "The Orbital Environment," technical symposiums, and publications.

American Institute of Aeronautics & Astronautics, 1801 Alexander Bell Drive, Suite 500, Reston, VA 22091, U.S.A. (703/264-7500). Largest and leading technical society devoted to science and engineering in aviation and space technology/systems. Special catalog of publications and services available. Aerospace America is their principal general interest magazine, plus many technical journals. Sponsors Technical Information Service, numerous international/national/regional conferences, and local sections. Have student programs and chapters, publishing *AIAA Student Journal.*

American Society of Civil Engineering/Aerospace Division, 345 East 47th St., New York, NY 10017, U.S.A. (1-800/548-ASCE;fax: 212/705-7975); division chair, Dr Stewart W. Johnson, 820 Rio Arriba, SE, Albuquerque, NM 87123, U.S.A. (505/298-2124; e-mail: StWJohnson@aol.com). Conducts annual conference and publishing proceedings on *Engineering, Construction, and Operations in Space*, including lunar construction. Publishes *Journal of Aerospace Engineering.*

Annals of Air and Space Law, Center for Research in Air and Space Law, McGill University, 3690 Peele Street, Montreal, H3A2K6, Canada. A scholarly journal published in conjunction with the Institute of Air and Space Law, which offers a doctoral degree.

ANSER, 1215 Jefferson Davis Highway, Ste. 800, Arlington, VA 22202, U.S.A. (703/416-2000; fax: 416-3389). Independent, non-profit research institute for public service and interest. Has a Center for International Aerospace Cooperation (CIAC) to facilitate information exchange and cooperative programs among nations and their aerospace systems.

Association for the Advancement of Science and Technology (ASCONT), 84/32 Profsoyunaza St., 117810, Moscow, Russia (7 095/333-2445; fax: 7 095/330-1200; e mail: Aberezbi@ESOC1.bitnet). Non-profit organization and clearinghouse for space sector in Russia and Commonwealth of Independent States. Produces *Space Bulletin* on the Russian space program for Gordon & Breach Science Publishers, as well as a directory of leading space organizations in the former U.S.S.R.

Association of Space Explorers, 35 White St., San Francisco, CA 94115, U.S.A. (415/931-0585). For those with spaceflight experience, this international organization holds annual congresses and promotes dialogue/publications toward cooperation in space missions and experimentation. Includes both astronauts and cosmonauts.
[NOTE: for a mailing list of 500 astronauts/cosmonauts, contact Jurgen Peter Esders, Avenue Huart-Hamoir 45, 1030 Brussels, Belgium (fax: +3.2.216.22.59) .... Bernice Scholl, P.O. Box 522579, Marthon Shores, FL 33042, U.S.A.]

Astronomy and Astrophysics Review, Springer-Verlag, 175 Fifth Ave., New York, N.Y. 10010, USA (212/460-1500). Reviews of scientific research on all aspects of astronomy and physics, including particle and cosmic ray. Request catalog of publisher's space books, which include *From Antarctica to Outer Space*.

Aviation Maintenance Foundation International, PO Box, 2826, Redmond, WA 98073, U.S.A. (1-800, 325-2637). In conjunction with the Russian Aeronautical Society and Novosti Press Agency, sponsoring a new Russian/English language magazine, *Samolyot*, on aviation and aerospace in the former U.S.S.R.

Aviation Space Education Association/Foundation, P.O. Box, 12603, Huntsville, AL 35815, U.S.A. (205/883-9922; fax: 205/650-0020). Promotes educational programs and newsletter, *Aerospace Ambassadors*, in conjunction with Citizen Ambassador Program of People to People International (tel: 509/534-0430). Arranges international tours of aerospace programs in other countries, such as **Cosmonaut** launch tours and training programs in Russia. Publishes *Aviation Space Magazine* and distributes *Cosmonautics* books on Russian space program for adults and children.

Aviation Week & Space Technology, McGraw-Hill Publishing, 1221 Avenue of the Americas, New York, NY 10020 (212/512-2000). Leading commercial weekly on the aerospace industry; also produces aerospace video cassettes and mailing lists. Same publisher issues *Aerospace Daily*, a newspaper for the industry (P.O. Box 5890, Cherry Hills, N.J. 08034, USA).

Aviation/Space Writers Association, 17 High St., Suite 1200, Columbus OH 43215, U.S.A. (614/221-1900). Professional society for aerospace writers, communication specialists, and publicists. Sponsors regional and annual conferences at which journalism awards for excellence are presented. Publishes newsletter, annual *Yearbook & Directory*, and special monographs.

Biosphere 2, PO Box 689, Oracle, AZ. 85623, U.S.A. (602/896-2108). A private, closed ecological system and habitat in the field of environmental sciences. This biosystem laboratory now operates in conjunction with Columbia University and its Lamont-Doherty Earth Observatory. Continuing scientific research on environmental systems, global climate change, and sustainable agriculture. Visitor center with hotel and conference facilities on 3.5 acre campus.

Boeing/Peat Marwick Commercial Space Group, Suite 455, 600 Maryland Ave., SW, Washington, DC, U.S.A. (202/479-4240). Under contract from NASA to provide professional support to companies interested in exploring potential benefits of space entrepreneurship. Provides briefings and consultation, technical planning and feasibility studies, design and financial analysis, market research and business planning, networking and linkage to academia, industry, and government.

British Interplanetary Society, The, 27/29 South Lambeth Rd., London, SW8 1SZ, U.K. (01-735-3160). World's oldest space-oriented organization. Publishes space books, as well as popular, *Spaceflight* magazine, *Journal of the BIS*, and *Space Education*. Sponsors resource library, local and international conferences.

British National Space Center (BNSC), 52 Horseferry Rd., London SW1P 1SZ, England. Coordinates United Kingdom space activities, and represents the nation within the ESA.

California Space Institute, UCSD/SIO A-021, 8605 LaJolla Shores Dr., LaJolla, CA 92093 (6l9/534-6381). Cal Space is a statewide research unit which supports and conducts space-related studies on all campuses of the University of California system. Seeks to identify and develop research for the benefit of the State of California citizens and economy, and facilitates communication between university researchers and members of the industrial and public sectors.

Canadian Aeronautics and Space Institute, 130 Slater St. #818. Ottawa, ON K1P 6E2, Canada (613/234-0191; fax: 234-9039). Professional society with annual conferences, proceedings, and publications, including *Canadian Aeronautics and Space Journal*.

Canadian Network for Space Researchers (CSNR), c/o Institute for Space and Terrestrial Science, 4850 Keele St., North York, ON M3J 3KI, Canada. A Centre of Excellence Program for a federally funded network of 16 universities, businesses, and research institutions throughout the nation; headquartered at the University of Calgary.

Canadian Publishing/Mclean Hunter Ltd., 777 Bay St., Toronto, ON M5W 1A7, Canada (416/596-5719). Publishes *Aviation and Aerospace*, *Canada's International Journal* combined with *Aerospace and Defense Technology*, *Aerospace Canada International*, and Aerospace *Canada*.

Canadian Space Agency, Place Air Canada, 500 Rene-Levesque Blvd. West, Montreal, Quebec H2Z 1Z7, Canada (514/496-4204). Manages Canada's civilian space program, including space science and technology projects, as well as astronaut program. Administers agreements with other space agencies, such as NASA, ESA, and NASDA, and promotes space projects that develop key technologies for Canadian industry. Primarily into robotic arm for space shuttle and station, as well as satellites. Publishes magazine, *Apogee*.

Canadian Tours of the Universe, CN Tower, 301 Front St. West, Tortono, Ontario, Canada M5V 2162 (416/363-8687). Provides 45 minute simulated Shuttle flight to spaceport in Jupiter in year 2019. Shows space films.

Centre for Industrial and Technical Development (CDTI), Edificio Cozco, Paseo de la Castellana 141, 28046 Madrid, Spain. Spanish delegation to ESA which facilitates technological developments.

Centre National d'Etudes Spatiales (CNES), 18 Avenue Edouard Belin, 31055 Toules Cedex, France ... . **CNES**, 2 Place Maurice Quentin 75039, Paris, France. The French space agency within ESA that studies the environment from space and manages projects such as, **Ariane, Topex-Poseidon**, and **Hermes**.

Challenger Center for Space Science Education, ll0l King St., Ste.100 Alexandria, VI 22314, U.S.A. (703/683-9740; fax: 683-7546). As a memorial to the late **Challenger Seven** astronauts, their families and friends are promoting nationwide network of affiliate media/educational centers for youth ... Prototype **Challenger Learning Centers** now functioning at: the Houston Museum of Natural Science (713/639-4635); Discovery Museum, Bridgeport, CT.(203/372-3521); Museum of Science & Industry, Tampa, FL (813/985-5531); H. B. Owens Science Center, Lanham-Seabrook, MD (301/577-8718); Kiser Middle School, Dayton, OH (513/224-7092); Mathematics & Science Center, Alexandria, VA (804/343-6525); Edmonton Space and Science Center, Edmonton, Alberta, Canada (403/452-5882); Reuben E. Fleet Space Theatre and Science Center (619/238-8971); Ontario Science Centre, Don Mills, Ontario, Canada (416/429-4100); Science Center of Iowa, Des Moines, Iowa (515/274-3404); Rochester Institute of Technology (716/232-4016); Discovery Place, Charlotte, North Carolina (704/372-6261) ... Publishes *Challenger Log* and special space science instructional materials; conducts interactive teleconferences for school children and workshops for teachers on space subjects.

China Great Wall Industry Corporation (CGWIC), 21 Huangsi Dajie, Xichengi Qu, Beijing 10011, China. Manages commercial space activities of the People's Republic of China, such as **Long March** launch vehicles; exhibits at international space conferences.

Civil Air Patrol/Aerospace Education, National Hdq./EDE, Maxwell AFB, AL 36112, U.S.A. In addition to Air Cadet program for youth and emergency services, provides speakers and instructional materials on aerospace opportunities. Conducts National Congress on Aviation and Space Education.

Committee on Space Research (COSPAR), International Council of Scientific Unions, 5l Boulevard de Montomorency, 750l6 Paris, France. Publishes *Advances in Space Research*, the Official Journal of COSPAR_and *COSPAR Information Bulletin*.

Contact, 1412 Potomac Ave., SE, Washington, DC 20003, U.S.A. (202/544-4984). Annual conference simulation for professionals from various disciplines, which is an action learning experience about making contact with alien cultures. Also has a Solar System Council or SOLSYS exercise to explore possible effects on human societies of expansion into our Solar System. Publishes proceedings and *Contact* newsletter.

Cosmos Center, The, 4200 Wisconsin Ave. NW, Washington, DC 20016, U.S.A. (202/828-1954; fax: 202/363-0958). Designated by Russian space officials as coordinators of travel and tour programs to Russian and CIS facilities, including Star City, NPO Energia, Baikonour, and **Mir** space station. Source of Russian space books and publications, including, *Cosmonautics,* as well as CD-ROMS, videos, slides, pictures, posters, and models.

CSPACE Press, P.O. Box 9331, Grand Rapids, MI 49509, U.S.A. (616/452-5500; fax: 616/452-5538). Publishes two unique space magazines—*Quest*, a quarterly dedicated to the **history** of spaceflight; *Countdown,* a bimonthly devoted to continuous coverage of every shuttle mission since 1983, has been discontinued in 1996.

East View Publications, 3020 Harbor Lane North, Minneapolis, MN 55447, U.S.A. (612/550-0961; fax: 559-2931; e-mail: eastview@skypoint.com). Extensive catalog of Russian aviation and space periodicals, as well as bibliographic materials.

Elsevier Science Ltd, The Boulevard, Lanford Lane, Kidlington, Oxford, OX5 1GB, UK (01865 843000; fax: 843010). Premier science publisher, including aerospace books and journals—e.g., *Acta Astronautica, Journal of the International Academy of Astronautica* in conjunction with Pergamon Press; *Space Policy*, in conjunction with Butter-worth–Heinemann Ltd. (See *Space Communications*.)

Energia USA, 2214 Rock Hill Road, Ste. 500, Herndon, VA, 22070, U.S.A. Represents NPO Energia, the Russian space organization responsible for **Energia** launch vehicle, **Buran** space shuttle, **Soyuz** spacecraft and **Proton** upper stage rocket, and **Mir** space station; organizes symposia in the U.S.A., symposia on these programs.

Euroconsult, 71 Boulevard Richard Lenoir, Paris 75011, FRANCE. Research on space economics and space industry, as well as consultancy for space companies. Some of their studies result in publications of Sevig Press (see separate entry).

European Space Agency, 8–10 Rue Mario Nikis, 75738 Paris Cedex 15 France (3314-273-7155) or U.S.A. office: 955 L'Enfant Plaza SW # 7800, Washington, DC 20024, U.S.A. (202/488-4158). Space agency for the European Community, and primarily into space transportation (**Ariane** rockets and **Hermes** orbiter), satellites, manned labora-tory modules for space shuttle (**Spacelab**) and station (**Columbus**), large orbiting struc-tures, and planetary exploration (**Ulysses** spacecraft). Request to be placed on mailing list of ESA News Releases, *ESA Newsletter, Fact Files, Annual Report*, and other special publications.

Extrasolar Planetary Foundation, The, Observatory Station, Pittsburg, PA. 15214. Scholarly organization devoted to search for and study of other planetary systems, espe-cially planets habitable by humankind. Publishes astronomically oriented magazine, *Cos-mic Search*.

Final Frontier, 1017 South Mountain Rd., Monrovia, CA 91016, USA (1-800-24-LUNAR, 818/932-1033; fax: 818/932-1036). Popular space magazine for national news-stand subscription. Also maintains a **space collectibles** service of books, videos, clothing, etc.

First Millennial Foundation, P.O. Box 81650, Rifle, CO 81650 (1-800/547-1277; fax: 970/625-5052; e-mail: mtsavage@pipeline.com. Organization promoting an eight-step plan to colonize the universe. Blueprint contained in book by Marshall Savage, *The Mil-lennial Project* (Little Brown, New York, 1994). Publishes magazine, *First Foundation News*. Conducts annual conference and has under way the building of an undersea space colony prototype in the Caribbean.

Foundation for Space Business Research, 1 Longfellow Place, Room 721, Boston, MA. 02114, U.S.A. (604/227-5993). Promotes research on space commerce and management, particularly among graduate students, such as at Harvard and Stanford universities. Acts as a funding mechanism for NASA and other government agencies. Provides a contact network for new space entrepreneurs.

German Space Agency (DARA), Konigswinterer Straße 522-524, D-5300, Bonn 3, Germany. Coordinates nation's space program and represents it in ESA.

Gordon and Breach Science Publishers/Harwood Academic, P.O. Box 90, Reading, Berkshire, RG1 8JL, U.K. or 820 Town Center Dr., Langhorne, PA 19047, U.S.A. (215/750-2642; fax: 6343). Publish excellent magazines, such as *Earth Space Review*, Russian *Space Bulletin*, and journals like *Earth Observation and Remote Sensing*, *Remote Sensing Reviews*, as well as books: *Remote Sensing for Geologists*, *Land Observation by Remote Sensing*, *Space Sailings* (solar sails technology), *Space Commerce '90* (proceedings of the Third International Conference on Commercial and Industrial Use of Outer Space). Also publishes the journal, *Space Commerce* (see separate listing below).

Gulf Publishing Space Science Series, 3301 Allen Parkway, Houston, TX 77019 (1-800, 231-6275 or 713/529-4301). Reference books on space technology and history, such as *Almanac of Soviet Manned Spaceflight*, *Space Almanac*, *Space Satellite Handbook*, *Introduction to Space Sciences*. Also a management book series, *Managing Cultural Differences* with content appropriate for multicultural space crews.

High Frontier, Inc., 2800 Shirlington Rd., Ste. 405, Arlington, VA 22206, U.S.A. (703/671-4111; fax: 931-6432). Project strategy for maximizing U.S. space technology in support of space policy/goals that are non-nuclear, technologically feasible, fiscally responsible, militarily sound, and politically practical. Space defense oriented, publishes newsletter, occasional reports, and video cassettes. Sponsor *Journal of Practical Applications in Space*, a quarterly and the **Space Transportation Association** which also issues a newsletter.

Icarus, Academic Press, Harcourt Brace Jovanovich Publishers, 1 East First St., Duluth, MN 55802 (Editor: 607/255-4873). International journal of Solar System studies published in affiliation with the Division of Planetary Science, American Astronomical Society.

Institute on Global Conflict and Cooperation, University of California—San Diego, Q-068, LaJolla, CA 92093 (619/534-3352). A University of California multicampus research unit supporting studies, teaching, conferences, and publications on topics related to international peace and conflict, including outer space. Publishes IGCC newsletter, biennial report, and *Research and Program Summary*.

Institute of Noetic Sciences, 475 Gate Five Road # 300, Sausalito, CA 94965, U.S.A. (415/331-5650). Founded by Apollo astronaut Edgar Mitchell, to encourage research in human consciousness. Publishes *Noetic Science Review*, bulletin, special reports and catalog of resources.

Institute of Space and Astronautical Science, 3-1-1, Yoshinodai, Sagamihara-shi, Kanagawa 229, Japan. Research into rocket and satellite models and panels.

Institute for Space Systems Operations, University of Houston, College of Natural Science & Mathematics, Houston, TX 77204 (713/743-9135: Fax:743-9134). Develops space-related research and academic programs, fosters graduate research opportunities in all phases of space science, engineering, and other fields related to NASA programs. Benefits from University's centers for Space Vacuum Epilaxi and Space Architecture, lunar and planetary geosciences. Director is co-inventor of Lunar Power System, and an administrator for the Texas Space Research Consortium.

Institute for the Social Science Study of Space, Fairmont State College, 1201 Locust Ave., Fairmont, WV 26554, U.S.A. (304/367-4674; fax: 366-4870). **ISSSS** concern is development of space social sciences as an interdisciplinary field. Publishes an occasional newsletter, *Space and Society*, particularly on Internet *Homepage*.

International Academy of Astronautics, BP 1268-16, F-75766, Paris Cedax 16, France (33/1 472382215; fax: 33/1 47238216). Publishes *Acta Astronautica Journal*, plus theme volume proceedings of annual International Astronautical Federation Congress.

International Association of Astronomical Artists, P.O. Box 1584, New York, NY 10011, U.S.A. (818/888-5709, Pierson or D. Davidson, e-mail: d.david@AMNH.org). Global organization of professional artists specializing in aerospace themes and exhibits; promotes exchanges between Russian and U.S.A space artists.

International Astronautical Federation (IAF), 3–5, Rue Mario Nikis, 75015, Paris, France (33 1 4574260). Global technical society that conducts annual conferences around the world in conjunction with national counterparts, then publishes proceedings and reprints of papers presented. Co-sponsor of ISY'92 World Space Congress.

International Lunar Exploration Conference, International Space Enterprise (ISE), 4909 Murphy Canyon Rd., Ste 330, San Diego, CA 92123, U.S.A. (619/637-5770; fax: 637-5776). Annual assembly of lunar development proponents who conduct action workshop and publish proceedings. ISE is conducting private enterprise lunar projects in conjunction with the Russian Lavochkin Association.

International Institute of Space Law, International Astronautical Federation, above address in France (IISL Secretariat fax: 42732120); U.S. Association, c/o Milton Smith, J.D., Sherman & Howard, L.L.C. 90 South Cascade Ave., Colorado Springs, CO 80903, USA (719/475-2440; fax:635-4596). Forum for lawyers, political scientists and others interested in issues of space and astrolaw. Promotes annual colloquia on Law in Outer Space; proceedings published by AIAA.

International Small Satellite Organization, 520 Huntmar Park Dr., Ste. 100A, Herndon, VA 22070 (1-800/636-ISSO; 703/709-9362; fax: 709-0790). Promotes small satellites, launch vehicles and their applications. Publishes *The ISSO Newsletter* six times a year.

International Space University, Communuaté Urbaine de Strasbourg, Blvd. Gonthier d'Andernach, 67400 Illkirch, France (33 88 65 54 45; fax: 33 88 65 54 47; e-mail: admissions@isu.isunet.edu); North American Office: 3400 International Dr., NW, 4M Ste.400, Washington, DC 20008, U.S.A. (1-202/237-1987; fax: 1202 237 8336). With worldwide campus at Strasbourg's Parc d'Innovation, and ISU 23 affiliate universities in 13 countires, this innovative institution focuses upon interdisciplinary, international space education. Residential Master of Studies program is at the main campus, plus ten-week Summer Session (SSP) for graduate degree at affiliate host campuses; produces and publishes an international space design project on a theme. Since 1987, graduate 1000 alumni from 59 nations taught by 450 experts from 25 nations. Publishes newsletter and *The Universe,* a quarterly report, and sponsors a satellite network linking its campuses—ISUnet is part of Internet. Promotes space research and sponsors an annual International Symposium.

Interstellar Propulsion Society, P.O Box 1292, LaJolla, CA 92038, U.S.A. (fax:619/457-9738; e-mail: ips@www.digimark.net). Promotes scientific and engineering advancements in space propulsion, leading to manned missions to other star systems. IPS had a digital library on World Wide Web, a member director, news group, research group, and publishes *IPS Newsletter* and *Interstellar Propulsion Journal.*

Jane's Information Group, l68 Brighton Rd. Coulsdon, Surrey CR5 2NH, U.K. (+44 1-08 763-l030) or l340 Braddock Pl. # 300, Alexandria, VA 22313, U.S.A. (703/683-3700). Among many aviation and space publications, *Interavia Space Directory* is a premier, annual source of information on aerospace firms, executives, organizations, products, and services.

Journal of Aviation/Aerospace Education and Research, 600 Clyde Morris Blvd., Daytona Beach, FL 32114 (904/239-6855). JAAER is a new semi-annual journal which focuses on aviation/aerospace history, educators, curriculum, and instructional innovations; intended for all levels of education.

Journal of Space Law, P.O. Box 308, University, Mississippi 38677, USA. Publishes articles by leading authorities on national and international legal and policy issues related to space activities.

Keidanren, 9-4 Ohtemachi 1-Chome, Chiyoda-Ku, Tokyo 100,
Japan (03/270=1441). This Federation of Economic Organizations publishes annually, *Year in Space—Japan.*

League of New Worlds, P.O. Box 542327, Merrit Island, FL 32954, U.S.A. Promotes R&D in habitats both underseas and in space, and publishes magazine, *New Worlds Exploration.*

Lunar and Planetary Science Conference Proceedings, Cambridge University Press. 32 East 57th St., New York, NY 10022, U.S.A. (1- 800, 872-7423). Proceedings publisher of annual conference for Lunar and Planetary Institute, Houston, Texas, beginning with l9th edition in l988 (see UNRA entry).

Lunar Enteprises Corporation, P.O. Box 4457, Burlingame, CA 94011, U.S.A. (408/996-9210; fax: 996-2125). Project to provide live video footage from the Moon beginning in 1997 via *Lunar Video Orbiter*. The LVO satellite is equipped with cameras using the latest optics technology to beam 24-hour coverage of the Moon's landscape, as well as views of Earth, and nearby asteroids/comets.

Lunar Power System Coalition, c/o NETROLOGIC # 201, 5080 Shoreham Pl., San Diego, CA 92122 (619/587-0970; Fax: 458-1624). A network of multidisciplinary professionals and activists interested in promoting lunar industrialization, especially solar power from the Moon, encourages research on Criswell/Waldron LSP patent (refer to Appendix B) ... . NETROLOGIC INC. applies neural networks to problem solving beyond conventional computers. Utilizing leading-edge technology, this scientific firm engages in Small Business Innovation Research contracts for NASA (robotic path planning, geographical information system, engine performance diagnostics, human machine interaction), as well as for the FAA, ONR, Air Force, Army, Navy, and other government agencies.

Moscow Aviation Institute/Cosmos International Center for Advanced Studies, Box 9, A-80, Moscow, 125080, Russia (7 095 158 58 65; fax: 7 095 229 3237; e-mail: cosmos@sovam.com). Large technical university with leading aerospace studies and research programs, such as College of Cosmonautics. COSMOS is a consortia of Russian space organizations that facilitates international agreements and programs.

National Academy of Public Administration, 1120 G Street, NW, Washington, DC 20005, U.S.A. (202/347-3190) Consultants and researchers for government agencies and officials, the Academy was founded through the inspiration of former NASA administrator James Webb and receives NASA fundings for many of its studies, some of which deal with space management and leadership. Its reports are available for purchase, such as *Encouraging Business Ventures in Space Technologies*.

National Aeronautics and Space Administration *(NASA)*, Headquarters, 300 E St SW, Washington, DC 20546 (202-358-0000). The NASA Library can be reached by telephone (202/358-0168); information on elctronic access, 202/488-2931. The External Relations Office, Code I (202/358-0900; fax: 358-3030), in charge of international technology policy, can provide a directory of Field Centers and special program offices, such as:

- Space Access and Technology (Code X: 202/358-4600);
- Human Resources and Education (Code F);
- Small & Disadvantaged Business Utilization (Code K);
- Space Flight (Code M);
- Aeronautics (Code R);
- Space Science (Code S);
- Life and Microgravity Sciences Applications (Code U);
- Mission to Planet Earth (Code Y)
- Policy & Plans (Code Z).

Also request a list of numerous NASA free space publications, such as—*SBIR Program Solicitation* (Code XC, Tel:202/358-46612/2320; 1-800/547-1181 (Electronic Bulletin Board); e-mail: mbattaglia@hq.nasa.gov)) ...

—**NASA Center for Aerospace Information**. 800 Elkridge Landing Road, Linthicum Heights, MD 21090, U.S.A. (410/859-5300, ext. 241). Coordinates information distribution on NASA reports, technology transfer, and industrial applications; publishes annual *Spinoff* magazine on space technology transfers, and sponsors *NASA Tech Briefs*, space technical magazine, a monthly published by Associated Business Publications (address below; 212/490-3999; fax: 986-7864) ... . *Beyond Low Earth Orbit*, a newsletter published by the Johnson Space Center (713/483-5899, e-mail: jconnolly@snmail.jsc.nasa.gov).

—**NASA Central Operations of Resources for Educators**, 15181 Route 58 South, Oberlain, OH 44074, U.S.A. (216/774-1051, Ext. 293/4). CORE distributes NASA-produced educational materials; request catalog. Also *NASA Technology Today* is a resource magazine for science and mathematics teachers published by Associated Business Publications (317 Madison Ave., Ste, 921, New York, NY 10017, U.S.A.; 212/490-3999; e-mail: ntb_edit@interramp. com.). The latter also publishes, *Russian Tech Briefs*, a bimonthly newsletter for technology transfer, including the Russian Space Agency.

—**NASA Office of Space Access and Technology** (Code X, Washington, DC 20546) also publishes an excellent bimonthly magazine, *Space Technology Innovation.*

—**NASA Field Centers/Facilities**—contact local Public Affairs Office concerning their Information Centers/Museums:

- AMES RESEARCH CENTER, Moffett Field, Ca. 94035 (415/604-0753)
- DRYDEN FLIGHT RESEARCH CENTER, PO Box 273, Edwards, CA. 93523 (805/258-33ll; 3802)
- GODDARD SPACE FLIGHT CENTER, Greenbelt Rd., Greenbelt MD. 20771 (301/286-5810)
- JET PROPULSION LABORATORY, 4800 Oak Grove Dr., Pasadena, CA. 91109 (8l8/354-7006; 2577)
- JOHNSON SPACE CENTER, 2101 NASA Road 1, Houston, TX 77058 (7l3/483-5lll; 0474)
- KENNEDY SPACE CENTER, Cape Canaveral, FL 32899 (407/867-2201; 2544)
- LANGLEY RESEARCH CENTER, 11 Langle Blvd, Hampton, VA 23681 (804/864-6005)
- LEWIS RESEARCH CENTER, 21000 Brookpark Road, Cleveland, OH 44135 (2l6/433-3484)
- MARSHALL SPACE FLIGHT CENTER, Huntsville, AL 35812 (800/633-7280 or 205/544-5418)
- MICHOUD ASSEMBLY FACILITY, New Orleans, LA 70189 (504/257-2601)
- STENNIS SPACE CENTER (NATIONAL SPACE TECHNOLOGY LABORATORIES), SSC, MS 39529 (601/688-3341 or 3757)
- SPACE TELESCOPE SCIENCE INSTITUTE, Baltimore, MD. 21218 (301/338-4757)
- WALLOPS FLIGHT FACILITY, Wallops Island, VA. 23337 (804/824-1579)

—**NASA Commercial Technology Network**, a virtual organization in 13 geographic centers located in universities, research institutes, contracting organizations and NASA Field Centers ... . Call National Technology Transfer Center, 1/800/678-6882, or Federal Laboratory Consortium, 1/800/472-6785 or e-mail: mosaic-x@nasa.uiuc.edu, or consult **homepage**: http:/nctn.oact.hq.nasa.gov. The U.S. regional technology transfer network locations and telephones are: RTTC

- Far West, Los Angeles, CA—213/743-6132; fax: 746-9043;
- Mid-West, Cleveland, OH—216/734-0094; fax: 734-0686;
- North-East, Westborough, MA—508/845-8762; fax: 366-0101;
- Mid-Continent, College Station, TX - 409/845-8762; fax: 845-3559;
- Mid-Atlantic, Pittsburgh, PA—412/648-7000; fax: 648-7003;
- South-East, Alachua, FL—904/462-3913; fax: 462-3898;
- Am Tech, Menlo Park, CA—415/325-5353; fax: 329-0320;
- NASA ARC/TCC, Sunnyvale, CA—408/734-4700; fax: 734-4946;
- NASA JSC/TCC, Houston, TX—713/335-1250; fax: 333-9285.

—**NASA Centers for the Commercial Development of Space (CCDS)**

- CENTER FOR BIOSERVE SPACE TECHNOLOGIES, University of Colorado, Box 429, Boulder, CO 80309 (303/492-3633; fax: 492-8883).
- CENTER FOR COMMERCIAL DEVELOPMENT OF SPACE POWER AND AD-VANCED ELECTRONICS, Auburn University, 231 Leach Center, Auburn, AL 36849 (334/844-5894; fax: 334-5900).
- CENTER FOR MAPPING, Ohio State University, 12l6 Kinnear Road, Columbus, OH 43212 (6l4/292-1600; fax: 292-8062).
- CENTER FOR MACROMOLECULAR CRYSTALLOGRAPHY, University of Alabama, THT-Box 76, UAB Sta., Birmingham, AL 35294 (205/934-5329; fax: 934-0480).
- CENTER FOR SATELLITE & HYBRID COMMUNICATIONS NETWORKS, Institute for Systems Research, University of Maryland, Rm. 3117/Williams Bldg., College Park, MD 20742 (301/405-7000; fax: 314/8586).
- CENTER FOR SPACE COMMUNICATIONS TECHNOLOGY, Florida Atlantic University Research Corporation, 777 Glades Rd., Boca Raton, FL 33431 (407/637-2343; fax: 367-3418).
- CENTER FOR SPACE POWER, Texas A&M University, Wisenbaker Rm. 223, College Station, TX 77543 (409/845-8768; fax: 847-8857).
- CONSORTIUM FOR MATERIALS DEVELOPED IN SPACE, University of Alabama, 4701 University Dr., Huntsville, AL 35899.
- (205/895-6620; fax: 205/895-6791).
- ITD SPACE REMOTE SENSING CENTER, NASA Stennis Space Center, Bldg. 1103, Ste. 118, NASA/SPC, MS 39529 (601/688-2509; fax: 688-2861).
- SPACE VACUUM EPITAXY CENTER, University of Houston, 4800 Calhoun Road, Houston, TX 77204 (713/743-3625; fax: 747-7724).
- WISCONSIN CENTER FOR SPACE AUTOMATION AND ROBOTICS, University of Wisconsin, 2348 Engineering Hall, 1415 Engineering Dr., Madison, WI 53706 (608/262-5526; fax: 262-9458).

—NASA Space Grant College and Consortia
(Includes the leading U.S.A. space studies and research programs in universities and national laboratories.)

- ALABAMA SPACE GRANT CONSORTIUM, University of Alabama, College of Science, Huntsville, AL 35899 (205/895-6538).
- ARIZONA SPACE GRANT COLLEGE CONSORTIUM, University of Arizona, Dept. of Planetary Sciences, Tucson, AZ. 85721 (602/621-6962).
- CALIFORNIA SPACE GRANT CONSORTIUM AND FELLOWSHIP PROGRAM, University of California—San Diego, Dept. of Chemistry B-017. LaJolla, CA 92093. (619/534-2998).
- COLORADO SPACE GRANT CONSORTIUM, University of Colorado, Laboratory of Atmospherics & Space Physics, Box 392, Boulder, CO 80309 (303/492=5300).
- CORNELL SPACE GRANT CONSORTIUM, Cornell University, 318 Space Sciences Bldg., Ithaca, NY 14853 (607/255-8544).
- FLORIDA SPACE GRANT CONSORTIUM, University of Florida, Dept. of Aerospace Engineering, 231 Aerospace Bldg., Gainesville, FL 32611 (604/392-0961).
- GEORGIA INSTITUTE OF TECHNOLOGY SPACE GRANT CONSORTIUM, School of Aerospace Engineering, Atlanta, GA 30332 (404/894-6812)
- UNIVERSITY OF HAWAII AT MANOA CONSORTIUM, Hawaii Institute of Geophysics, 2525 Correa Road, Honolulu, HI 96822 (808/948-6490).
- AEROSPACE ILLINOIS SPACE GRANT CONSORTIUM, Argonne National Laboratory, 9700 South Case Ave., Argonne, IL 60439 (708/972-7357).
- IOWA SPACE GRANT COLLEGE CONSORTIUM, Iowa State University, College of Engineering, 101 Marston Hall, Ames, IA 50011 (515/294-6617).
- THE JOHNS HOPKINS SPACE GRANT CONSORTIUM, Dept. of Physics and Astronomy, Baltimore, MD 21218 (301/338-7350).
- MASSACHUSETTS INSTITUTE OF TECHNOLOGY CONSORTIUM, Dept. of Aeronautics/Astronautics, 37–441, Cambridge, MA 02139 (617/253-0906).
- MICHIGAN SPACE GRANT COLLEGE PROGRAM, University of Michigan, Dept. of Aerospace Engineering, 2508 Patterson Pl., Ann Arbor, MI 48109 (313/764-3344).
- NEW MEXICO SPACE GRANT CONSORTIUM, New Mexico State University, Dept. 3-0, Box 30001, Las Cruces, NM 88003 (505/646-5870).
- OHIO AEROSPACE INSTITUTE CONSORTIUM, 2100 Aerospace Parkway, Brook Park, OH 44142 (216/891-2100).
- THE PENNSYLVANIA STATE UNIVERSITY CONSORTIUM, 114 Kern, University Park, PA 16802 (814/865-2407).
- ROCKY MOUNTAIN SPACE GRANT CONSORTIUM, Utah State University, Rm. 324A, SER Bldg., Logan, UT 84322 (801/750-3554).
- TENNESSEE VALLEY AEROSPACE CONSORTIUM, Vanderbilt University, Dept. of Mech. Engineering, Box 1612, Sta. B, Nashville, TN 37203 (615/322-7062).
- TEXAS SPACE GRANT CONSORTIUM, University of Texas, Center for Space Research, WRW 402, Austin, TX 78712 (512/471-1356).
- VIRGINIA SPACE GRANT CONSORTIUM, University of Virginia, Office of Govt. Relations, Madison Hall, University & Rugby, Charlottesville, VA 22906 (804/924-3337).

- UNIVERSITY OF WASHINGTON CONSORTIUM, College of Arts and Sciences, Seattle, WA 98195 (206/545-3815).

—**NASA Graduate Students Researchers Program** in Aerospace Science and Technology; Life Sciences. Contact NASA Hdq. Educational Affairs, Washington, DC 20546, U.S.A. (202/453-1000).

National Air & Space Museum, Smithsonian Institution, Washington, DC 20560, U.S.A. (202/357-1300). Major exhibits and audio-visual presentations on aviation and space in several locations. Conducts public symposia and publishes magazine, *Air & Space* (370 L'Enfant Promenade, SW 10th Fl., Washington, DC 20024). Request Smithsonian Institution Press catalog which contains space-related volumes (e.g. *Envoys of Mankind*). NASM also provides leadership to a network of local aerospace museums, many of which publish newsletters and have gift shops which sell aerospace books, such as:

- CALIFORNIA MUSEUM OF SCIENCE & INDUSTRY'S AEROSPACE HALL 700 State Dr., Los Angeles, Ca 90037 (213/744-7483).
- CERNAN EARTH & SPACE CENTER, Triton College, 2000 5th Ave. River Grove, IL. 60171 (708/456-5815).
- CHICAGO MUSEUM OF SCIENCE AND INDUSTRY/HENRY CROWN SPACE CENTER, 57th St. & Lakeshore Dr., Chicago, IL 60637 (312/684- 1414).
- FLORENCE AIR AND SPACE MUSEUM, U.S. Highway 301, North Airport Entrance Rd., Florence, SC 29503 (803/665-5118).
- HANGER 9/EDWARD H. WHITE MEMORIAL MUSEUM, Brooks Air Force Base, TX 78235 (512/536-2203).
- KANSAS COSMOPHERE AND SPACE CENTER, 100 N. Plum St., Hutchinson, KS 67501 (316/662-2305).
- MICHIGAN SPACE CENTER, Jackson Community College, 2111 Emmons Road, Jackson, MI 49201 (517/787-0800).
- MUSEUM OF FLIGHT, 9404 E. Marginal Way So., Seattle, WA 98108 (206/767-7373).
- NEIL ARMSTRONG AIR AND SPACE MUSEUM, Interstate Highway 75, Wapakoneta, OH 45895.
- KIRKPATRICK CENTER MUSUEM COMPLEX, 2100 NE 52nd St., Oklahoma, OK 73111 (405/427-5461).
- MANDELL FUTURES CENTER, THE FRANKLIN INSTITUTE, Benjamin Frankin Parkway at 20th St., Philadelphia, PA 19103 (215/448-1208).
- MUSEUM OF FLIGHT, 9404 E. Marginal Way 5, Seattle, WA 98108 (206/764-5720).
- ROSWELL MUSEUM, North Main & 11th Sts., Roswell, NM 88201 (505/622-4700).
- RUBEN H. FLEET SPACE THEATER AND SCIENCE CENTER, Balboa Park, PO Box 33303, San Diego, CA 92103 (619/238-1168).
- SAN DIEGO AEROSPACE MUSEUM, Balboa Park, 2001 Pan American Plaza, San Diego, Ca. 92103 (519/234-8291).
- SPACE CENTER, PO Box 533, Alamorgordo, NM 88311 (505/437-2840).

- U.S. AIR FORCE MUSEUM, Wright Patterson Air Force Base, Dayton, OH 45433 (513/255-3284).
- U.S. NAVAL AVIATION MUSEUM, Naval Air Station, Pensacola, FL 32508 32508 (904/452-3606).
- VIRGIL I. GRISSOM STATE MEMORIAL, Spring Mill State Park, Mitchell, IN 47446 (812/846-4129).

[Also see entry above for CHALLENGER CENTER.]

National Space Club, 655 Fifteenth NW, Suite 300, Washington, DC 20005, U.S.A. (202/639-4210). Membership limited to those affiliated with government, military industrial complex, and public press. Politically oriented toward peaceful advancement of aerospace technologies and utilization of space resources. Publishes newsletter and presents awards/grants to deserving citizens. Conducts annual National Space Outlook conference to provide an overview and information to industry on national space programs to aid in future business planning.

National Space Development Agency of Japan, World Trade Center Bldg. 2-4-1 Hamamatsucho, Minato-ku, Tokyo l05, Japan (03/435- 6111); in U.S.A.: 300 South Grand Ave., Ste. 2760, Los Angeles, CA. 90071 (213/660-2216). NASDA promotes space applications and coordinates space agreements with other national space agencies. Primarily into rockets, space observations and satellites, advanced space communications and broadcasting, planetary exploration; developing special module (JEM) for international space station—a manned multipurpose space experiment laboratory .... A separate R&D entity, **Institute of Space and Astronautical Science** operates under the Ministry of Education and Culture—3-l-l Yunodai, Sagamihara, Kanagawa Pref. 229 , Japan (0427/51-0141).

National Space Society (NSS), 922 Pennsylvania Ave. SE, Washington, DC 20003, U.S.A. (202/543-1900; 1-800/543-1280; e-mail: nsshdq@nss.org). Aimed to promote a spacefaring civilization, beginning with a Return to the Moon permanently. Largest U.S. space activist organization (26 000) with extensive chapter network (112) at the local levels, both nationl and international (request directory—708/758-3537). Publishes newsletter and monthly magazine, *Ad Astra*; electronic newsletter *SpaceViews* (request@world.std.com), **Homepages** (http://**www.global.org/bfreed/nss/nss-home.html**), *Return to the Moon* (http://**www.ari.net/back2moon.html**), and Internet **NSS Space Forum** (log on, "go to", keyword "space" and select "Lifestyles and Interests"). Sponsors annual "Space Development Conference" in month of May, as well as special events such as the International Lunar Exploration Conference. During shuttle missions, provides "Dial-a-Shuttle" information (1-900, 909-NASA). Distributes position papers, audio/video cassettes, and other services. HEADQUARTERS for

—**Spacecause**, non-profit, national lobbying network on behalf of the civilian space movement. Coalition of public, academic, and civilian space activities. Affiliated with Spacepac (listed below under "s"), and public *Voters Guide to the Presidential Candidates* ... .

—**NASA Alumni League** for those former employers and contract workers of NASA, NACA, or JPL who support nation's civil aeronautics and space program. Provides independent communications to those organizations and government on aerospace programs and policies. Provides space program updates, quarterly newsletter, and invitations to space conferences and exhibits; available for expert consultation. Above address or call 202/543-3587; 1/800/343-7793; fax: 202/546-4189.

Orbit Book Company, Krieger Publishing, PO Box 9542, Melbourne, FL 32902 (407/724-9542; fax: 951-3671). Request catalog of extensive offering of titles on space technology, spaceflight, and space environment.

Organization for the Advancement of Space Industrialization and Settlement (OASIS), PO Box 1231, Santa Monica, CA 90278, U.S.A. (310/364-2290). Affiliated with NSS's L5 Chapter of Southern California and the California Space Development Council, it promotes space development regionally through programs, such as Space Week, Careers in Space, seminars for educators, etc.

Pasha Publications Inc. PO Box 9188, Arlington, VA 22219 or 1616 N. Fort Meyers Dr., Ste. 1000. Arlington, VA 22209, U.S.A. (1-800/424-2908; fax: 703/528-3742). Publishes annual *NASA Contracts and Procurements Directory ... Space Business News*, a commercial newspaper of 26 issues per year (1-800/424-2908) ... *Military Space,* a newsletter.

Pergamon Press, Fairview Park, Elmsford NY l0523, U.S.A. (914/592- 7700) or Headington Hill Hall, Oxford 0X3 OBW, U.K. Publishes six major international space periodicals/volumes. In addition to those mentioned above (*Acta Astronautica, Advances in Space Research*, and *COSPAR Information Bulletin*), also publishes *Space Technology, Progress in Aerospace Sciences, Automatic Control in Space (IFAC Symposium Proceedings*). A subsidiary of Elsevier Science Ltd.

Phillips Business Information Inc., PO Box 61110, Potomac, MD 20859, U.S.A. (1-800/777-5006; fax: 301/309-3847). Publishes annually *The World Satellite Directory* and *Via Satellite*, as well as *Satellite News, GPS Report, The Spectrum Report*, and *Mobile Satellite News.*

Planetary Society, The, 65 North Catlina Ave., Pasadena, CA 9ll06, U.S.A. (8l8/793-5700). Promotes realistic continuing program of planetary exploration and the search for extraterrestrial life. Publishes *The Planetary Report,* provides book service on the Solar System, and funds research, as well as public information programs.

Russian Space Agency (RKA), 42 Shchepkina ul., 12987, Moscow, Russia (7 095 288-9905; fax: 7 085 251-8702, 883-5672). Federal institution coordinating all space activities in Russian and Commonwealth Independent Nations.

Sevig Press, 6. rue Bellart, F-75015, Paris, France (+33 1 42 73 28 37; fax: 42 73 20 95). Publishes reference works, *The World Space Industry Survey,_A* ten-year outlook based on their EUROSPACE data base; *European Space Directory; Space Directory of Russia.*

Society of Aerospace Communicators, Inc., PO Box 15748, Chevy Chase, MD 20825, U.S.A. (301/652-3381; fax: 656-5371). Forum for responsible and informed journalists and public relations executives in the world aerospace field. Provides newsletter, directory, and briefings.

Space Access Society, 4855 E. Warner Rd. #24-150, Phoenix, AZ 85044, USA (602/431-9283; e-mail: hvanderbilt@bix.com). Dedicated to promoting affordable acess to space for all. Conducts annual conference and publishes newsletter, *Space Access Update.*

Space Age Publishing Corporation, 75–5751 Kaukini Hgy, Ste. 209, Kailua-Kona, Hawaii 96740, U.S.A. (808/326-2014; fax: 808/326-1825). Useful publications include *Space Fax Daily* (available in paper and electronic editions), *Space Calender* (weekly), and special reports, such as *Pacific Alhoa Spaceport Study.* Conducts periodic Space Enterprise Conference, for example on Pacific Rim Opportunities. Affiliated project, Lunar Enterprises Corporation (see listing under "L").

Space Agency Forum (SAF), contact any national space agency, or ESA Washington Office, 955 L'Enfant Plaza, SW, Ste 7800, DC 20024, U.S.A. (202/488-4158; fax: 488-4930). The SAF Secretariat is managed by IAF in Paris (see above listing under "I"). An outcome of ISY '92, the Forum acts as an informal, coordinating entity for the world's space agencies and affiliated members, such as the International Space University. Provides agency representatives with a mechanism for discussion of common space issues and exchange of information. Usually meets in conjunction with the annual Congress of the International Astronautical Federation.

Space Business Roundtable originated in Houston and currently operates in Dallas, Texas; Washington, D.C.; New York, NY; South Florida; Seattle, WA; Los Angeles, CA. [In Southern California, sponsor is WORLD SPACE FOUNDATION, P.O. Box Y, South Pasadena, CA 91030, U.S.A. (818/357-2878), which also publishes newsletter, *Under the Stars*; promotes special space projects and products, including participation in Earth–Moon Race and the Columbus 500 Sail Cup Race.] Local roundtables provide information and discussion on space business opportunities to business executives, assist in start-up of commercial space technology companies, foster excellence in space education, and accelerate space research to reduce risk and barriers to capital investment in space. (See **The Space Foundation**).

Space Catalog, The, Aerospace Resources International (1514 Vivian Ct., Silver Springs, MD 20902, USA). Carries space books and collectibles from various publishers.

Space Commerce, edited by The Egan Group, 1701 K St. NW, l2th Fl., Washington, DC 20006 (202/755-0720). Published by Gordon and Breach Science/Horwood Academic (P.O. Box 786 Cooper Station, New York, NY 10276, U.S.A.; 1-800/545-8398; fax 212/645-2459—see separate listing above under "G"). This journal is dedicated to the commercial development of space, and provides a forum for the international space business community and academia.

Space Commerce Corporation, 6900 Texas Commerce Tower, Houston, TX 77002 (713/227-9008; FAX:227-9000). Attorneys who represent Russian space interests in North America. Negotiate contracts for commercial users of Russia's space hardware/software and expertise. Also offer an IBM-compatible computer game on **Soviet Buran Pilot,** which allows player to pilot the automated Russian shuttle.

Space Communications, Elsevier Science Publishers, PO Box 1991, 1000 BZ Amsterdam, the Netherlands (Editor: Box 530, Engineering Center, University of Colorado, Boulder, CO 80308, USA). An international journal in the field of space satellite communications and policy.

Space Forum: The International Journal of Space Policy, Science, and Technology for Industrial Applications, Earth Space Institute, 12 Cour Saint-Eloi, 75012, Paris, France. Multidisciplinary journal to promote communications between those in space activities and those who benefit from such participation. Aimed at academic, public, and industrial sectors to discuss issues related to building infrastructure and transportation both on the ground and in space, as well as legal, political, and financial dimensions of practical space applications.

Space Frontier Foundation, Intrepid Sea Air Space Museum, 46th St. at 12th Ave., New York, NY 10036 (212/757-7780); l6 First Ave., Nyack, NY 10960. Grass-roots organizations of American citizens dedicated to the opening of the high frontier for space exploration and settlement, protection of the Earth's biosphere through space technology, and application of free enterprise system to utilization of space resources for the benefit of Earth's peoples.Publishes *Space Front*, a quarterly newsletter.

Space Foundation, The, Houston, TX. National, non-profit organization of business and industrial leaders working in the space industry. Conducts monthly Space Business Roundtables nationally (see entry above), and co-sponsors the annual conference, Space Technology, Commerce & Communications [c/o T. F. Associates, 79 Milk St., Suite 1108, Boston, MA 02l09, U.S.A. (6l7/292-6480).]

Space Information Laboratories, Inc., P.O. Box 5090, Vandenberg AFB, CA 93437, U.S.A. (805/347-3075; fax: 337-3081). Non-profit corporation dedicated to space education and research; conduct **The Endeavor Center** and feature NASA Teacher Resource Center.

Space Network Television, The, TSN Communications, P.O. Box 2778, Rancho Mirage, CA. 92270, U.S.A. (619/770-3474). Cable, 24-hour television coverage of space-related programs, including features, documentaries, and children's shows. Emphasis is upon NASA activities, aerospace industry, space technology, and future projects.

Space News, 6883 Commercial Dr., Springfield, VA 22159, U.S.A. (1-800/368-5713; fax: 703/642-7352). Excellent weekly newspaper of international information related to the space field in general, and the satellite and remote sensing industries in particular.

Space Nursing Society, 3053 Rancho Vista Blvd., Ste 8377, Palmdale, CA 93551 (805/266-2832; fax: 266-2891). Professional society of nurses who are educators, researchers, or practitioners in fields of space medicine and life sciences. Publishes newsletter, *Expanding Horizons*, and SNS is a test site for **Spaceline**, a computerized data base that contains references to life science research related to space.

Spacepac, 3435 Ocean Park Blvd, Suite 210S, Santa Monica, CA 90405 (202/543-1900). Promotes space development through political action, particularly lobbying for legislation. Affiliated with National Space Society, its chapters, and Spacecause (see listing under "N").

Space Policy, Butterworth–Heinemann, Ltd., Linacre House, Jordan Hill, Oxford, OX2 8DP, U.K. (0462/67255; fax: 0462/480947); U.S.A.: 80 Montvale Ave., Stoneham, MA 02180 (617/438-8464; fax: 438-1479). Leading international quarterly journal on space policy—uses, activities, and developments in industrial, economic, legal and social contexts ... . Butterworth-Heinemann also publishes related journals on *Telecommunications Policy* and *Futures*.

Space Policy Institute, The George Washington University, Washington, DC 20052, U.S.A. (202/994-7292; Fax: 994-1639). Academic program and research in space policy, including data base, projects and reports (e.g., *NASA Documentary History of the Space Age* and *Emerging Policy Issues for Long-Duration Human Space Exploration)*. Director is North American editor of journal, *Space Policy*.

Space Publications, P.O. BOX 5752, Bethesda, MD. 20824, U.S.A. (301/718-9603). Publishes *1995–1996 U.S. Space Directory* with information on companies, government agencies, universities, educational institutions, and associations.

Spaceport Systems International, L.P., 3769-C Constellation Road, Lompoc, CA 93436, U.S.A. (805/733-7370; fax: 733-7372; e-mail: calspace.com). Commercial polar launch site at Vandenberg Air Force Base managed by a partnership of California Commercial Spaceport, Inc. and ITT Corporation ... . The **California Spaceport Authority (CSA)** is that state's non-profit entity to promote such commercial programs in the $2.5 billion Federal facility ... **California Spaceport Tours** (1-800/666-6302) arranges for tours of the remodeled infrastructure where the Space Shuttle sometimes lands and where a variety of the nation's ballistic missiles have been and are launched. The base serves as home to the USAF 30th Space Wing (805/734-8232, Ext.6-7363). See listing above for **Space Information** Laboratories.

Spaceport, USA, Kennedy Space Center, FL 32899 (407/442-2121). Operated privately for NASA; provides large visitor complex and exhibits, as well as IMAX Theater. Nearby in Titusville is the U.S. Astronaut Hall of Fame (407/269-6l00) and the U.S. Space Camp (1-800, 63-SPACE), summer training center for students and teachers.

Space Power, Council for Economic and Social Studies, 1133 13th St. NW, Ste 2-C, Washington, DC 20005, U.S.A. (202/371-2700; fax: 371-1523); Editor—phone/fax: 607/785-6499; e-mail: canough@bingvaxa.CC.Binghamton.Edu = **Space Power Net-**

work). Sponsored by the SUNSAT Energy Council, a journal primarily devoted to space-based energy, power-relay satellites, power beaming, etc. International quarterly for discussion and analysis of advanced concepts, initial treatments, and basic research on technical, economic, and societal aspects of: large-scale, space-based solar power, space resource utilization, space manufacturing, space colonization, and other areas related to the development and use of space for the benefit of humanity.

Space & Rocket Center, One Tranquility Center, Huntsville, AL 35807, U.S.A. (1-800, 633-7280). Founded by Werner von Braun, this is the Official Visitor Center for NASA Marshall Flight Center with major exhibits. Sponsors the U.S. Space Camp and Space Academy programs for youth and adults.

Space Studies Institute, PO Box 82, Princeton, NJ 08542, U.S.A. (609/921-0389). Conducts and supports leading-edge research for opening up the resources of space. Publishes book, *The High Frontier* by Gerard O'Neil; *SSI Update*, a quarterly newsletter; distributor for Russian *NPO Energia Guide to Products and Services* and *MIR 1 Space Station: A Technical Overview*. Conducts annual Princeton Conference on Space Manufacturing and publishes proceedings through UNIVELT (see listing under "u"). Among innovative space technology projects, it funds the SSI Lunar Prospector Probe, a private spacecraft with a mission to map the Moon, especially for its mineral and water potential.

Space Technology, Quelle-Wisstech Communications, Co., PO Box 1272, Boston, MA 02254, U.S.A. (617/899-3868). Commercial update newsletter and publications on space experiments, propulsion, imaging, planetary exploration, space station, and commercialization. Also produces monographs, such as *Space Information Sourcebook*; *Space Directory* of manufacturers and suppliers; *Space Physiology & Medicine*; and *Space Resources and Space Settlements.*

Spaceflight Television, PO Box 2119, Eugene, OR 97402, U.S.A. Provides electronic books and videos about space.

Students for the Exploration and Development of Space, MIT Rm. W20, 445-77 Massachusetts Ave., Cambridge, MA 02139 (617/253-8897). Promotes space science educational programs internationally, and actively seeks to educate youth about space opportunities and careers. SEDS has high school, college, and university chapters throughout the world which conduct annual conferences; UKSEDS provides scholarship to two-week summer European Space School: SEDS—USA sponsors space design contests.

Transquest Corporation, P.O. Box 30326, Cleveland, OH 44130 (1-800/929-8953; fax: 216/888-3992; e-mail: rsn@transquest.com). Distributor for Videocosmos Company of Moscow of their English language edition **Novosti Kosmonautiki**. *Russian Space News*, the English edition, is a bi-monthly, 40-page magazine of Russian and foreign developments in space, including daily report from **Mir**.

United Nations/Office of Outer Space Affairs, Rm. F-832, UN Vienna International Centre, P.O. Box 500, Vienna A-1400, Austria (43 1 211-31-4951; fax: 43 1 21-34-558-30). Implements UN space policies and treaties, assists Committee on Peaceful Use of

Outer Space (COPUOS), and issues special publications ... . Other UN space documents are available through Publications Sales Section, United Nations, NY 10017, U.S.A. (Rm. DC2-853/Dept 701—212/963-8302). Titles include: *Maintaining Outer Space for Peaceful Uses* (#E85.III.A.11), *The United National Treaties on Outer Space* (#E84.I.10), and *International Cooperation in the Peaceful Uses of Outer Space* (#94-43650 E 111194).

United Societies in Space, Inc., 6841 South Yosemite, Englewood, CO 80112, U.S.A. (1-800, 895-META; fax: 303/721-1182; e-mail: usis@usa.net). USIS is a non-profit, educational corporation dedicated to fostering legal, financial, and governance systems in space which encourage investment, industrialization, and settlement aloft. Publishes biyearly journal, *Space Governance*, conducts annual conference, co-sponsors with the World Bar Association an annual *Global Space Essay Contest*, as well as a **Homepage** on the Internet. Proposes to become an UN/NGO to facilitate the creation of a **Lunar Economic Development Authority**, and foundation of space **Metanation** on behalf of the "common heritage of humankind" (see Appendix D).

U.S. Department of Commerce/Office of Air & Commercial Space Programs, Rm. 5027, 14TH & Constitution Ave., Washington, DC 20230 (202/482-6125). Formulates government commercial aerospace policy, acts an an industry advocate and does market research, as well as provide information and publications (e.g., *Introduction to Space Commerce; Space Commerce, An Industry Assessment; Commercial Space Ventures*).

U.S. Department of Energy/Office of Technology Policy, Rm. 4639, 600 Independence Ave. SW, Washington, DC 20230. Sets national technology policy on energy, including space-based. DOE has a National Renewable Energy Laboratory which engages in solar energy research. The department also has an Office of Space and Defence Power Systems concerned with nuclear energy (301/903-3321).

U.S. Department of Transportation/Office of Commercial Space Transportation, 400 7th St., SW, Rm. 5415 (S-50), Washington, DC 20590 (202/366-5770). Formulates government policy on space transportation and facilitates private sector access to STS and launch services. OCTS mission is to promote development of a safe and competitive U.S. commercial space transportation. Request 1995 publications, *Office of Commercial Space Transportation, OCST/DOT Strategic Plan* and *Commercial Spacecraft Mission Model Update*.

U.S. Government General Accounting Office, PO Box 6015, Gaitherburg, MD 20877, U.S.A. (202/275-6241). Issues monthly and annual Index of reports and testimonies on general affairs of the Federal government; includes section on Science, Space, and Technology. You may then request free up to 5 copies of reports dealing with all aspects of NASA, the U.S. Space Program, and the U.S. Ballistic Missile Agency.

U.S. Government Printing Office, Superintendent of Documents, Washington, DC 20402 (202/783-3238). Request to be placed on special mailing list for notification about all NASA and space-related publications. Publishes *Space Resources* (NASA SP-509), 5 volumes.

U.S. Space Foundation, 2860 South Circle Dr., Colorado Springs, CO 80906, U.S.A. (719/576-8000; fax: 576-8801). Promotes international education, understanding, and support of space, as well as dialogue on public space policy issues. Publishes *Spacewatch* magazine and sponsors annual National Space Symposium.

UNIVELT, PO Box 28130, San Diego, Ca. 92l28, U.S.A. (619/746-4005), Publisher of extensive titles in space science and technology, largely proceedings from conferences of the American Astronomical Society, the Goddard Memorial Lecture Series, National Space Society's International Space Development Conferences, and the Space Studies Institute. Also distributes space books of other publishers; catalog available upon request.

University Corporation for Atmospheric Research, The (UCAR) PO Box 3000, Boulder, CO. 89397 (303/497-1662). Non-profit consortium of more than 50 U.S. research universities with doctoral programs in atmospheric sciences and related fields. Manages the National Center for Atmospheric Research (NCAR) under sponsorship of the National Science Foundation. Organizes and manages cooperative national and international programs that bring together the resources of the Federal government and the university scientific community. Its UCAR Foundation mission enhances industry competitiveness by academic technology transfer, and uses revenues from these commercial applications to support ongoing research and academic programs. The UCAR Foundation is majority owner of the for-profit External Tank Corporation (ETCO), 1777 Broadway, Ste. 405, Boulder, CO 80302 (303/444-6221).

Universities Space Research Association, HDQ. American Cith Building, 10227 Wisconsin Circle, Columbia, MD 21044, U.S.A. (410/730-2656; fax: 730-3496); DC office: 202/479-2609; fax: 479-2613. Private non-profit corporation organized under the National Academy of Sciences. Consists of a consortium of 62 U.S. universities and research institutions who cooperate with the U.S. Government and other organizations to further space science and technology. USRA develops and operates laboratories and other R&D facilities for this purpose, such as Lunar and Planetary Institute (established by NAS and NASA) 3303 NASA Road One, Houston, TX 77058 (713/486-2172) [Request to be placed on mailing list for *Lunar and Planetary Information Bulletin* and to receive announcements of annual conferences/proceedings.] USRA also manages: the Institute for Computer Applications in Science and Engineering at NASA Langley Research Center in Hampton, Virginia; Research Institute for Advanced Computer Science at NASA Ames Research Center in Moffett Field, California; Division of Space Biomedicine in NASA's Space Biomedical Research Institute at Johnson Space Center, Houston, Texas; Microgravity Science and Applications Program at NASA's Marshall Space Flight Center; Atmospheric Sciences in Washington, DC and Huntsville, Alabama; Space Astronomy Program in Washington, DC; University Advance Design Programs at various NASA centers; Centers for Excellence in Space Data and Information Sciences Program at College Park, Maryland. Request to be placed on mailing list for free newsletter, *USRA Quarterly,* PO Box 391, Boulder, CO 80306, U.S.A. (303/440-9160).

Warren Publishing Inc., 2115 Ward Court NW, Washington, DC 20037 (202/872-9200; Fax 293-3435). Publisher of *Satellite Week*, *Mobile Satellite Reports*, and *Space Commerce Bulletin.*

World Space Foundation, P.O. BOX Y, South Pasadena, CA 91031, U.S.A. Organization devoted to building and testing solar sailing spacecraft. (See **Space Business Roundtable**.)

2.   ELECTRONIC SPACE DIRECTORY *

The Internet provides many space resources, such as **Home Pages. Mosaic** permits you to browse through data organized on the **World Wide Web** ... . E-mail can be used to connect with other space organizations and publications, so some of these addresses on the Information Highway were listed above. But through electronic mail, you can subscribe to space mailing lists, roam data bases and obtain other services related to your space interests. For example:

American Institute of Aeronautics and Astronautics—
　　Customer Service—deustserd@AIAA.org
　　Aerospace Access—dispatch@ebscodoc.com

Astronomy Web Page—
　　http://info.er.usgs.gov/network/science/astronomy/index.html

Educational Tours of the Solar System—
　　http://www.c3.lanl.gov:1331/c3/people/calvin/homepage.html

European Space Agency Data Dissemination Network
　　listserv@esoc.bitnet
　　ESA NEWS—esrin.esa.it (access code, **Telenet**)
Global Network Navigator
　　http://gnn.com/gnn.html

Internet by e-mail—
　　listserv@ubvm.cc.buffalo.edu

NASA Press Release Mailing List, etc.—
　　domo@Hq.NASA.gov. Access to data center, Log in as **nodsi**
　　(National Space Science Data Center):
　　nssdc.gsfc.nasa.gov (access code = **Telnet, Register** ... .
　　Spacelink access to NASA data base of news =
　　spacelink.msfc.nasa.gov (**Telnet, Register**) ...

---

*The author acknowledges assistance in preparing this section to Joe Rudich's "The Electronic Frontier," *Ad Astra*, September/October 1995, pp. 32–36. Helpful *references* are:

Hahn, H. and Stout, R. (1995) *The Internet Yellow Pages*. New York, NY: McGraw-Hill.
Young, M. L. and Levine, J. R. (1995) *The Internet Starter Kit for Dummies*. New York, NY: IDG Books.

Earth View (from Space) =
http://images.jsc.nasa.gov/html/home.htm

Extragalactic Database = ned:ipac.caltech.edu.Login:ned

National Space Society—
nsshq@nss.org
American Online = "NSS"
Home Pages =
http://www.global.org/bfreed/nss/nss-home.html
http://www.ari.net/back2moon.html
NSS Messages/Announcements =
MajorDomo@ARL.net

Nontechnical Issues in Space Exploration—
sci.space.policy

Planetary and Solar System Science—
sci.space.science

Planetary Discussion—
alt.sci.planetary

Politics of Space Programs—
talk.politics.space

Satellite Science Data/Fife—
Log in as **fifeuser** then address:
pldsg3.gfsc.nasa.gov (access **Telenet**)

SEDS (Students for Exploration and Development of Space)—
bit.listserv.sedsnews

SCAN: Selected Current Aerospace Notices—
listserv@sti.nasa.gov
SETI Institute—Extraterrestrial Intelligence
http://www.metrolink.com/seti/SETI.html

Shuttle Mission Guide—
http://www.ksc.nasa.gov/shuttle/missions.html

Space Calendar—Worldwide
file://explorer.arc.nasa.gov/pub/SPACE/FAQ/calendar

Spacecraft Design—
   sci.space.tech
   http://www.ksc.nasa.gov./shuttle/technology/sts-newsref/
    stsref-toc.html

Space Discussion Group—
   sci.space

Space Environment—
   http://satori2.lerc.nasa.gov/
   envet.gsfc.nasa.gov Login:envnet (Password: henniker)

Spaceflight and Astronomy Developments—
   sci.space.news

Space Movie Archive—
   http:/www.univ-rennes.fr/ASTRO/anim-e.htmlt

Starship Design Home Page—Interstellar Travel
   http://128.194.15.32/~dml601a/ssd/sdhp.html

Sustainable Global Systems—Mission Earth project of Society for Computer Simula
   tions International and its *Simulations in Service to Society Newsletter*: major
   domo@scs.org., then type subscribe mission earth ... . To post messages:
   mission-earth@scs.org or mcleod@sdsc.edu

Timely Astronautics Updates—
   clari.tw.sp

United Nations Office of Outer Space Affairs—
   ftp://ecf.hq.eso,org/un/un-homepage.html

United Societies in Space Homepage
   http://www.patagsystem.com (choose "Technology", search USIS
or http://www.dworld.con/USIS

Wiley–Praxis Catalogue—*Expanding the Universe*
   praxis@pavilion.co.uk (C. J. Horwood)

Exhibit 113—Space Exploration and Governance Challenges. To explore other worlds will require **innovative governance systems.** For understanding of the present inadequate arrangements, refer to: Reijnen, B. C. M. *The United Nations Space Treaties Analyzed.* Paris, FR: Editions Frontières, 1992; and Benko, M. and Schrogi, K.-U. (eds), *International Space Law in the Making—Current Issues in the UN Committee on the Peaceful Use of Outer Space,* Vol. 1 of Forum for Air and Space Law. Paris, FR: Editions Frontières, 1993.

Appendix D
The Model Treaty on Jurisdiction in Outer Space

(This is a Model Treaty first proposed in 1992 as Appendix No. 7, *Cosmic Man Coming*, by Declan J. O'Donnell, privately published. It was revised and republished in 1994 in Space Governance Journal, VI., No. 1, June, 1994, pp. 16 and 17. It is amended herein for republication in this form for *Living and Working in Space* in 1996. It will be reviewed, updated, and placed on the agenda for adoption by delegates at the Metanation Convention, Denver, Colorado, August 4, 2000 AD. In final form at that date it will be opened for signature until August 4, 2010 AD.) Refer to Appendix C, United Societies in Space, Inc, to address commentary.

PREAMBLE

WHEREAS, humankind has a shared vision of living and working in outer space and would like to focus that vision into a common plan with social, financial, and political fairness for the benefit of mankind, and,

WHEREAS, The Outer Space Treaty of 1967 (a/k/a the Treaty on Principles Governing the Activities of States in the Exploration and Use of Outer Space, including the Moon and other celestial bodies) was signed by 120 nations and ratified by 93 such nations and is regarded internationally as the consensus Treaty on Principles of Space Policy for purposes of International Law and as the progenitor of the underlying concept of developing space for "the benefit of Mankind," and,

WHEREAS, the Moon Agreement of 1979, also called the Moon Treaty, was adopted by the United Nations and signed by eight nations with the stated concept of dedicating space and all celestial bodies, including the Moon, to the *Common Heritage of Mankind*, without defining that phrase in the body of that Treaty, and,

WHEREAS, the 1982 United Nation's Convention on the Law of the Sea, adopted by the General Assembly of the United Nations in 1982 and signed by most of the developed nations, including the United States of America in 1995, has defined the phrase *The Common Heritage of Mankind*, (CHOM) as follows:

- Neither the area nor its contents can be appropriated by anyone for any purpose,
- All nations must have an active sharing in the management of the area and its contents,
- All benefits reaped from natural resources in the area must be shared by all nations in the sense of being actually distributed to them,
- The area and its contents must be preserved for future generations of mankind, and,
- The area and its contents must be dedicated to peaceful purposes exclusively, and,

WHEREAS, there is a potential tension in space policy between the consensus for space development for *the Benefit of Mankind* in the 1967 Outer Space Treaty and the apparently anti-development statement for *The Common Heritage of Mankind* as used in the 1979 Moon Agreement and the 1982 United Nations Agreement on the Law of the Sea, which should be extended to outer space under principles of maritime and International Law, and,

WHEREAS, this Treaty is intended to reconcile that tension by creating a new regime in space, called the Metanation, under the logo *United Societies in Space*, which can manage the venue *for the benefit of mankind* while administrating same in a manner not inconsistent with the principles of CHOM and be the Trustee for humankind in space, the ultimate beneficiary under both competing policies and principles, and,

WHEREAS, other fundamental principles of Space Policy (such as the absolute right of innocent passage in space, the risk allocation of damages sustained by all nations to the one sponsoring any damaging launch, and the free use and exploration of space for all nations without any discrimination) shall be regulated by the Metanation as part of our heritage in space and representing our interplanetary commons, and,

WHEREAS, the Metanation shall be established by world citizen convention on August 4, 2000 AD., at Denver, Colorado, or at any adjourned meeting thereafter, by Delegates representing six billion people on Earth and in space, with ratification by this Treaty as to fundamental jurisdictional authority not inconsistent with the 1967 Outer Space Treaty and in furtherance of the Common Heritage of Mankind (CHOM), so our extended estate offworld may be developed, managed, and maintained for all nations and all people fairly, and,

WHEREAS, no sentient being shall be denied representation as a citizen of the Metanation, whether living on Earth or in space, and no person shall be denied free access to space, nor innocent passage through space, for peaceful purposes, now, therefore:

TREATY

The States Party hereto agree on the following Articles for the benefit of mankind, consistent with the principles of the United Nations Charter, and in order to advance the Common Heritage of Mankind.

ARTICLE I: Recognition

1.1   United Societies in Space. States Party hereto recognize the Metanation as a free state created by world citizen convention under the name United Societies in Space. It is recognized that jurisdiction in the territory of outer space as defined herein shall be attributed to this entity as asserted by it subject to the provisions of this Treaty.

1.2   Citizenship. All citizens of the world and in space shall be deemed citizens of the Metanation and each individual and private enterprise is accorded free access to outer space and innocent passage through space, subject to reasonable regulation by Metanation and by his or her original or residential state. Each Nation's rights under the 1967 Outer Space Treaty is also recognized.

1.3   Full Faith and Credit. Recognition herein is self-executing as limited herein and States Party hereto acknowledge the entity and its Constitution and its right to exist as a nation in the territory as described below. Full faith and credit are extended to its official acts and its money under international comity.

ARTICLE II: Jurisdiction.

2.1   In Space. United Societies in Space and its official subdivisions of authority shall have and hold the following jurisdiction in outer space.

(a) *Exclusive jurisdiction* and management authority in deep space beyond the Moon and including the side of the Moon away from Earth, and all planetary commons in this galaxy and beyond.

(b) *No jurisdiction* within 200 miles of Earth, consistent with the 1982 UN Convention on the Law of the Sea which provides nations with a 200 mile exclusive economic zone off of their shores and into the seas, (and this is extended upwards into space). It is specifically recited that no principle nor any precedent of Aviation Law shall be altered or amended solely by the adoption of this Treaty.

(c)*Secondary jurisdiction* from the 200 mile low Earth orbit to the Moon, including the near side of the Moon facing towards Earth, subject to all existing satellite treaties and regional organization agreements, UN Conventions and Agreements, and the 1967 Outer Space Treaty and its progeny, subject only to the principle that each individual and private enterprise is accorded free access to space and innocent passage through space, under reasonable regulation by the Metanation.

2.2   Conflicts. Nothing herein shall be deemed to change, alter, amend, or conflict with any Aviation Law where it is rightfully applied to non-outer space activity, provided that the Metanation shall be presumed to operate under space law while flying its vessels to and from Earth.

ARTICLE III: Trusteeship

3.1   Trusteeship. Chapter XI of the United Nations Charter, a Declaration regarding non-self governing territories, is ratified as it declares at Art. 73 that "Members of the United Nations which have or assume responsibilities for the administration of territories whose people have not yet attained a full measure of self-government recognize the principle that the interest of the inhabitants of these territories are paramount, and accept as a sacred trust the obligation to promote to the utmost, within the system of international peace and security established by the U.N. Charter, the well being of the inhabitants of these territories." Outer space beyond the 200 mile low Earth orbit is deemed to be such a territory, the Metanation is deemed to be the Trustee Manager of that territory, and those spacefaring nations and others who sign or accede to this Treaty are settlers who accept as their sacred duty an obligation to promote to the utmost space as a territory.

3.2   Interplanetary Commons. The term *Interplanetary Commons* shall mean and refer to space and space resources, such as the Sun, the Moon, the planets and the asteroids, which are in the Metanation jurisdiction of outer space. The Metanation shall manage our interplanetary commons for the benefit of mankind, as well as for the common heritage of mankind, and may create a synergy in space (rather than a policy tension as to these competing principles) by executing its Trusteeship over Interplanetary Commons to include the following:

(a) Regulate uses of space resources so that there is no appropriation by nations, but, instead, only relevant and material uses for limited times and in appropriate ways under leases and sub-leases,

(b) Include all nations and/or representatives of all nations in the legislature of Metanation so that an active sharing in the management of our Interplanetary Commons is substantially effected,

(c) Require users, lessees, and sub-lessees to actually distribute to the Metanation a fair fee not only for rent, but also for an amount called a Developmental Royalty in lieu of taxation, in order to pay for the orderly development of the venue for all peoples and to be shared equally by all nations in this venue,

(d) Police the premises and clear debris and restore damaged areas so that our Interplanetary Commons, as reasonably developed for users, will be preserved for future generations of mankind, and

(e) Protect the sentient beings residing in outer space and those guests traveling through space from the perils of warfare by keeping the peace in space for all times and in all places in space, and minimally guaranteeing safety and security for 100 years.

3.3   UN Trustee System. In the event that the UN does not sponsor a specific Trusteeship Agreement for Metanation, the States Signatory hereto shall operate like such a system in principle, and as provided from time-to-time by Metanation itself through legislative, judicial and executive governance activity not inconsistent with that UN Trusteeship system. Specifically, there shall be cooperation among all participating nations to effect the following UN principles:

(a) Metanation shall be the sole entity for the administration and supervision of the territory and each nation shall be represented by it in this management. (See Art. 75.)

(b) The objectives of the Metanation Management shall include:

(i) To further international peace and security.

(ii) To promote the political, economic, social, and educational advancement of tenants and settlers and users of space resources and our Interplanetary Commons.

(iii) To encourage recognition of the interdependence of Earth-kind and Space-kind in a setting of respect for human rights by nations and fundamental freedoms for individuals without discrimination as to race, sex, religion, and culture.

(iv) To guarantee equal treatment under all laws of the Metanation for sentient beings in its territories as to governance activity in respect to societal, economic, and free enterprise and commercial matters. (See Art. 76, a, b, c, and d.)

ARTICLE IV: Friendship

4.1   *War Outlawed.* During the 100 year development program from 2001 AD to 2101 AD it shall be illegal to instigate, execute, or maintain any warfare in the territory known as outer space.

4.2   *Friendship.* Each nation signatory hereto shall extend peaceful relations and international friendship to all other nations signatory hereto and will not engage in any acts of war towards each other in space, nor towards Metanation. This covenant will operate independently of trading sanctions, most favored nation statuses, and shall relate only to activities involving outer space and its Phased Development.

4.3   *No Army.* Neither Metanation nor any other entity using space shall maintain any army in that territory. However, the United Nations and Metanation may agree from time to time to employ a peacekeeping force when needed to police the provisions of this Article.

ARTICLE V: International Participation

5.1   *Project.* Metanation shall sponsor the Space Development Phased Project commencing at a time within one decade after its organizational Convention at Mile High Stadium in Denver, Colorado, on August 4, 2000 AD, (i.e. not later than August 4, 2010 AD when this Treaty is closed for signatures). It shall create an offworld estate for humanity with a design capacity for at least one billion people in twelve or more settlements in orbits and on the Moon with a scheduled transport system for carrying goods and people, which system may include facilities on Earth. The budget shall not exceed one quadrillion space dollars.

5.2   *Participation.* Each signatory nation and all acceding nations shall be included in the Space Development Phased Plan (SDPP), for goods and services and for settlers and workers and human resources of every kind that it may nominate for the project, and as may be selected by Metanation. International participation by nations shall be attended by international cooperation among them as a condition to that participation.

5.3   *Project Financing.* The standard for money in space shall be quid pro quo, or fair consideration of this for that. Space money shall be at a value at all times exactly equivalent to what it has purchased for the project during the term provided. Each participating nation shall peg its money value at an appropriate percentage of Metanation money stated value for project purposes. All contributions by nations, individuals, corporations and consortia to the project, whether for services, products, equipment, capital assets or software, shall be compensated by Metanation in space dollars as pegged. Each participating nation shall receive and accept such space dollars as regulated by Metabank, the central bank of the Metanation, with full faith and credit not only of said Metanation space dollars, but, also, of each participating nation's money as so pegged. Similarly, in order to control world inflation otherwise exacerbated by the Project (i.e. the SDPP), all participating nations' central banks shall cooperate with the Metabank on fiscal and monetary policies and their administration under this international regime.

5.4   *State's Rights to Projects.* Nothing herein shall prevent, hinder or delay the national, regional and international space projects of each nation from going forward as financed by those nations, provided that a synergy shall be sought with the Project authorized herein. Metanation shall be authorized to permit, and assist as needed, every peaceful space project of every participating nation whether or not it is tied to SDPP, or totally unrelated. Likewise, all nations shall be enabled to lease space resources from Metanation for money as pegged or for services and products as may be agreed from time-to-time at stated fair values within the budget of Metanation and its SDPP In this regard Metanation will serve as the developer of the venue for the benefit of participating members of the international community, consistent with SDPP purposes.

5.5   *Authorities.* Space Port Authorities and Economic Development Authorities in the space venue shall be hosted by Metanation as the host nation and financial guarantor of their financial instruments to the extent they are approved by the Central Bank of the Metanation, thereby reinsuring the guarantees of other host nations and taking out their debt and equity positions in respect to such authorities. In this regard the following three entities are sponsored by Metanation at a minimum:

(a) The Lunar Port Authority for development of a fleet of single state to orbit vessels (SSTO) and other transport vessels to and from the moon and otherwise.

(b) The Lunar Economic Development Authority for the reasonable and consensus planning and development of the whole Moon as a venue for private enterprise and national research and exploration, and,

(c) The Gerard K. O'Neil Orbital Authority for the eventual development of city ships in orbits about the Moon, Mars, and at Earth geosynchronous and wider orbits for settlement by citizens of all nations without discrimination, and as may be apportioned through reasonable selection procedures with passport and visa standards to be published by Metanation.

ARTICLE VI: International Comity in the Space Venue.

The Metanation shall seek to promote space as a place for synergy. It is authorized and directed by the participants to host space-related activities and maintain municipal services for the usage of all nations at the Moon and elsewhere as needed in our interplanetary commons in order to facilitate space travel by mankind and all nations. Specifically, Metanation may maintain residential facilities for settlers; hospitals; rescue services; schools; social security and retirement funds for workers; insurance and re-insurance agencies; common industrial, commercial and recreational areas; a consensus zoning map; a building code; a licensing and inspection department; a police force; and a site-specific court system relevant to space-related dispute resolution and general jurisdiction over crimes and punishments in outer space for the equal protection of residents and guests.

6.2 Comity with other Principles of Astro Law. Pursuant to this Trusteeship for Metanation the following Amendments are made to the 1967 Outer Space Treaty:

(a)*Article I* is amended to authorize and direct the existence and maintenance of Metanation as Trustee to effect the principles of these treaties for mankind as a beneficiary. In addition, Article I is amended to provide authority in United Societies in Space to regulate in a reasonable manner the exploration and usage of space by nations, free enterprise companies, and individuals. There shall be freedom of scientific investigation in space and a sharing of information with Metanation as Trustee for all nations and residents and guests in space, subject to the right of reasonable inspection, rents and royalties in lieu of taxation, and appropriate environmental regulation.

(b) *Article II* is amended to assert that all nations and all persons shall subordinate their free use of space to the entity *United Societies in Space* as a governance authority and a nation in space exempt from the sanctions otherwise against national appropriations in order to regulate the usage of interplanetary commons more effectively for mankind.

(c)*Article IV* is amended to provide that USIS is authorized and directed to regulate power in space, including solar and chemical and nuclear power, as well as the maintenance of oxygen plants and hydrogen processing plants, and may exclude nuclear weapons and all other weapons of mass destruction and its Courts shall have primary jurisdiction over these and similar municipal matters that will affect the venue and its development for all nations.

(d) *Article V* is amended to provide that all vessels in space shall have minimal safety features and conform with common standards for docking and entry for rescues. Every vessel shall have an affirmative duty to assist others in distress, but Metanation shall maintain specialized facilities for such assistance. Astronauts shall be deemed envoys of mankind. Article V and its consequent convention on rescue and return of astronauts shall be amended to permit the return of rescued astronauts to an

USIS Port of safe harbor, hospital, or rescue facility pending final return to a home port at the expense of the rescued astronauts and their sponsor nation, or home port.

(e) *Article VI* is amended to provide that residents and guests in space undertake personal responsibility for their contracts, negligence, torts and criminal activities, as well as cause indirect liability for their principals such as the nation, company or other individual so authorizing and directing them in space. This liability includes responsibility for adhering to all reasonable regulations and development codes of the Metanation.

(f) *Article VII* is amended to provide that all nations maintain exclusive jurisdiction over their vessels and agents in space subject to the common law and regulations of Metanation, such as municipal codes and traffic laws and the like, and any conflicts in law shall be resolved so that nation laws apply inside the nation's vessel and the conflicting Metanation law will apply outside of the vessel and as to common usage of our interplanetary commons.

(g) *Leasehold Estates* in space shall not exceed 99 years and shall not exceed in scope that which reasonably relates to the purpose or business of the tenant. A rental fee for the use and a royalty fee based on commercial values may be assessed in lieu of taxation. Metanation shall have no authority to assess any individual tax in respect to income or otherwise, but shall assess a use fee on all space resources and space territory transactions as may be associated with a leased premise. The regulatory scheme is centered around assessing commercial centers of activity rather than personal taxation.

(h) *That quid pro quo* is the primary standard for monetary value in space for the Metabank and for its participating national central banks, provided that reasonable regulations may be made on a common and fair basis. Worldwide inflation, deflation and fiscal policies are authorized and directed for the benefit of mankind during the SDPP and until 2010 AD.

ARTICLE VII: Applicability.

The provisions of this Treaty on jurisdiction in Space shall apply to the states signatory hereto and their activities in and about space and it shall apply to such other states which are deemed to accede to its terms in policy and practice and to the General Assembly of the United Nations, all subject to the following provisions:

(a) That the prime sponsor, the United States of America, and the Co-sponsors, the European Community, Russia, Japan, China, Canada, Mexico, Australia, India and others, shall sign and ratify this treaty during or before the year 2010 AD.

(b) That the Trust Agreement for Administration and Supervision under Article 75 *et seq.*, Chapter VII, UN Charter, shall not diminish the sovereign nature and wide-scope of authority granted herein.

(c) That the United States shall be the sole depository Nation and the United Nations shall be the sole Agency for administration of signatures.

(d) That English shall be the official language and the official measuring system shall be the metric system.

(e) That effectiveness by signature of the sponsors and co-sponsors shall be required by the close of the year 2001 AD, pending final ratification permitted no later than 2010 AD.

(f) That nations may accede hereto by formal agreement and the consent of the United Societies only.

(g) That this Treaty shall be registered by the depository, Co-sponsors, and accessory nations pursuant to Article 102 of the Charter of the United Nations.

ARTICLE VIII: Amendments.

No amendments shall be made to this treaty without the prior consent of United Societies in Space. Withdrawal or expulsion of a nation after signature, except United Societies in Space, may be made as an amendment hereto with majority approval of nations signatory hereto.

THIS BOOK'S LAST WORDS

European astronaut Ulf Merbold's recalling his first spaceflight in October 1983 when he was a payload specialist on the Shuttle carrying the ESA **Spacelab**:

That is something which I cannot express—it's incredibly spectacular because the Earth is such a beautiful planet. There are northern lights, sunrises, and sunsets in every orbit, the mountains and the oceans ... .

It's almost a shock to realize how thin the atmosphere is, how fragile it is. You truly can see the pollution over the industrial areas. You realize that Earth is a spaceship for 5 billion passengers who have to travel together. It almost makes you obliged to protect the environment for those who will be born after us!

—*ESA Newsletter,* March 1991, #2, p. 2

FINIS

References

PROLOGUE

1. Criswell, D. R. (1981) Human Roles in Future Space Operation, *Acta Astronautica.* vol. 8: 9–10, pp. 1161–1171.
2. Gibson, E. (ed.) (1995) *The Greatest Adventure: Stories and Photographs by Men and Women who have Flown in Space.* Emeryville, CA: Pierson/Publishers Group West (4065 Hollis, zip code 94608, U.S.A.).
3. Isakowitz, S. J. (ed.) (1995) *International Reference Guide to Space Launch Systems,* second edition. Washington, DC: American Institute of Aeronautics and Astronautics.
4. Cheston, T. S., Chafer, C. M., and Chafer, S. R. (eds) (1984) *Social Sciences and Space Exploration* (NASA SP-192) ... Connors, M. M., Harrison, A. A., and Akins, F. R. (1985) *Living Aloft—Human Requirements for Extended Spaceflight* (NASA SP-483). Washington, DC: U.S. Government Printing Office ... Harrison, A. A., Clearwater, Y. A., and McKay, C. P. (eds) (1991) *From Antarctica to Outer Space—Life in Isolation and Confinement.* New York: Springer-Verlag.
5. National Commission on Space (1986) *Pioneering the Space Frontier.* New York, NY: Bantam Books.
6. Chaikan, A. (1994) *A Man on the Moon: The Voyages of the Apollo Astronauts.* New York, NY: Penguin Books.
7. McKay, M. F., McKay, D. S., and Duke, M. B. (eds) (1992) *Space Resources* (NASA SP509). Washington, DC; U.S. Government Printing Office, 5 vols. (Note volume 4, *Social concerns.*)
8. Wiley-Praxis (1995) *Expanding the Universe,* a catalog of titles on astronomy & astrophysics, space science & technology, remote sensing, and atmospheric physics. Available from John Wiley & Sons, Inc. (Baffins Lane, Chichester, West Sussex, PO19 1UD, U.K.; tel: 1-800/979-4539; fax: 1-202.320-2300).

CHAPTER 1

1. Matsunaga, S. (January 21, 1985) An Introduction Statement for the U.S. Senate Joint Resolution 46 Relating to NASA and Cooperative Mars Exploration.
2. McLucas, J. and Myerson, H. (1988) "Planning the International Space Year," *Space Policy*. Vol. 4:2, pp. 108–111.
3. The ISY World Space Congress Abstracts (1992), a 716 page volume of summaries of 2700 technical papers presented, is available from the America Institute of Aeronautics and Astronautics (1801 Alexander Bell Av., Ste. 500, Reston, VA 22091, U.S.A.). Since the WSC was also the site for the 43rd Congress of the International Astronautical Federation (Aug. 28–Sept. 5, 1992), copies of the numerous papers and proceedings may be obtained from IAF Headquarters (3–5, Rue Mario-Nikis, 75015, Paris, France).
4. Boorstein, D. "Realms of Discovery Old and New," August 31, 1992, Inaugural Address to World Space Congress (IAF-92-1000) ... Goldin, D. S. "The Light of a New Age," address to the Association of Space Explorers, August 24, 1992.
5. Walker, C. "President's Message," *Ad Astra*, the Magazine of the National Space Society, May/June 1995, page 2.
6. Reich, R. (1987) *Tales of a New America*. New York, NY: Times Books.
7. McKay, M. F., McKay, D. S., and Duke, M. B., eds (1992) *Space Resources*. Washington, DC: U.S. Government Printing Office (NASA SP-509, 6 vols).
8. For these four cases, the author acknowledges information from these sources: Ojalehto, G. D. and Hertzfeld, H. R. (1995) "International Beat—Nations Aim High Despite Falling Budgets," July, pp. 4–7, 40–41; Ojalehto, G. D. and Vondrak, R. R. (1991) "National Scene—A Look at the Growing Civil Space Club," February, pp.12–16, *Aerospace America*, journal of the American Association of Aeronautics and Astronautics (1801 Alexander Bell Av., Ste. 500, Reston, VA 22091, U.S.A.).
9. National Commission on Space (1986) *Pioneering the Space Frontier*. New York, NY: Bantam Books ... . Ride, S. (ed) (1987) *NASA Leadership and America's Future in Space*. Peoria, Il.: *Aviation Week* and *Space Technology*.
10. Advisory Committee on Future US Space Programs (1990, December) *Report of the Advisory Committee on the Future of the U.S. Space Program*. Washington, DC: U.S. Government Printing Office ... Air War College-Center for Aerospace Doctrine, Research and Education (1990, April). *Building a Consensus Toward Space*. Washington, DC: U.S. Government Printing Office ... National Space Council (1990) *Report to the President*. Washington, DC: U.S. Government Printing Office.
11. Michaud, M. A. (1987) *Reaching for the High Frontier: the American Pro-space Movement, 1972–1984*. New York, NY: Praeger.
12. *Presidential Commission on Space Shuttle Challenger Report* (1986) Washington, DC: U.S. Government Printing Office.
13. Simon, M. C. (1987) *Keeping the Dream Alive: Putting American and NASA Back into Space*. San Diego, CA: Earth–Space Operations Press (International Space Enterprises, 4909 Murphy Canyon Rd., Ste. 330, San Diego, CA 92123, U.S.A.) ... . Hurt, H. (1988) *For All Mankind*. New York, NY: Atlantic Monthly Press.
14. Synthesis Group (T. P. Stafford, Chairman) (1991) *America at the Thresh-*

old—America's Space Exploration Initiative. Washington, DC: U.S. Government Printing Office.

15. National Space Society's Return to Moon and Stay issue (1995), "NSS Hosts Space Policy Seminar on Capitol Hill," *Ad Astra, May/June*, pp. 6–9 ... Twelve position papers are available upon request from the NSS (922 Pennsylvania Ave. SE, Washington, DC 20003, U.S.A.).

16. Harris, P. R. (1993) "Space Policy for the new US Administration—a White House Conference on Space Enterprise," *Space Policy*, May, Vol. 9:2, pp. 82–84.

17. Logsdon, J. M. (1990) "The Decision to Go to the Moon: The Pathway to Moon and Mars Leads Straight through Capitol Hill," *Ad Astra*, December ... Bormanis, A. and Logsdon, J. M. (eds) (1992) Emerging Policy Issues for Long-duration Human Space Exploration. Washington, DC (20052): Space Policy Institute/The George Washington University.

18. Logsdon, J. M. (1970) *The Decision to Go to The Moon: Project Apollo and the National Interest.* Cambridge, MA: MIT Press ... (1986) "The Space Shuttle Program: A Policy Failure?," *Scientific American*, May 30.

19. For information on the organization and its publications, contact Canadian Aeronautics and Space Institute (130 Slater St., Suite 818, Ottawa, ON K1P 6E2, Canada—tel: (613) 234–0191; fax: 234–9039).

20. Newkirk, D. (1990) *Almanac of Soviet Manned Space Flight.* Houston, TX: Gulf Publishing ... . Harvey, B. (1988) *Race into Space—the Soviet Space Programme.* Chichester, UK: Ellis Horwood.

21. Harris, P. R. and Moran, R. T. (1996) *Managing Cultural Differences.* Houston, TX: Gulf Publishing, 4th Edition (refer to Chapter 13, pp.332–340).

22. *Space Bulletin*, published by Gordon & Breach Science Publishers (PO Box 90, Reading, Berkshire RG1 8JL, U.K. Editorial office at ASKONT (84/32 Profsoyuznay St, 117, Moscow, Russia; fax: {7 095} 330-1200) ... . *Russian Aerospace Market* and *Space School* Learning System, Technoex-Conversion Research Centre (197342 C. Herepoypr,a/r 170, St. Petersburg, Russia; fax: {812} 275-87-21) ... . For information on *Samolyot*, contact Aviation Maintenance Foundation International (PO Box 2826, Redmond, WA 98073, U.S.A.; tel: 1-800, 325-2634).

23. 1993 Maryland–Moscow Conference Proceedings available from University of Maryland, College of Behavioral and Social Science (Dr Warren R. Phillips, Chairman, LeFrak Hall # 2182, College Park, MD 20742, U.S.A.; tel: 301/405-4157; fax: 405-4137) ... . The 1993 Altai Conference Proceedings, from Dr Sergei A. Zhukov, President, Moscow Space Club—tel: (095) 291-3901; fax: 324-7101.

24. To keep up on developments in the space program of the Russian Federation, we recommend the annual *Space Directory of Russia* ... . The same publisher also produces a *European Space Directory*, our next topic—both available from Sevic Press (6 rue Bellart, F-75015, Paris, France; tel: +33 {1} 42 73 28 37; fax: 42 73 20 95).

25. Lytkin, V. Finney, B., and Aleptko, L. (1995) "The Planets are Occupied by Living Beings: Tsiolkovsky, Russian Cosmism, and ETI," *Quarterly Journal of the Royal Astronautical Society* ... . Dudenkov, V.N. (1992) *Russian Cosmism: Philosophy of Hope and Salvation* (in Russian). St. Petersburg, RF: Sintez Publishers ... . Brand, P. (ed.) (1984) *Cosmos: The USSR in Space.* Victoria: Michael Edgley International

Pty. (Available from Astratech, PO Box 351, Mulgrave North, Victoria 370, Australia ... . For additional titles on Soviet space history, see Clarke, P. (1988) *The Soviet Manned Space Program*. London, U.K.: Salamander Books ... . Johnson, N. L. *The Soviet Space Programme*, this annual issued through Teledyne Brown Engineering (1250 Academy Loop, Colorado Springs, CO 80910, U.S.A.). Oberg, J. "Russian Space Program: Running on Empty," IEEE Spectrum, December 1995, pp. 18–34.

26. Freeman, M. (1994) *How We Got to the Moon—the Story of the German Space Pioneers*. Washington, DC: 21st Century Science & Technology (PO Box 16285, zip code 20041).

27. Curien, H. and Geiss, J. (co-chairmen) (1994), *International Lunar Workshop*. Paris, France: European Space Agency (ESA SP-1170).

28. United Nations Office for Outer Space Affairs (PO Box 500, A 1400, Vienna, Austria; Tel: 43 1 21131-4951; Fax: 43 1 232156). Ask for list of publications available, such as *UN Report of the Committee on the Peaceful Uses of Outer Space* (1994, Supplement No. 20 - A/49/20).

29. Kingwell, J. (1994) "Viewpoint—The Rise of the Asian Space Dragon," *Space Policy*, Vol. 10:3, August, pp.185–6.

30. Lunar Energy Park Study Group (1995), "A Proposed Plan for a Lunar-based Power Plant—Space Technology Utilized for Global Energy Supply in the Future," *Science and Technology in Japan*, Vol. 14:53, April, pp. 54–59.

31. Kong, D. "A Chinese Perspective on Space Development" (1995), *Space Governance*, Vol.2:1, June, pp.18–20. The journal of United Societies in Space, Inc. (6841 South Yosemite #3-C, Englewood, CO 80112, USA) ... . For further information, contact the Chinese Academy of Space Technology (PO Box 2417, Beijing, PRC); China Great Wall Industry Corporation (PO Box 847, Beijing, PRC; fax: 8311808).

32. Tefft, S. (1988) "The Chariot of Indra," *Air & Space*, April May, pp.33–42 ... . (1986) "India—The Way Forward—Space Centres," *Spaceflight*, December, Vol. 28:12, pp. 431-436 ... . Rao, R. (1986) "Asia into Space," *L5News*, January, pp.10–12.

33. White, F. (1987) *The Overview Effect: Space Exploration and Human Evolution*. New York, NY: Houghton Mifflin.

34. Robinson, G. S. and White, J. M. (1986) *Envoys of Mankind: Declaration of First Principles for the Governance of Space Societies*. Washington, DC: Smithsonian Institution Press.

CHAPTER 2

1. Astronauts/Lattimer, D. (1988) *All We Did was Fly to the Moon*. Gainesville, FL: The Whispering Eagle Press ... . Hurt, H. (1988) *For All Mankind*. New York, NY: Atlantic Monthly Press ... . STS/Apollo 11 (1994) *Apollo 11—First Lunar Landing—25th Anniversary Special Edition*. (Available from STS/Apollo 11, PO Box 751387, Memphis, TN 38175, U.S.A.)

2. Finney, B. K. and Jones, E. M. (eds) (1985) *Interstellar Migration and the Human Experience*. Berkeley, CA: University of California Press.

3. Salk, J. (1972) *Man Unfolding*. New York, NY: Harper & Row ... . (1973) *The Survival of the Wisest*. New York, NY: Harper & Row ... . (1983) *Anatomy of Reality*. New York, NY: Columbia University Press.

4. Harris, P. R. (1996) *The New Work Culture and HRD Transformational Management*, Amherst, MA: Human Resource Development Press ... with R. T. Moran and W. G. Stripp (1993) *Developing the Global Organization*. Houston, TX: Gulf Publishing.

5. Tanner, N. P. in Finney and Jones, op. cit.

6. Von Braun, W., Ordway, F. I., and Dooling, D. (1985) *Space Travel—A History*. New York, NY: Harper & Row ... . Burdett, G. L., Hearth, D. and Soffen, G. A. (1987) *The Human Quest in Space*. San Diego, CA: UNIVELT, AAS Vol. 65 ... . Cosmos Books (1994) *Cosmonautics* (Available from Cosmos Books, 106-38ID, 4200 Washington, DC 20016, U.S.A.; Tel 1-800/819-8051 ... . Dudenkov, V.N. (1992) *Russian Cosmism: Philosophy of Hope and Salvation* (in Russian). St. Petersburg, RF: Sintez Publishing ... Freeman, M. "Krafft Ehricke's Extraterrestrial Imperative," *Space Governance* Journal, Vol. 2:2, Dec. 1995, pp. 20–23.

7. As an example of such books printed in English within the U.S.A., consider the following: O'Neill, G.K. (1989) *The High Frontier*. Princeton, NJ: Space Studies Institute and (1981) *2081—A Hopeful View of Human Future*, New York, NY: Touchstone/Simon & Shuster ... Heppenheimer, T.A. (1979) *Toward Distant Suns* and (1977) *Colonies in Space*, Harrisburgh, PA: Stackpole Books ... Stine, H. (1985) *Handbook for Space Colonists*. Holt, Rinehart & Winston ... Oberg, J. E. (1981) *New Earths: Restructuring Other Planets*. Harrisburg, PA: Stackpole Books; Oberg, A. R. (1985) *Spacefarers of the 80s and 90s: The Next Thousand People in Space*. New York, NY: Columbia University Press; Oberg, J. E. and Oberg, A. R. (1986) *Living on the Next Frontier: Pioneering in Space*. New York, NY: McGraw-Hill ... Savage, M. T. (1994) *The Millennial Project—Colonizing the Galaxy in 8 Easy Steps*. New York, NY: Little Brown Publishers, 2nd Edition.

8. Hardy, D. A. (1994) "Picture This—the Best Interstellar Probe Might Just be an Airbrush," *Ad Astra*, March/April, pp. 29–34; (1990) *Visions of Space—Artists Journey Through the Cosmos*. New York, NY: Gallery Books ... . For information about the International Association for Astronomical Arts, see Appendix C or contact David A. Hardy (99 Southam Road, Hall Green, Birmingham B28 OAB, U.K.).

9. Ordway, F. I. and Liebermann, R. (1992) *Blueprint for Space—Science Fiction to Science Fact*. Washington, DC: Smithsonian Institution Press ... Gump, D. P. (1990) *Space Enterprise Beyond NASA*. New York, NY: Praeger ... Goldman, N. G (1985) *Space Commerce: Free Enterprise on the High Frontier*. Cambridge, MA: Ballinger ... Soffen, G. A. (ed.) (1987) *Visions of Tomorrow: A Focus on National Transportation Issues*. San Diego, CA: UNIVELT, AAS. Vol. 69.

10. Lakeoff, S. and York, H. F. (1989) *A Shield in Space—Technology, Politics, and Strategic Defense Initiative*. Berkley, CA: University of California Press ... Bowman, R. M. (1985) *Star Wars: Defense or Death Star?* Available from the Institute for Space and Security (5115 S. AIA Hgy., Melbourne Beach, FL 32951, U.S.A.) ... National Defense University Press (1986/4) *America's Plans for Space*—Symposium Proceedings. Washington, DC: U.S. Government Printing Office.

11. Butler, G. V. (ed.) (1990) *The 21st Century in Space*. San Diego, CA: UNIVELT, AAS Vol. 70 ... Graham, D. O. (1994) *Space Policy 2000—Double Prime*. Available from High Frontier Inc. (2800 Shirlington Rd, Ste. 405, Arlington, VA 22206, U.S.A.).

12. Contact the National Space Society (922 Pennsylvania Ave, SE, Washington, DC 20003-2140, U.S.A.) for copies of their Position Papers on such subjects as Space Development, Space Research, Space Commercialization, Space Station, Space Materials, Return to Moon/Mars, Outer Space and The Global Environment, Making Orbit, Space Settlement, etc.

13. NASA Advisory Council (1987) Report of the Task Force on NASA Space Goals ... NASA (1986) *America's Plans for Space* (S/N 008-020-01068-5) and *Planetary Exploration Through Year 2000* (S/N 033-000-00987-9) ... An example of a NASA special publication is: Ride, S. (1987) *Leadership and America's Future in Space*, which became available through the Superintendent of Documents ... Request a list of past and present reports and special publications from the data base of NASA's Center for Aerospace Information, Linthicum, MD (1/301/621-0390, Ext. 391), or the Public Information Office, NASA Headquarters, Washington, DC 20546 (1-202/358-1600), or the U.S. Government Printing Office, Washington, DC 20402, U.S.A.).

14. As of 1995, the United States Space Foundation has sponsored eleven symposia with published reports averaging between 200 and 300 pages ... In that same year, the National Space Society was holding its fourteenth International Space Development Conference. Information on the proceedings' volumes for these organizations, as well as the American Astronautical Society, may be obtained from UNIVELT, Incorporated (PO Box 281340, San Diego, CA 92198, U.S.A.).

15. Meyers, T. (ed.) (1996) *The Case for Mars IV* ... . Stocker, C. (ed.) (1989) *The Case for Mars III*, AAS Vols 74 & 75 ... . McKay, C. (ed.) (1985) *The Case for Mars II*, AAS Vol. 62 ... . Boston, P. J. *The Case for Mars I*. San Diego, CA (92198): UNIVELT, Incorporated (PO Box 28130).

16. Clark, P. (1988) *The Soviets Manned Space Program*. New York, NY: Salamander/Crown ... Johnson, N. L. (1987) *Soviet Space Programs, 1980–1985*, Vol. 66; (1980) *Handbook of Soviet Manned Spaceflight,* Vol. 48; (1979) *Handbook of Soviet Lunar and Planetary Exploration*. San Diego, CA: UNIVELT Inc.

17. Sarkisyan, S.A., *et al.* (1985) "Socio-Economic Benefits Connected with the Use of Space Power and Energy Systems," (Paper presented at International Astronautical Federation Congress, 1985, and now available in proceedings published by Pergamon Press, New York).

18. Chinese Society of Astronautics (1987) *Proceedings of the Pacific Basin International Symposium on Advances in Space Science Technology and its Applications.* San Diego, CA: UNIVELT Inc ... Chiu, H. H. (ed.) (1994) *Proceedings of the Pacific International Conference on Aerospace Science and Technology*. Tainan, Taiwan: Institute of Aeronautics and Astronautics, National Cheng Kung University.

19. United Societies in Space, Inc. (6841 South Yosemite, Englewood, CO 80112, U.S.A.). Its journal, *Space Governance*, published the two winning essays: Von Puttkamer, J. (1994) "Space Humanization: Always a Mission to Planet Earth," Vol.

1:1, June, pp. 18–27; Foulkes, R. A. (1994) "Why Space? An Anthropologist's Response," Vol. 1:2, pp. 22–31.

20. Toffler, A. (1992) "The Space Program's Impact on Society," in Korn, P. (ed.) *Human and Machines in Space: The Payoff*. San Diego, CA: UNIVELT/American Astronautical Society, pp. 77–106.

21. Harris, P. R. (1989) "Human Dimensions in Space Policy," *Space Policy*, May, Vol. 5:2, pp. 147–154; "The Influence of Culture on Space Development," *Behavioral Science*, Jan., Vol. 31:1, pp. 12–28.

22. Woodcock, G. R. (1991) *Space Exploration*, Vol. 1: *Mission Engineering*; Vol. 2: *Systems Engineering and Design*. Melbourne, FL: Orbit Book/Krieger Publishing ... . Hoard, M. (1985) *Handbook of Space Technology: Status and Projections*. Boca Raton, FL: CRC Press.

23. Surkov, Y. *Exploration of Terrestrial Planets from Spacecraft*. Chichester, UK: Ellis Horwood ... Baker, D. (1981) *The History of Manned Space Flight*. New York, NY: Crown Publishers ... . American Astronautical Society (1977–1994) *History of Rocketry and Astronautics*, 15 Volumes. San Diego, CA: UNIVELT Inc.

24. *Life Support and Habitability Series*—Sulzman, F. M. and Genin, A. M. (eds) (1994) *Space Biology and Medicine*, Vol. II ... Rummel, J. D., Kpte; Nikov, V. A. and Ivanov, M. V. (eds) (1993) *Space Biology and Medicine*, Vol. I. Washington, DC: American Institute of Aeronautics and Astronautics ... NASA Life Sciences Division (1988) *Exploring the Universe—A Strategy for Space Life Sciences*: (1987) *Advanced Missions with Humans in Space*. Washington, DC: U.S. Printing Office.

25. Tascone, T. F. (1987) *Introduction to the Space Environment*; (1991) *Space Environmental Hazards*. Melbourne, FL: Orbit Book/Krieger Publishing.

26. Cheston, T, S., Chafer, C. M., and Chafer, S. B. (eds) (1984) *Social Sciences and Space Exploration*. Washington, DC: U.S. Government Printing Office, NASA-SP 192 ... Anadejiona, P., Goldman, N. C. and Meeks, P. J. (eds) (1984) *Space and Society: Challenges and Choices*. San Diego, CA: UNIVELT Inc., AAS Vol. 59.

27. Grigoriev, A. I., Kozerenko, O. P., Myasnikov, V. I., and Egorov, A. D. (1986) "Ethical Problems between Ground-Based Personnel and Orbital Station Crewmembers." Moscow: Institute of Biomedical Problems, Paper presented at the XXXVII Congress of International Astronautical Federation, Innsbruck, Austria, October 4–11, 1986 ... Connors, M. M. (1991) "Communication Issues of Spaceflight" in Harrison, A. A., Clearwater, Y. A., and McKay, C. P. (eds) *From Antarctica to Outer Space*, Ch. 24, pp. 267–280.

28. Lorr, D. B., Garshnek, V., and Cadoux, C. (eds) (1989) *Working in Orbit and Beyond: The Challenge for Space Medicine*. Melbourne, FL: Orbit Books/Krieger Publishing.

29. Ballyn, B. (1987) *Voyagers of the West: A Passage in the Peopling of America on the Eve of Revolution*. New York, NY: Alfred Knopf.

30. Piradov, A. S. (1988) "Creating a World Space Organization," Space Policy, May, Vol. 4:2, pp. 112–114 ... O'Donnell, D. J. (1994) "Overcoming Barriers to Space Travel," Space Policy, November, Vol. 10:4, pp. 252–255.

31. O'Donnell, D. J. and Harris, P. R. (1995) "Strategies for Lunar Development and Port Authority," *Space Governance*, June, Vol. 2:1, pp. 6–13 ... Bingham, J.,

Gilbreath, W., and O'Leary, B. (eds) (1979) *Space Resources and Space Settlements*. Washington, DC: U.S. Government Printing Office, NASA SP-428.

32. Harris, P. R. (1992) "The Influence of Culture on Space Development," *Social Concerns,* Vol. 4, pp. 189–217, in McKay, M. F., McKay D. S. and Duke, M. B. (eds) *Space Resources*. Washington, DC: U.S. Government Printing Office, NASA SP-509, 6 volumes.

33. Robinson, G. S. and White, H. M. (1986) Envoys of Mankind—*A Declaration of First Principles for the Governance of Space Societies*. Washington, DC: Smithsonian Institution Press ... . See also *Space Governance*, the journal of United Societies in Space, Inc. (6841 South Yosemite, Englewood, CO 80112, U.S.A.; tel: 1/800-895-META; fax: 1-303/721-1182).

34. International Astronautical Federation (1994) *36th Colloquium on the Law of Outer Space*. Washington, DC: American Institute of Aeronautics and Astronautics ... . Annual proceedings are published of the International Institute of Space Law, IAF, 3-5 Rue Mario Nikis, Paris, France.

35. Refer to *Space Communications*, an international journal published by Elsevier Science Publishers (PO Box 1991, 1000 BZ Amsterdam, the Netherlands, or Journal Information Center, 655 Avenue of the Americas, New York, NY 10010, U.S.A.).

36. Johnson, S. W. and Johnson, M. A. (1990) "The Civil Engineer and Space." (Unpublished paper available from the authors at 820 Rio Arriba SE, Albuquerque, NM 87123, U.S.A.).

37. Harris, P. R. (1996) *The New Work Culture and HRD Transformational Management*. Amherst, MA: Human Resource Development Press.

38. Moran, R. T. and Harris, P. R. (1982) *Managing Cultural Synergy*. Houston, TX: Gulf Publishing.

39. Johnson-Freese, J. (1990) *Changing Patterns of International Cooperation in Space*. Malabar, FL: Orbit Books ... Paine, T. (ed.) (1991) *Leaving the Cradle: Human Exploration in Space in the 21st Century*. San Diego, CA: UNIVELT, Goddard Memorial Symposium, Vol.78.

40. The two 1993 reports cited were: *Partners in Space—International Cooperation in Space: Strategies for a New Century* available from U.S.-CREST (1840 Wilson Blvd., Ste 204, Arlington, VA 22201, U.S.A.); *International Space Cooperation—Learning from the Past, Planning for the Future*, from AIAA (370 L'Enfant Promenade SW, Washington, DC 20024, U.S.A.).

41. Clarke, A. C. (1986) *July 20, 2019—Life in the 21st Century*. New York, NY: Omni Books/Macmillan (ch. 2/7); (1987) "Peacesat," *Spaceflight*, May, Vol. 29:5.

42. Harvey, B. (1988) *Race into Space—the Soviet Space Programme*. Chichester: Ellis Horwood, UK; (1996) *The New Russian Space Programme: From Competition to Collaboration*. Chichester, UK: Wiley–Praxis.

43. McDougall, W. A. (1985) *The Heavens and the Earth—A Political History of the Space Age*. New York, NY: Basic Books.

44. Cummings, D. (1995) "Editorial" in the Spring issue of *USRA Quarterly,* available upon request from the publisher—Universities Space Research Association (PO Box 391, Boulder, CO 80306, U.S.A.).

45. Jones, E. M., Quigg, P. W., and Gabrynowicz, J.I. (eds) (1985) *The Space Settlement Papers*. Los Alamos, NM: Los Alamos National Laboratory (LA-UR-85-3874).

CHAPTER 3

1. Robinson, G. S. (1995) "Natural Law and a Declaration of Humankind Interdependence," *Space Governance*, Part 1, June, Vol. 2:1, pp. 14–17; Part 2, Vol. 2:2, pp. 32–35. (Journal of United Societies in Space, Inc., 6841 South Yosemite, Englewood, CO 80112, U.S.A.).
2. Bullock, A. and Stallybrass, O. (eds) (1977) *The Harper Dictionary of Modern Thought*. New York, NY: Harper & Row.
3. Miller, J. G. (1956) "Editorial, Behavioral Science, A New Journal." *Behavioral Science*, Vol. 1:1, January, pp.1–5. (Published by the International Society for Systems Science, PO Box 40, Fallbrook, CA 92088, U.S.A.) ... . See also by same author (1978), *Living Systems*. New York, NY: McGraw-Hill, pp. XV-XVI.
4. Bonting, S. L. (1991, Vol. 1; 1992, Vol. 2) *Advances in Space Biology and Medicine*. Greenwich, CT: JAI Press (PO Box 1678, zip code 06836, U.S.A.) ... .Austrian Society for Aerospace Medicine (eds) (1992) *Health from Space*. New York, NY: Springer-Verlag.
5. Connors, M. M., Harrison, A. A., and Akins, F. R. (1985) *Living Aloft—Human Requirements for Extended Spaceflight*. Washington, DC: U.S. Government Printing Office (NASA SP-483).
6. Pitts, J. A. (1985) *The Human Factors: Biomedicine in the Manned Space Program to 1980*. Washington, DC: U.S. Government Printing Office (NASA SP-4213).
7. AIAA five-volume series on Space Biology and Medicine (1993/1994/1995): *Space and Its Exploration*, vol. I; *Life Support and Habitability*, vol. II; *Humans in Spaceflight*, vol. III (2 books); *Crew Health, Performance, and Safety*, vol. IV; *Reference Material*, vol. V. Available with catalog from the American Institute of Aeronautics and Astronautics (PO Box 753, Waldorf, MD 20604, U.S.A.; FAX: 301/843-0159).
8. National Research Council/Space Science Board (1987). *A Strategy for Space Biology and Medical Science for the 1980s and the 1990s*. Washington, DC: National Academy Press ... See also NASA Life Science Working Group (1987) *Advance Missions with Humans in Space*, January; Life Sciences Division (1987) *Humans in Space*, April. Washington, DC: NASA HDQ. Office of Space Science and Applications/Life Sciences Division.
9. Niccogossian, A. E., Huntoon, C. L., and Pool, S. L. (1989) *Space Physiology and Medicine*. Philadelphia, PA: Lea & Febiger.
10. To access existing NASA/RECON database, contact the Center for Aerospace Information (800 Elkridge Landing Road, Linthicum Heights, MD 21090, U.S.A.; Tel: 301/621-0l00; Fax: 301/621-0134; E Mail: eroundtree@sti.nasa.gov. Inquire about categories such as: space habitats, stations, spacecraft environments, law, human behavior, adaptation, colonies, industrialization, bases, commercialization.

11. Bluth, B. J. and Helppie, M. (1986) *Soviet Space Stations as Analogs.* Washington, DC: NASA Headquarters (NAGW-869) ... . Harris, P. R. (1992) "The Influence of Culture on Space Developments" in *Social Concerns*, Vol. 4, pp. 189–219 of Space Resources, edited by McKay, M. F., McKay, D. S., and Duke, M. B. Washington, DC: U.S. Government Printing Office (NASA SP-509).

12. Harrison, M. H. (1987) "Space Station: Opportunities for Life Sciences," *Journal of the British Interplanetary Society*, March Vol. 40. pp. 117–124.

13. Cheston, T. S. (1986) *The Human Factors in Outer Space Production.* Boulder, CO: Westview Press ... Connors, M. M. (1989) "Human Aspects of Mission Safey," in *The Case for Mars III.* C. Stoker (ed.) San Diego, CA: UNIVELT, Vol. 74, ch. 7, pp. 203–214 ... Lorr, D. B., Garshnek, V, and Cadoux, C. *Working in Orbit and Beyond—The Challenges of Space Medicine.* San Diego, CA: UNIVELT, Vol. 72 ... Hansson, P. A. (1989) "Work on Human Adaptation to Long-Term Space Flight in the U.K.," in *The Case for Mars III.* C. Stoker (ed.) San Diego, CA: UNIVELT, Vol. 75, pp. 151–160 ... Wilford, J. N. (1990) *Mars Beckons: The Mysteries, the Challenges, The Expectations of our Next Great Adventure in Space.* New York: Alfred A. Knopf.

14. Harrison, A. A., Clearwater, Y. A., and McKay, C. P. (eds) (1991) *From Antarctic to Outer Space—Life in Isolation and Confinement.* New York, NY: Springer Verlag ... Rummel, J. D., Kotelnikov, V. A., and Ivanov, M. V. (eds) (1993–95) *Space and Its Exploration.* Washington, DC: AIAA Publications, 5 volumes, including the Russian experience ... Allen, J. (1991) *Biosphere 2: The Human Experiment.* New York, NY: Penguin Books ... Osland, J. S. (1995) *The Adventure of Working Abroad—Hero Tales from the Global Frontier.* San Francisco, CA: Jossey-Bass Publishing.

15. Brady, J. V. (1990) "Toward Applied Behavior Analysis of Life Aloft," *Behavioral Science* January, Vol. 35:1, pp. 11–13 ... (1988) *Human Behavior in Space Environments: A Research Agenda.* Baltimore, MD: Johns Hopkins University School of Medicine.

16. Angelo, J. and Easterwood, G. W. (1991) *Lunar Base Concepts, Technologies, and Applications* ... Finney, B. R. (1985) "Lunar Bases: Learning to Live in Space," in *Lunar Bases and Space Activities of the 21st Century.* W. W. Mendell (ed.) Houston, TX: Lunar and Planetary Institute, pp. 731–756 ... Funaro, J. J. (1989) "Anthropologists as Designers of Offworld Colonies," in *The Case for Mars III.* C. R. Stoker (ed.). San Diego, CA: UNIVELT, Vol. 75, pp. 201–216.

17. Connors, M. M. and Harrison, A.A. (1994) "Crew Systems: Integrating Human as Technical Subsystems for the Exploration of Space," *Behavioral Science*, Vol. 39:3, pp. 183–212 ... . Harrison, A. A. & Connors, M. M. (1985) "Psychological and Interpersonal Adaptations to Mars Missions," in *The Case for Mars II* C. McKay (ed.) San Diego, CA: UNIVELT, Vol. 62, pp. 643–654.

18. Finney, B. R. and Jones, E. M. (eds) (1985) *Interstellar Migration and the Human Experience*, Berkely, CA: University of California Press ... Finney, B. R. (1987) "Anthropology and the Humanization of Space," *Acta Astronautica.* Oxford, U.K.: Pergamon Press, Vol. 15, pp. 189–194.

19. Bullock, A. and Stallybrass, O. (eds) (1977) *The Harper Dictionary of Modern*

Thought. New York, NY: Harper & Row.

20. Harrison, A. A. (1988) "Beyond Earthnocentrism: Anthropology on the High Frontier," *Space Power*, Vol. 7:3/4, pp. 345–352.

21. Cheston, T. S. (1984) "Orbital Human Factors," in *Social Sciences and Space Exploration*, Cheston, T. S., Chafer, C. M., and Chafer, C. B. (eds) Washington, DC: U.S, Government Printing Office, NASA EP-192, pp. 65–71; 113–114; 121.

22. Maruyama, M. and Harkins, A. (1975) *Cultures Beyond Earth.* New York, NY: Random House.

23. Harrison, A. A. and Elms, A. C. (1990) "Psychology and the Search for Extraterrestrial Intelligence," *Behavioral Science,* July, Vol. 35:3, pp. 207–218. [Dr Harris has a book on this subject to be published by McGraw-Hill in 1996.]

24. *Contact—Cultures of the Imagination Proceedings* have been published annually since 1985. For information on their purchase, direct inquiries to Prof. James J. Funaro, Cabrillo College, Aptos, CA 95004, U.S.A.

25. Foulkes, R. A. (1994) "Why Space? An Anthropologist's Response," *Space Governance*, Journal of United Societies in Space Inc., December, Vol. 1:2,pp. 22–27.

26. Finney, B. R. (1992) "Space Migrations: Anthropology and the Humanization of Space," *Social Concerns*, Vol. 4, pp. 164–188, in *Space Resources.* McKay, M. F., McKay, D. S., and Duke, M. B. (eds) Washington, DC: U.S. Government Printing Office, NASA SP-509 ... (1988) "Will Space Change Humanity," in *Frontiers and Space Conquest* Schneider, J. and Leger-Orine, M. (eds) Hingham, MA: Kluwer Academic Press, pp. 155–172.

27. Raby, N. (1984) "Creating Space Culture: Issues for Consideration," a paper circulated at the NASA Summer Study at the California Space Institute and incorporated into the proceedings report, *Space Resources*, Vol. 4, cited previously in endnote #24.

28. Atkinson, J. B. and Shafritz, J. M. (1985) *The Real Stuff—a History of NASA's Astronaut Recruitment Program.* New York, NY: Praeger ... . Johnson, N. (ed.)(1993) *Cosmonautics—a Colorful History, Vol. 1; Facts & Figures, Vol. 2.* Washington, DC: Cosmos Books (4200 Wisconsin Ave. NW, Ste. 106-381).

29. Helmreich, R. L. (1983) "Applying Psychology to Outer Space: Unfilled Promises Revisited," *American Psychologist*, Vol. 38, pp.445–450 ... with Wilhem, J. A. and Runge, T. E. (1983) "Psychological Factors in Future Space Missions." in *Human Factors in Outer Space Production.* Cheston, T. S. and Winters, D. L. (eds). Boulder, CO: Westview Press.

30. Harrison, A. A. and Summit, J. (1991) "How Third Force Psychology Might View Humans in Space," *Space Power.* Vol. 10:2, pp. 185–203.

31. Connors, M. M., Harrison, A. A., and Akins, F. R. (1985) *Living Aloft—Human Requirements for Extended Spaceflight.* Washington, DC: U.S. Government Printing Office, (NASA SP-483).

32. Cheston, T. S. (1986) "Space and Society," in *Beyond Spaceship Earth*, Hargrove, E. C. (ed.) San Francisco, CA: Sierra Book Club, pp. 20–46.

33. Christensen, J. M. and Talbot, J. M. (1986) "Psychological Aspects of Space Flight," *Aviation, Space and Environmental Medicine*, March, pp. 203–212 (NASA Contract # 3924).

34. Kanas, N. (1987) "Psychological and Personal Issues in Space," *The American Journal of Psychiatry*, June, Vol. 144:6, pp. 703–709.

35. White, F. (1987) *The Overview Effect: Space Exploration and Human Evolution.* New York, NY: Houghton Mifflin ... Harris, P. R. and Moran, R. T. (1996) "Managing Relocations and Transitions," Ch. 7 in *Managing Cultural Differences*, 4th Edition. Houston, TX: Gulf Publishing/Book Division.

36. Oberg, J. E. and Oberg, A. R. (1986) *Pioneering Space: Living on the Next Frontier.* New York, NY: McGraw-Hill.

37. Harrison, A. A., Sommer, R., Struthers, M. J., and Hoyt, K. (1986) "Privacy and Space Habitat Design," Davis, CA: University of California, Psychology Dept (NASA Grant. Report NAG-237).

38. Foley, T. (1995) "The Space Frontier Mentality," *Ad Astra*, July/August, pp. 53–56.

39. Carr, G. (1981) "Comments of a Skylab Veteran," *The Futurist*, Vol. 15, p. 38.

40. Bluth, B. J. (1984) "Sociology and Space Development," in *Social Science and Space Exploration.* Washington, DC: U.S. Government Printing Office, NAS EP-192, pp. 72–79 ... . "The Human Use of Outer Space," *Society*, Jan–Feb., pp. 31–36.

41. Bluth, B. J. and Helppie, M. (eds) (1986) *Soviet Space Stations as Analogs.* Washington, DC: NASA Headquarters (NAGW-659).

42. MacDaniel, W. E. (1988) "Scenario for Exraterrestrial Civilization," *Space Power*, Vol. 7:3/4, pp. 365–381 ... (1985) "Free Fall Culture," *Space Manufacturing*, Vol. 5. Washington, DC: American Institute of Aeronautics and Astronautics ... (1984) *Space Utilization, Sociocultural Problems Within Extraterrestrial Populations—A NASA Report.* Niagara University, NY: Space Studies Settlement Project/Dept. of Sociology ... (1983) "The Future, Extraterrestrial Space Humanization, and Sociology," *Space Journal*, Vol. 1:1, pp. 1–19. Niagara University, NY: Space Studies Settlement Project/Dept of Sociology.

43. Ortner, D. J. (ed) (1983) *How Humans Adapt: A Biocultural Odyssey.* Washington, DC: Smithsonian Institution Press.

44. Rudoff, A. (1977) "Space Sociology—A Terrestrial Perspective: Public Perceptions of the Space Program," unpublished paper presented at the Twenty-third Annual Meeting of the American Astronautical Society, October.

45. Mauldin, J. H. (1995) "Reflections on the Sociology of Interstellar Travel," *Ad Astra*, magazine of the National Space Society, July/August, pp. 48–52.

46. Miller, J. G. (1994) *Living Systems.* Niwot, CO: University of Colorado (Paperback edition—PO Box 849, zip code 80544) ... Dr Miller and his wife, Jessie, have a new book under way entitled, *Greater Than the Sum of its Parts*, which is being published also as a series in *Behavioral Science*, journal of the International Society of Systems Science (PO Box 40, Fallbrook, CA 92088, U.S.A.). To date, these have been released under that same title: I. Subsystems Which Process Matter, Energy and Information (Vol. 37:1, Jan. 1992, pp. 1–38; II. Matter–Energy Processing Subsystems (Vol. 38:1, Jan. 1993, pp. 1–73); III. Information Processing Subsystems (Vol. 40:3, pp. 171–270).

47. Finney, B. R. (1985) "Lunar Bases—Learning to Live in Space," in *Lunar Bases and Space Activities of the 21st Century.* W. W. Mendell (ed.). Houston, TX: Lunar and Planetary Institute, pp. 731–756.

48. Miller, J. G. (1992) "Applications of Living Systems Theory to Life in Space," in *Social Concerns* (Vol. 4, pp. 231–259) of *Space Resources.* M. F. McKay, McKay, D. S., and Duke, M. B. (eds) Washington, DC: U.S. Government Printing Office (NASA SP-506, 6 vols).

49. Harris, P. R. (1992) "The Future of Management: The NASA Paradigm;" "The Influence of Culture on Space Development" in *Social Concerns* (Vol. 4, pp. 120–142; 189–219) of *Space Resources.* M. F. McKay, McKay, D. S. and Duke, M. B. (eds). Washington, DC: U.S. Government Printing Office (NASA SP-509, 6 vols).

50. Miller, J. G. principal investigator (1998/9) *Living Systems Applications to Space Habitation* (NASA Proposal/UCSD # 886049). LaJolla, CA: University of California—San Diego/School of Medicine.

51. Miller, J. G. (1991) "Applications of Living Systems Theory to Life in Space," A. A. Harrrison *et al* (eds.) *From Antarctica to Outer Space.* New York, NY: Springer-Verlag, pp. 177–198.

52. Miller, J. G. and Miller, J. L. (1992) "Introduction: the Nature of Living Systems," *Behavioral Science*, Vol. 35:3. July, pp. 157–163.

53. Ehricke, K. A. (1957) "The Anthropology of Astronautics," *Astronautics*, Vol. 2:4, November.

CHAPTER 4

1. Launius, R. D. (1994) NASA: A History of the U.S. Civil Space Program. Melbourne, FL: Krieger Publishing ... . Neal, V., Lewis, C.S., and Winter, F. H. (1995) Spaceflight: A Smithsonian Guide. New York, NY: Macmillan Publishing ... . Gibson, E. (1995) The Greatest Adventure. Emeryville, CA: Association of Space Explorers/Publishers Group West (4065 Hollis, zip code 94608).

2. Finney, B. R. and Jones, E. M. (eds), *Interstellar Migration and the Human Experience.* Berkley, CA: University of California Press ... . Robinson, G. S. and White, H. M. *Envoys of Mankind—A Declaration of First Principles for the Governance of Space Societies.* Washington, DC: Smithsonian Institution Press, 1986 ... Robinson, G. S. "Natural Law and a Declaration of Human Interdependence," *Space Governance*, journal of United Societies in Space, Inc., June 1995, Vol. 2:1, Part I; Dec. 1995,Vol. 2:2, Part II.

3. Harris, P. R. and Moran, R. T. (1996) *Managing Cultural Differences, 4th Edition.* Houston, TX: Gulf Publishing).

4. Sulzman, F. M. and Genin, A. M.(eds) (1994) *Life Support and Habitability—Space Biology and Medicine.* Washington, DC: American Institute of Aeronautics and Astronautics, Vol. II ... Tascione, T. F. (1988) *Introduction to the Space Environment.* Melbourne, FL: Orbit Book/Krieger Publishing, ... Connors, M. M., Harrison, A. A., and Akins, F. R. (1985) *Living Aloft—Human Requirements for Extended Spaceflight.* Washington, DC: U.S. Government Printing Office, (NASA SP-483).

5. Atkinson, J. D. and Shafritz, J. M. (1985) *The Real Stuff—a History of NASA's Astronaut Recruitment Program.* New York, NY: Praeger Publishers ... Harvey, B. (1988) *Race into Space—the Soviet Space Programme.* Chichester: Ellis Horwood,

UK ... O'Berg, A. R. (1985) *Spacefarers of the '80s and '90s—the Next Thousand People in Space.* New York, NY: Columbia University Press ... Global Space Productions (1995) *Women Space Pioneers.* Videotape distributed by Afterburner Enterprises (PO Box 7310, Beverly Hills, CA 90212, U.S.A.).

6.   Harris, P. R. (1994) *High Performance Leadership.* Amherst, MA: Human Resource Development Press, Ch. 5 ... .Harris, P. R. (1992) "The Future of Management: the NASA Paradigm," *Social Concerns* (Vol 4, pp. 120–142) in *Space Resources* edited by M. F. and D. S. McKay, and M. B. Duke. Washington, DC: U.S. Government Printing Office, NASA SP-509.

7.   Hall, S. B. (ed.) (1985) *The Human Role in Space.* Park Ridge, NJ: Noyce Publications ... Smolders, P. (1986) *Living in Space.* Shrewsbury, U.K.: Airlife Publications ... Bluth, B. J. and Helppie, M. (1986) *Soviet Space Stations as Analogs.* Washington, DC: NASA Headquarters, NAGW-659.

8.   Simons, G. F., Vazquez, C., and Harris, P. R. (1993) *Transcultural Leadership* ... Elashmawi, F. and Harris, P. R. (1993) *Multicultural Management* ... Moran, R. T., Harris, P. R., and Stripp, W. G.(1993) *Developing the Global Organization*—three volumes in the Managing Cultural Differences Series. Houston, TX: Gulf Publishing Company.

9.   Harrison, A. A., Clearwater, Y. A., and McKay, C. P. (eds) (1991) *From Antarctica to Outer Space—Life in Isolation and Confinement.* New York, NY: Springer-Verlag ... . Bell, L., *et al.* (eds) (1991) *Proceedings Report—the First nternational Design for Extreme Environments Assembly*, 1991, Houston, Texas (for information on this and subsequent reports—IDEEA Two, 1993 in Montreal, Canada, and IDEEA Three, 1996 in Oxford England—contact IDEEA/USA, College of Architecture/SICSA, University of Houston, Houston, TX 77204 (fax: 713/743-2356).

10. Rogers, T. F. "Space Tourism: The Perspective from Japan and Some Implications for the United States," *The Journal of Practical Applications in Space*, Winter 1995, Vol. VI:2, pp. l09–149 ... Other citations, see *Ad Astra*, journal of the National Space Society (Feb. 1991, pp. 24–15); *Space World*, (June 1988, p. 22).

11. *Earth Space Review*, January–March 1995, Vol. 4:1, Human Factors Issue, published by Gordon and Breach Science Publishers (PO Box 90, Reading, Berkshire RG1 8JL, U.K.).

12. Harris, P. R. (1995) "The Impact of Culture on Human and Space Development—New Millennial Challenge," *Acta Astronautica*, Journal of International Academy of Astronautics, October, Vol. 36:7, pp. 399–408 ... .Also Finney, B. R. "Space Migrations: Anthropology and the Humanization of Space," pp. 164–188. Both in *Space Resources*, NASA-SP 509, cited above in ref. #6.

13. O'Donnell, D. J. (1994) "Founding a Space Nation Utilizing Living Systems Theory," *Behavioral Science*, journal of International Society for the Systems Sciences, April Vol. 39:2, pp. 93–116 ... O'Donnell, D. J. and Harris, P. R. (1994) "Space-based Energy Needs a Consortium and a Revision of the Moon Treaty," *Space Power,* journal of SUNSAT Energy Council, Vol. 13:1/2, pp. 121–l34 ... O'Donnell, D. J. and Harris, P. R. (1996) "Strategies for Lunar Development and Port Authority," *Space Governance*, journal of United Societies in Space, June 1996, Vol.2:1 (USIS address: 6841 South Yosemite, Englewood, CO 80112, U.S.A.; fax: 303/721-1182).

14. O'Neill, G. K. (1989) *The High Frontier—Human Colonies in Space.* Princeton, NJ: Soace Studies Institute (PO Box 82, zip code 08542 ... Mendell, W. W. (ed.) (1985), *Lunar Bases and Space Activities of the 21st Century.* Houston, TX: Lunar and Planetary Institute ... Savage, M. T. (1994) *The Millennial Project—Colonizing the Galaxy in Eight Easy Steps.* New York, NY: Little Brown & Co.

15. Koelle, H. H. IAA's *Lunar Base Quarterly,* April 1995 (available from Dr Koelle at Aerospace Institute, Techniche Universität, Marchstr 14, D-10587, Berlin, Germany ... Benaroya, H. "Lunar Industrialization," *The Journal of Practical Applications in Space,* Fall 1994, Vol. VI:1, pp. 85–94 ... . Criswell, D. R. and Harris, P. R. "An Alternate Solar Energy Source," *Earth Space Review,* April–June 1993, Vol. 2:1, pp. 11–15 ... Lunar Study Steering Group, *Mission to the Moon—Europe's Priorities for Scientific Exploration and Utilization of the Moon.* Paris, France: European Space Agency, June 1992 (ESA SP-1150).

16. White, F. (1987) *The Overview Effect: Space Exploration and Human Evolution.* New York, NY: Houghton-Mifflin.

17. Connors, M. M. (1991) "Communication Issues of Spaceflight," Chapter 24 in *From Antarctica to Outer Space* edited by A. A. Harison, Y.A. Clearwater, and C.P. McKay. New York, NY: Springer-Verlag.

18. Springer, S. P. and Deutsch, G. (1993) *Left Brain, Right Brain*, Fourth Edition. New York, NY: W. H. Freeman and Company ... Iaccino, J. F. *Left Brain—Right Brain Differences.* Hillsdale, NJ: Lawrence Erlbaum Associates, 1993.

19. Harris, P. R. *The New Work Culture and HRD Transformational Management.* Amherst, MA (22 Amherst Rd., zip code 01002): Human Resource Development Press, 1995 ... Criswell, D. R. (ed.) *Automation and Robotics for the National Space Program.* San Diego, CA: California Space Institute/University of California, 1985 (NASA Grant NAGW829/Cal Space Report CSI/85-01).

20. Ehricke, K. A. "Lunar Industrialization and Settlement—Birth of a Polyglobal Civilization," in *Lunar Bases and Space Activities of the 21st Century* edited by W. W. Mendell. Houston, TX: Lunar and Planetary Institute, l985, pp. 827–855.

21. Peters, T. J. and Waterman, R. H. (1982) *In Search of Excellence.* New York, NY: Harper & Row ... Also see, Schein, E. (1985) *Organizational Culture and Leadership*; Guzzo, R. A. and Salas, E. (1995) *Team Effectiveness and Decision Making in Organizations.* San Francisco, CA: Jossey Bass Publishing.

22. Gushin, V. I. (1995) "Problems of Psychological Control in Prolonged Spaceflight," *Earth Space Review,* Jan–Mar., Vol. 4:1, pp. 28–31.

23. Moran, R. T., Harris, P. R., and Stripp, W. G. (1993) *Developing Global Organizations.* Houston, TX: Gulf Publishing.

24. Harrison, A. A. "Beyond Earthnocentrism: Anthropology on the High Frontier" *Space Power (*1988), Vol. 7:3/4, pp. 345–352 ... Harris, P. R. and Moran, R. T. (1996) *Managing Cultural Differences.* Houston, TX: Gulf Publishing.

25. Sherman, P. J. and Wiener, E. L. "At the Intersection of Automation and Culture," *Earth Space Review,* Jan–Mar. (1995), Vol. 4:1, pp. 11–13.

26. Blagonravov, A. A. (ed.) (1965) *Collected Works of K. E. Tsiolkovsky* (NASA-TT-F-236, Vol. 1 and 237, Vol. 2 and 238, Vol.3); Tsiolkovsky, K.E./Tikonravov, M. K. (ed.) *Works on Rocket Technology* (NASA-TT-F-243). These informative, 30-

year-old translations are still available from NASA Center for Aerospace Information (Document Request, 800 Elkridge Landing Road, Linthicum Heights, MD 21090, U.S.A.; fax: 301/621-0134 - Order # N65-21736/32975/35965 and N65-36875 required).

CHAPTER 5

1. Surkov, Y. (1990) *Exploration of Terrestrial Planets from Spacecraft—Instrumentation, Investigation, and Interpretation.* Chichester, UK: Ellis Horwood.

2. Harris, P. R. (1996) *The New Work Culture and HRD Transformational Management.* Amherst, MA: Human Resource Development Press.

3. Harris, P. R. (1994) *High Performance Leadership.* Amherst, MA: Human Resource Development Press.

4. Moran, R. T., Harris, P. R., and Stripp, W. G. (1993) *Developing Global Organizations.* Houston, TX: Gulf Publishing.

5. Hall, S. B. (ed.) (1985) *The Human Role in Space—Technology, Economics, and Optimization* (THURIS Study). Park Ridge, NJ: Noyes Publications.

6. Pitts, J. A. (1985) *The Human Factors: Biomedicine in the Manned Space Program to 1980.* Washington, DC: U.S. Government Printing Office, NASA SP-4213.

7. Harrison, A. A. and Connors, M. M. "Human Factors in Spacecraft Design," *Journal of Spacecraft and Rockets*, AIAA, Vol. 27:2, p. 479 ... . Harrison A. A., Clearwater, Y. A., and McKay, C. F. (1991) "Recommendations for Future Research" in *From Antarctica to Outer Space.* New York, NY: Springer-Verlag ... Harrison, A. A. (1994) "Humanizing Outer Space: Some Suggestions for Metanation, *Space Governance*, Journal of United Societies in space, Inc., Dec., Vol. 1:2, pp. 11–13, 20.

8. Connors, M. M., Harrison, A. A., and Akins, F. R. (1985) *Living Aloft—Human Requirements for Extended Spaceflight.* Washington, DC: U.S. Government Printing Office, (NASA SP-483), Ch. 24.

9. Harrison, A. A. (1985) "A Model of Space Station Decor and Crew Performance," (Unpublished abstract of NASA proposal, 14 pp.) ... Harrison, A. A., Sommer, r., Struthers, N., and Hoyt, K. (1987) "Privacy and Space Habitat Design," NASA Grant NAG-237) ... Harrison, A. A. and Connors, M. M. (1984) "Groups in Exotic Environments," in L. Berkowitz (ed.) *Advances in Experimental Psychology.* New York, NY: Academic Press, Vol. XVIII.

10. Baker, D. (1985) *The History of Manned Space Flight.* New York, NY: Crown Publishers ... . Clark, P. (1988) *The Soviet Manned Space Program.* New York, NY: Orion Books/Crown Publishers (Salamander Books).

11. Sulzman, F. M. and Genin, A. M. (eds) (1994) *Life Support and Habitability.* Washington, DC: American Institute of Aeronautics and Astronautics, Vol. II ... Nicogossian, A., Huntoon, C. L., and Pool, S. L. (eds) (1989) *Space Physiology and Medicine.* Philadelphia, PA: Lee & Febiger, 2nd Edition ... Briarity, L. G. (1989) *Biology in Microgravity—A Guide for Experimenters.* Noordwijk, the Netherlands: European Space Agency.

12. Harvey, B. (1988) *Race Into Space—-the Soviet Space Programme.* Chichester,

UK: Ellis Horwood; (1996) *The New Russian Space Programme: From Competition to Collaboration.* Chichester, UK: Wiley–Praxis.

13. Reichardt, T. and Rhea, J. (1987) "The Human Factor: a Conversation with Dr Arnauld Nicogossian," *Space World,* March, pp. 8–13.

14. Foley, T. M. (1995) "Learning from Living in Space," *Aerospace America,* April, pp. 24–33.

15. Schaefer, R. (1991) "Moon Docs Among Moon Rocks," *Ad Astra,* Feb. Vol. 2:3, pp. 8–11.

16. Douglas, W. K. (1984) *Human Performance Issues Arising from Space Station Missions.* Huntington Beach, CA: McDonnell Douglas Astronautics Company (NAS2-11723 contract) ... (1991) "Psychological and Sociological Aspects of Manned Spaceflight," in *From Antarctica to Outer Space* edited by Harrison, A. A., Clearwater, Y. A., and McKay, C. R. New York, NY: Springer-Verlag.

17. Hall, S. B. (1985) *The Human Role in Space—Technology, Economics, and Optimization.* Park Ridge, NJ: Noyes Publications.

18. Helmreich, R. L., Wilheim, J. A., and Runge, T. E. (1983) Human Factors in Outer Space Productions. Boulder, CO: Westview Press.

19. Furnham, A. and Bochner, S. (1987) *Culture Shock—Psychological Reactions to Unfamiliar Environments.* New York, NY: Methuen.

20. For proceedings, refer to Harrison, A. A., Clearwater, Y. A., and McKay, C. F. (1991) *From Antarctica to Outer Space—Life in Isolation and Confinement.* New York, NY: Springer-Verlag.

21. Harrison, A. A. and Connors, M. M. (1991) "Crew Systems: Theoretical, and Practical Issues of Human and Technology." Paper presneted at American Group Psychotherapy Association, San Antonio, Texas, February, 10 pp.

22. Presidential Commission on the Space Shuttle Challenger (1986) *Report of the Presidential Commission on the Space Shuttle Challenger,* Vol. 2/3. Washington, DC: U.S. Government Printing Office.

23. Oberg, A. (1985) *Spacefarers of the 80s and 90s: the Next Thousand People in Space.* New York, NY: Columbia University Press.

24. Friedmann, L. "Frangos Goal: Reducing Bone Loss on Earth and in Space," *UC San Diego Engineering* (LaJolla, California 92093, U.S.A.), Spring 1995, p. 5.

25. Hurt, H. (1988) For All Mankind. New York, NY: Atlantic Monthly Press ... . White, F. (1987) *The Overview Effect—Space Exploration and Human Evolution.* New York, NY: Houghton Mifflin ... Ride, S. with Okie, S. (1986) *To Space & Back.* New York, NY: Lee & Shepard Books.

26. Pitts, J. A. (1985) *The Human Factor: Biomedicine in the Manned Space Program to 1980.* Washington, DC: U.S. Government Printing Office, NASA SP-4213.

27. Ockels, W. J. (1989)"Liftoff—Why Conduct Life Science Research in Space?" *As Astra,* May, Vol. 1:5, pp. 3/22.

28. Colon, A. R. and P. A. (1992) "The Psychosocial Adaptation of Children in Space: a Speculation," *Journal of Practical Applications in Space,* Spring, Vol. 3:3, pp. 5–22.

29. Nicholas, J. M. (1989) "Interpersonal and Group Behavior Skills for Crews on Space Stations," *Aviation and Environmental Medicine,* January, 22 pp ... (1987)

"Small Groups in Orbit: Group Interaction and Crew Performance on Space Station," *Aviation and Environmental Medicine*, October, pp. 1009–1013 ... with Fouchee, H. C. and Ulschak, F. L. (1988) "Crew Productivity Issues in Long Duration Space Flight," paper presented at the AIAA Aerospace Science Meeting, January, 12 pp.

30. Kanus, N. (1987) "Psychological and Interpersonal Issues in Space," *The American Journal of Psychiatry*, June, Vol. 144:6, pp. 703–706.

31. For further information, request catalogs of Center for Study of Work Teams (University of North Texas, PO Box 13587, Denton, TX 76203, U.S.A.) ... . Human Resource Development Press (22 Amherst Rd., Amherst, MA 01002, U.S.A.) ... Pfeiffer & Company International Publishers (8517 Production Ave., San Diego, CA 92121) ... Gulf Publishing Company—Managing Cultural Differences Series (3301 Allen Parkway, Houston, TX 77019, U.S.A.) ... Crisp Publications (1200 Hamilton Ct., Menlo Park, CA 94205, U.S.A.).

32. Harris, P. R. (1994–6) *The New Work Culture Series*, 3 volumes. Amherst, MA: Human Resource Development Press.

33. Bellevag, I. (1995) "A Western View of Life on Mir," *Ad Astra*, May/June, p. 21.

34. Francis, B. and Young, D. (1980) *Improving Work Groups*. San Diego, CA: University Associates/Pffeifer & Co.

35. Harris, P. R. and Moran, R. T. (1996) *Managing Cultural Differences*. Houston, TX: Gulf Publishing, Ch. 5 ... Harris, P. R. *High Performance Leadership*. Amherst, MA: HRD Press, Ch. 5.

36. Hooper, G. R. (1986) *The Soviet Cosmonaut Team*. Woodbridge, Suffolk, U.K.: GRH Publications (36, Bury Hill, Melton) ... Bluth, B. J. and Helppie, M. (1986) *Soviet Space Stations as Analogs*. Washington, DC: NASA Headquarters (Grant NAGW-659).

37. Criswell, D. R. (ed.) (1985) *Automation and Robotics for the National Space Program*. LaJolla, CA: California Space Institute/University of California—San Diego (NASA Grant NAGW629) ... Redmond, C. (ed.) (1994) *Robotics Handbook: Version 1*. Washington, DC: NASA HDQ/Office of Advanced Concepts and Technology (tel: 202/358-1757).

38. Jackson, J. (MSIS manager) (1994) *Man–Systems Integration Standards (MSIS)*. Houston, TX: NASA/Johnson Space Center, NASA-STD-3000, Vol. III; Vol. I, Rev. B.

39. Skaar, S. B. and Ruoff, C. F. (eds) (1994) *Teleoperations and Robotics in Space.* Washington, DC: American Institute of Aeronautics and Astronautics ... Also, see Bode, M. E. (ed.) (1995) *Robotic Observatories*. Chichester, UK: Wiley–Praxis.

40. For information on activities and pubications of the American Society of Training and Development, write to their headquarters (1640 King St. Alexandria, VA 22313, U.S.A.; Tel: 703/683-8100 ... Another helpful professional organization is the Society for Education, Training, and Research International (SIETAR, 808 17th St., NW, Washington, DC 20006, U.S.A.: Tel; 202/466-7883).

41. Hancock, P. A. (1985) "The Principles of Maximal Adaptability in Setting Stress Standards" in Eberts, R. E. and Eberts, C. G. (eds) *Trends in Ergonomics: Human Factors II.* North-Holland, the Netherlands: Elsevier Science Publishers BV.

42. Bullock, A. and Stallybrass, O, (eds) (1977) *The Harper Dictionary of Modern Thought*. New York, NY: Harper & Row, pp. 186–7; 212.

43. Gregory, W. H. (1986) "Researchers at M.I.T. Say It's a Matter of Ease," *Commerical Space*, Summer, Vol. 2:2, pp. 58–60 ... . For current information on International Space Station, contact NASA Headquarters, Washington, DC 20546, Attn. Office of Advanced Concepts and Technology (tel. 202/358-4434), as well as Office of Life and Microgravity Sciences and Applications.

44. Clearwater, Y. A. (1991) "Functional Esthetics to Enhance Well-being in Isolated Confined Environments" in *From Antarctica to Outer Space* edited by A. A. Harrison, Y.A. Clearwater, and C. P. McKay. New York: Springer-Verlag, Ch. 30, pp. 331–348 ... Clearwater, Y. A. and Harrison, A. A. (1990) "Crew Support for Initial Mars Expedition," *Journal of the British Interplanetary Society*, Vol. 43, pp. 513–518.

45. Schaefer, K. E. *Bioastronautics*. New York: Macmillan.

46. *Earth Space Review* devoted its whole July-September 1995 issue (Vol. 4:3) to the subject of space debris, describing different types, means of detection, and current research. For information, contact Gordon & Breach Science Publishers (PO Box 90, Reading, Berkshire RG1 8JL, U.K. or 820 Town Center Drive, Longhorne, PA 19047, U.S.A.).

47. Butler, G. V. (1990) *The 21st Century in Space*. San Diego, CA: UNIVELT, AAS #70 ... Beckey, I. and Herman, D. (eds) (1985) *Space Stations and Space Platforms*. Washington, DC: American Institute of Aeronautics and Astronautics.

48. McKay, C. F. (ed.) (1985) *The Case for Mars II*. San Diego, CA: UNIVELT, AAS #62.

49. Altman, I. (1973) "An Ecological Approach to the Functioning of Isolated and Confined Groups," in J. E. Rasmussen (ed.) *Man in Isolation and Confinement*. Chicago, IL: Aldine, pp. 241–270.

50. Hargrove, E. C. (ed.) (1986) *Beyond Spaceship Earth—Environmental Ethics in the Solar System*. San Francisco, CA: The Sierra Club.

51. Johnson, R. G. (ed.) (1990) *Global Environmental Change—the Role of Space in Understanding Earth*. San Diego, CA: UNIVELT, AAS #76.

52. Benson, J. "Conversations with Charles Kennel, NASA Associate Administrator for Mission to Planet Earth," *Aerospace America*, August 1995, pp. 10–12.

53. MacDaniel, W. E. (1988) "Scenario for Extraterrestrial Civilization," *Space Power*, Vol. 7:3/4, pp. 365–381

54. Powers, R. M. (1985) *Mars: Our Future on the Red Planet*. New York, NY: Houghton Mifflin.

CHAPTER 6

1. Campbell, J. (1968) *Hero with a Thousand Faces*. Princeton, NJ: Princeton University Press.

2. Cordell, B. M. (1989) "Manned Mars Overview." Paper delivered at the 25th Joint Propulsion Conference, American Institute of Aeronautics and Astronautics, Mon-

terey, California, July 10–11; (1987) "Human Survival on Manned Mars Missions," *Space Reflections*. San Diego Hall of Science, May–June, pp. 5–7.

3. Finney, B. R. and Jones, E. M. (eds) (1984) *Interstellar Migration and the Human Experience*. Berkeley, CA: University of California Press ... . Finney, B. R. (1991) "Scientists and Seamen" in Harrison/Clearwater/McKay (eds), *From Antarctica to Outer Space*. New York, NY: Springer-Verlag, Ch. 9, pp. 89–102.

4. Harrison, A. A. and Connors, M. M. (1984) "Groups in Exotic Environments" (NASA Ames Research Contract NCA2-OR-180-803). In L. Berkowitz (ed.) *Advances in Experimental Social Psychology*. New York, NY: Academic Press.

5. Gunderson, E. K. E. (1963) "Emotional Symptoms in Extremely Isolated Groups," *Archives of General Psychiatry*, 9: 362–368; (ed.) (1974) *Human Adaptability to Antarctic Conditions*. Washington, DC: American Geophysical Union ... . Palinkas, L. A. (1991) "Group Adaptations and Individual Adjustment in Antarctica: A Summary of Research," in Harrison/Clearwater/McKay (eds) *From Antarctica to Outer Space*. New York, NY: Springer-Verlag, Ch. 21.

6. Harris, P. R. and Moran, R. T. (1996) *Managing Cultural Difference—High Performance Strategies for the Global Marketplace*. Houston, TX: Gulf Publishing, 4th Ed., Ch. 7.

7. Harrison, A. A., Clearwater, Y. A., and McKay, C. P. (eds) (1991) *From Antarctica to Outer Space—Life in Isolation and Confinement*. New York, NY: Springer-Verlag ... . Refer to Ch. 10, Weybrew, B. B. "Three Decades of Nuclear Submarine Research: Implications for Space and Antarctic Research" ... Ch. 11, Valen, R. J. and Caldwell, B. S."National Park Service Areas as Analogues for Antarctic and Space Environments" ... Ch. 4, Lugg, D., "Current International Human Factors Research in Antarctica" ... Ch. 5, Taylor, A. J. "The Research Program of the International Biomedical Expedition in the Antarctic and its Implications for Outer Space."

8. Osland, J. S. (1995) *The Adventures of Working Abroad—Hero Tales from the Global Frontier*. San Francisco, CA: Jossey-Bass Publishers.

9. Black, J. S., Gregersen, H. B., and Mendenhall, M. E. (1992) *Global Assignments—Successfully Expatriating and Repatriating International Managers*. San Francisco, CA: Jossey-Bass Publishers.

10. Tung, R. L. (1987) *The New Expatriates—Managing Human Resources in an International Context*. Cambridge, MA: Ballinger.

11. Torbiorn, I. (1982) *Living Abroad: Personal Adjustment and Personal Policy in Overseas Settings*. New York, NY: John Wiley ... . Brewster, C. (1991) *The Management of Expatriates*. London, U.K.: Kogan Page.

12. *Mobility, the Magazine of the Employee Relocation Council* (1720 N. St., NW, Washington, DC 20036, U.S.A.; tel: 202/857-0857).

13. Harris, P. R. (1995) Relocation Preparation Index in *Twenty Reproducible Assessment Instruments*. Amherst, MA: Human Resource Development Press ... Kelley, C. and Meyers, J. (1993) *The Cross-cultural Adaptability Inventory*. Yarmouth, ME: Intercultural Press/NCS Assessment Inc ... .Wederspan, G. M. (1995) *Overseas Assignment Inventory*. Boulder, CO: Prudential Relocation Services (2555 55th St., Ste 201D, zip code 80301) ... Dunbar, R. L., Bird, A. and Gudelis, R. (1992) *Expatriate Profile—Computer-based Training Program*. New York, NY: Dunbar, Bird & Gudelis, New York University Stern School of Business.

14. Palinkas, L. A. (1985) *Health Performance of Antarctic Winter-over Personnel.* San Diego, CA: Naval Health Research Center, Reports No. 85-18; 85-48; 85-49. The last-mentioned report is on Sociocultural Influences on Psychological Adjustment in Antarctica.

15. Bluth, B. J. and Helppie, M. (1986) *Soviet Space Station as Analogs.* Washington, DC: NASA Headquarters (Grant NAGW-659).

16. Furnham, A. and Bochner, S (1987) *Culture Shock—Psychological Reactions to Unfamiliar Environments.* New York, NY: Methuen.

17. White, F. (1987) *The Overview Effect—Space Exploration and Human Evolution.* New York, NY: Houghton Mifflin.

18. Austin, C. N. (ed.) (1987) *Cross-cultural Re-entry—A Book of Readings.* Yarmouth, ME: Intercultural Press.

19. Levine, A. S. (1991) "Psychological Effects of Long-duration Space Missions and Stress Amelioration Techniques," Ch. 23 in Harrison/Clearwater/McKay (eds), *From Antarctica to Outer Space.* New York, NY: Springer-Verlag.

20. Helmreich, R. L. Wilheim, J. A., and Runge, T. E. (1983) "Psychological Considerations in Future Space Missions," in S. Cheston and D. Winters (eds) *Human Factors in Outer Space Production.* Boulder, CO: Westview Press.

21. Brady, J. V. (1991) "Small Groups and Confined Microsocieties," in Harrison/Clearwater/McKay (eds), *From Antarctica to Outer Space.* New York, NY: Springer-Verlag.

22. Rivolier, J., et al (eds) (1987) *Man in Antarctica.* Philadelphia, PA: Taylor & Francis, IPS ... . Barabrasz, A. E. (1991) *A Review of Antarctic Behavioral Research,* Ch. 3 in Harrison/Clearwater/McKay (eds) *From Antarctica to Outer Space.* New York, NY: Springer-Verlag.

23. Kelly, T. (1986) "Frozen Assets," *Canada Today/d'aujourd'hui.* Canadian Embassy (Rm. 30, 1771 N. St. NW, Washington, DC 20036, U.S.A.).

24. Carrere, S., Evans, G. W., and Stokols, D. (1991) "Winter-over Stress: Physiological and Psychological Adaptation to an Antarctic Isolated and Confined Environment," Ch. 20, in Harrison/Clearwater/McKay, *From Antarctica to Outer Space.* New York, NY: Springer-Verlag.

25. Collins, P. (1990) "Choosing the Right Stuff," *Space Policy,* Aug., Vol. 6:3, pp. 281–282.

26. For information on the findings from the ISEMI, EXEMSI, and EUROMIR studies, contact Long-Term Programme Office, European Space Agency (8/10 rue Mario Nikis, 75738 Paris Cedax 12, France) or ESA Washington Office (955 L'Enfant Plaza SW, Ste. 7899, Washington, DC 20024, U.S.A.)

27. Allen, J. and Nelson, M. (1987) *Space Biospheres.* Melbourne, FL: Orbit Books/Krieger Publications.

28. Crawford, B. (1990) "Inside a Bonsai World (Biosphere 2)," *American Way* March, pp. 34–6, 107.

29. Shanks, C. (1991) "Genesis of a Space Colony—Biosphere 2 is Breaking Ground for Future Space Settlements," *Ad Astra,* Jan., Vol. 3:1. pp. 30–34.

30. Walford, R. L., *et al.* (1992) "The Calorifically Restricted, Low-fat Nutrient Dense Diet in Biosphere Significantly Lowers Blood Glucose, Total Leukocyte, Cholesterol, and Blood Pressure in Humans," *Proceedings of the National Academy of*

Science, Vol. 89: 11533-37 ... David, L. (1994) "Biosphere 2 Dwelling on the Future—The World's Largest Test Tube Yields Clues About Living on Earth and in Space," *Final Frontier*, January–February, pp. 41–45.

31. For information on proceedings for *IDEEA ONE/TWO*, contact IDEEA USA, Sasakawa International Center for Space Architecture, University of Houston (Houston, TX 77204, U.S.A.; tel: 713/743-2350; fax: 713/743-2358) ... for *IDEEA THREE*, contact IDEEA UK Secretariat, Oxford School of Architecture (Gipsy Lane, Oxford OX3 OBP, U.K.; tel: +44 01865 484876; fax: +44 01865 484883; E-mail: dbibby@brookes.ac.uk).

32. Harris, P. R. (1991) "Personnel Deployment Systems: Managing People in Polar and Outer Space Settings," Chapter 7 in Harrison/Clearwater/McKay (eds) *From Antarctica to Outer Space.* New York, NY: Springer-Verlag; (1996) "Challenges in the Space Environment," *Proceedings of 26th International Conference on Environmental Systems*, Warrensdale, PA: SAE International, (400 Commonwealth Drive, zip 15096).

33. Garshneck, V. (1990) "Humans in Space from Day One," *Space Policy*, Aug., Vol. 6:3. pp. 262–269 ... . Atkinson, J. D. and Shafritz, F. M. (1985) *The Real Stuff: A History of the Astronaut Recruitment Program*, New York, NY: Praeger Publications.

34. Piradov, A. S. (1988) "Creating a World Space Organization," *Space Policy*, May Vol. 4:2. pp. 112–114 ... . O'Donnell, D. J. (1994) "Founding a Space National Utilizing Living Systems Theory: New Research Challenge," *Behavioral Science*, April, Vol. 39:2, pp. 93–116. (See Appendix D in this book, *Living and Working in Space*, 2nd edn.)

35. European Space Agency (1990) "The European Astronaut Centre," *ESA Fact File No. 3*, July 10; "ISEMI: The Human Factor," *ESA Fact File No. 4*, Nov. 14 ... (1991) "Astronauts Wanted," *ESA Newsletter*, Jan., pp. 1–4. See ESA listing, Appendix C.

36. Manzey, D., Horman, H-J, Fassbender, C., and Schiewe, A. (1995) "Implementing Human Factors Training for Space Crews," *Earth Space Review*, Vol.4: 1, Jan–Mar., pp. 24–27.

37. Barclay, J. (1991) *Psychological Assessment*. Melbourne, FL: Krieger Publishing ... Black, J. S., Gregersen, H. B., and Mendenhall, M. E. (1992) "Selecting the Right People," Ch.3 in their *Global Assignments*. San Francisco, CA: Jossey-Bass Publishers ... Harris, P. R. and Moran, R. T. (1996) "Managing Transitions and Relocations," Ch. 7 in their *Managing Cultural Differences*. Houston, TX: Gulf Publishing.

38. Robinson, G. S. and White, H. M. (1986) *Envoys of Mankind—a Declaration of First Principles for the Governance of Space Societies*. Washington, DC: Smithsonian Institution Press (refer to Parts 1 through 5).

39. Harvey, B. (1996) *The New Russian Space Programme: From Competition to Collaboration*. Chichester: UK: Wiley–Praxis.

40. Hall, S. B. (ed.) (1985) *The Human Role in Space—Technology, Economics and Optimization*. Park Ridge, NJ: Noyes Publications.

41. Filbert, H. E. and Keller, D. J. (1989) "Astronaut Interdisciplinary, Medical and Dental Training for Manned Mars Missions," in C. Stoker (ed.) *Case for Mars III—Strategies for Exploration*. San Diego, CA: UNIVELT, pp. 161–171.

42. Swanson, G. E. (ed.) (1995) "STS-71 Preview—Shuttle/Mir Docking Mission," *Countdown*, May/June, Vol. 13: 3, pp. 44–47 (Crew Biographies). Published by

CSPACE Press, Inc. (123 32nd St., Grand Rapids, MI 49509).

43. Cooper, H. S. (1987) *Before Lift-Off—The Making of a Space Shuttle Crew*. Baltimore, MD: Johns Hopkins University Press ... . Harvey, B. (1996) *The New Russian Space Programme—From Competition to Collaboration*. Chichester, UK: Wiley/Praxis..

44. Manzey, D. and Schiewe, A. (1992) "Psychological Training of German Science Astronauts," *Acta Astronautica*, Vol. 27, pp. 147–154 ... Manzey, D. Schiewe, A., and Fassbender, C. (1995) "Psychological Countermeasures for Extended Manned Space Flights," *Acta Astronautica*.

45. Nicholas, J. M. (1989) "Interpersonal and Group Behavior Skills Training for Crews on Space Station," *Aviation Space and Environmental Medicine*, Vol. 60, pp. 603–608.

46. Harris, P. R. (1994) *High Performance Leadership—HRD Strategies for the New Work Culture* ... Schneier, C. E. *et al.* (eds) (1995) *The Training and Development Sourcebook—Second Edition* ... Rothwell, W. J. and Sredl, H.J. (1993) *The ASTD Reference Guide to Professional Human Resource Development Roles and Competencies*. Amherst, MA: Human Resource Development Press.

47. Ortner, D. J. (ed.) *How Humans Adapt: a Biocultural Odyssey*. Washington, DC: Smithsonian Institution Press ... Sheffield, C. and Rosen, C. (1984) *Space Careers*, New York, NY: William Morrow.

48. Baker, D. (1985) *The History of Manned Space Flight*. New York, NY: Crown Books.

49. Hall, E. T. and Hall, M. E. (1987) *Hidden Differences*. Garden City, NY: Anchor Press/Doubleday.

50. Oberg, A. R. (1985) *Spacefarers of the '80s and '90s*. New York, NY: Columbia University Press.

51. Clearwater, Y. A. and Harrison, A. A. (1990) "Crew Support for an Initial Mars Mission," *Journal of the British Interplanetary Society*, Vol. 43, pp. 513–518.

52. National Commission on Space (1986) *Pioneering the Space Frontier*, New York, NY: Bantam Books, p. 175.

53. Cordell, B. M. (1989) "Manned Mars Overview," Paper presented to 25th Joint Propulsion Conference of the American Institute of Aeronautics and Astronautics, Monterrey, CA, July 10–11.

54. Schaefer, C. (1991) "Moon Docs Among Moon Rocks," *Ad Astra*, Feb.,Vol. 3:2, pp. 8–11.

55. Klaus, P. N. (1991) "Decreasing Stress through the Introduction of Microenvironments," Ch. 13 in Harrison/Clearwater/McKay (eds), *From Antarctica to Outer Space*. New York, NY: Springer-Verlag.

56. Jones, E. M. (ed.) (1985) *The Space Settlement Papers*. Los Alamos, NM: Los Alamos National Laboratory (Report # LA-UR-85-3874), p. 13.

57. Miller, J. G. (1991) "Applications of Living Systems Theory to Life in Space," Ch. 16, in Harrison/Clearwater/McKay (eds), *From Antarctica to Outer Space*. New York, NY: Springer-Verlag.

58. Harrison, A. A., Sommers, R., Struthers, N., and Hoyt, K. (1986) *Privacy and Habitat Design*. Moffet Field, CA: NASA Ames Research Report #NAG 2-357.

59. Hoffman, J. A. (1986) *An Astronaut's Diary*. Montclair, NJ: Caliban Press.

60. National Aeronautics and Space Administration (1995) *Man–Systems Integration Standards*. Houston, TX: NASA/Johnson Space Center, July (Revision B, #NASA-STD-3000).

61. Austin, C. D. (ed.) (1987) *Cross-cultural Re-entry—a Book of Readings*. Yarmouth, ME: Intercultural Press. A volume for space re-entry is needed comparable to this one on return from foreign deployment.

62. Harvey, B. (1988) *Race into Space—the Soviet Manned Space Programme*. Chichester, UK: Ellis Horwood, pp. 305–307.

CHAPTER 7

1. McFarland, D. E. (1985) *The Managerial Imperative—The Age of Macromanagement*. Cambridge, MA: Balliger/Harper & Row, p. 137.

2. McKay, M. F., McKay, D. S., and Duke, M. B. (eds) (1992). *Space Resources,* Washington, DC: U.S. Government Printing Office (NASA SP-509). Refer to volume 4, Social Concerns, chapter II on "New Space Management and Structure" by K. J. Murphy; "The NASA Paradigm" by P. R. Harris, pp. 16–141.

3. Davidson, F. P., Meador, C. L., and Salkeld, R. (eds) (1980), *How Big and Still Beautiful? Macroengineering Revisited.* Washington, DC: American Association for the Advancement of Science, Vol. 40, p. 78.

4. Davidson, F. P. with J. B. Cox (1983) *MACRO*. New York, NY: William Morrow and Company.

5. Davidson, F. P. and Meador, C. L. (eds) (1992) *Macroengineering—Global Infrastructure Solutions*. Chichester, UK: Ellis Horwood.

6. Harvey, B. (1988) *Race into Space—the Soviet Space Programme*. Chichester, UK: Ellis Horwood.

7. Newkirk, D. (1990) *Almanac of Soviet Manned Space Flight*. Houston, TX: Gulf Publishing ... . (1994) *Cosmonautics—A Colorful History.* Washington, DC (4200 Wisconsin Ave., zip code 20016): Cosmos Books.

8. Salkeld, R. "Space: Macro-arena for Macro-engineering," pp. 131–138 in Davidson, Giacoletto, and Salkeld (eds) *Macro-Engineering and the Infrastructure of Tomorrow*. Washington, DC: American Association for the Advancement of Science, Vol. 23, 1978 ... Moore, P. (1976) *The Next Fifty Years in Space*. New York, NY: Taplinger Publishing.

9. Surkov, Y. (1990) *Exploration of Terrestrial Planets from Spacecraft.* Chichester, UK: Ellis Horwood ... Curtis, A. R. (1994; 1992). *Space Satellite Handbook*, Third Edition; *Space Almanac*, Second Edition. Houston, TX: Gulf Publishing.

10. Glaser, P. E., Davidson, F. P., and Csigi, K. I. (eds) (1993). *Solar Power Satellites—The Emerging Energy Option*. Chichester, UK: Ellis Horwood.

11. Bonnet, R. M. "Taking the Next Step: the European Moon Program," *The Planetary Report*, Vol. XV, Jan. 1995, pp.8–11; *Mission to the Moon—Europe's Priorities for the Scientific Exploration and Utilization of the Moon*. Paris, France: European Space Agency, June 1992 (ESA SP-1150), 190 pp ... . For further information, contact European Space Agency headquarters (8–10, Rue Mario Nikis 75738, Paris

Cedex 15, France, or ESA, 955 L'Enfant Plaza, SW, Ste. 7800, Washington, DC 20023, U.S.A.—tel: 202/488-4158 or fax: 488-4930).

12. O'Donnell, D. J. "Founding a Space Nation Utilizing Living Systems Theory," *Behavioral Science*, Vol. 39:2, April 1994, pp. 93–117; "Overcoming Barriers to Space Travel," *Space Policy*, Vol. 10:4, Nov. 1994, pp. 252–255 ... with P. R. Harris, "Is it Time to Replace the Moon Agreement?" *The Air and Space Lawyer,* Vol. 9:2, Fall 1994, pp. 3–9; "Space-based Energy Needs a Consortium and a Revision of the Moon Treaty", *Space Power*, Vol. 13/1&2, pp. 121–134 ... with P. R. Harris, "Strategies for Lunar Development and Port Authority," *Space Governance*, Vol. 2:1, June 1995, pp. 6–13 ... .For further information, consult the journal of United Societies in Space, *Space Governance*, Vol. 1: 1&2, 1994; Vol. 2: 1&2 (available from USIS, 6841 South Yosemite, Englewood, CO 80112, U.S.A.; tel: 1/800/895-META; fax: 1/303/721-1182).

13. Savage, M. T. (1994) *The Millennial Project—Colonizing the Galaxy in Eight Easy Steps*. New York, NY: Little Brown & Co. ... "The Millennial Project Strategy," *Space Governance journal*, Vol. 2:1, June 1995, pp. 26–31 ... For further information—the First Millennial Foundation, PO Box 347, Rifle, CO 81650 (tel: 1/303/625-2815; e-mail: mtsavage@pipeline.com.

14. DOE Office of Policy, Planning & Analysis, *The Potential of Renewable Energy, An Interlaboratory White Paper*, March 1990, SERI-DE #AC0E-83CH10093). Available from the Solar Energy Research Institute (1617 Cole Blvd., Golden, CO 8040l, U.S.A.).

15. Sagan, C. (1994) *The Pale Blue Dot.* New York, NY: Random House.

16. Ehricke, K., "The Anthropology of Astronautics," *Astronautics* 2:4, 1957, pp. 26–27 ... *Colonize Space! Open the Age of Reason*, Proceedings of the Krafft A. Ehricke Memorial Conference, June 15–16, 1985. New York, NY: New Benjamin Franklin House.

17. Toffler, A. (1992) "The Space Program's Impact on Society," in *Human and Machines in Space: the Payoff*, edited by Paul Korn. San Diego, CA (PO Box 28130, zip code 92198): American Astronautical Society/UNIVELT, Inc.

CHAPTER 8

1. National Aeronautics and Space Administration (1989) *Orders of Magnitude: A History of NACA & NASA, 1915–1990*; Launius, (1995), NASA: *A History of the U.S. Civil Space Programme*. Washington, DC: U.S. Government Printing Office ... . Levine, A. S. (1982) *Managing NASA in the Apollo Era*. Washington, DC: NASA Scientific and Technical Information Branch (MNASA-SP-4102) ... Mark, H. and Levine, A. S. (1984) *The Management of Research Institutions: A Look at Government Laboratories*. Washington, DC: U.S. Government Printing Office (NASA SP-481) ... Harvey, B. (1995) *The New Russian Space Programme: From Competition to Collaboration*. Chichester, UK: Wiley–Praxis.

2. Johnson, W. L. (ed.) (1967) *The Management of Aerospace Programs*. San Diego, CA: UNIVELT Inc. (AAS Vol. 12) ... Johnson, N. (1985) *Soviet Space Program: 1980–1985*. San Diego, CA: UNIVELT Inc. (AAS Vol. 66).

3.   Connors, M. M., Harrison, A. A., and Akins, F. R. (eds) (1985) *Living Aloft: Human Requirements for Extended Space Flight*. Washington, DC: U.S. Government Printing Office (NASA SP-483).

4.   Davidson, F. P. and Meador, C.L. (eds) (1992) *Macroengineering—Global Infrastructure Solutions*. Chichester, UK: Ellis Horwood ... *Technology in Society, An International Journal* (Polytechnic University, 333 Jay Street, Brooklyn, NY 11201).

5.   MacDaniel, W. E. (1985) "Intellectual Stimulant," *Extraterrestrial Society Newsletter*, Fall, Vol. 7:2 (Sociology Dept., Niagara University, New York, 14109, U.S.A.).

6.   Harris, P. R. and Moran, R. T. (1996) *Managing Cultural Differences*, Ch. 5 ... Elashmawi, F. and Harris, P. R. (1993) *Multicultural Management*. Houston, TX: Gulf Publishing Co.

7.   Harris, P. R. (1996) *The New Work Culture and HRD Transformational Management*. Amherst, MA: Human Resource Development Press ... (1996) "Diversity in the Global Work Culture," *Equal Opportunities International*, Vol. 15:2, pp. 36–51 (Barmarick Publications, Enholmes Hall, Parington, East Yorkshire, HU12 0PR, U.K.).

8.   Compton, W. D. (1989) *Where No Man Has Gone Before: A History of Apollo Lunar Exploration Missions* (NASA SP-4214) ... . Fries, S. D. (1992) *NASA Engineers and the Age of Apollo* (NASA SP-4104). San Diego, CA: UNIVELT INC. (distributor PO Box 28130, zip code 92198).

9.   Seamans, R. C. and Ordway, F. I. (1977) "The Apollo Tradition: An Object Lesson for the Management of Large-scale Technological Endeavors," *Interdisciplinary Science Reviews*, Vol. 2:4, pp. 270–303.

10. McFarland, D. E. (1985) *Managerial Imperative: Age of Macromanagement*. Cambridge, MA: Ballinger/Harper & Row.

11. O'Toole, J. (1987) *Vanguard Management*. New York, NY: Berkley Publishing Group.

12. Baugh, J. (1981) "A Study of Decision Making Within Matrix Organizations." Unpublished doctoral dissertation, United States International University, San Diego, California. (Available from University Microfilms International, 300 N. Zeeb Rd., Ann Arbor, MI 48106, U.S.A.)

13. Harris, P. R. (1994) *High Performance Leadership—HRD Strategies for the New Work Culture*. Amherst, MA: Human Resource Development Press.

14. Sayles, L. R. and Chandler, M. K. (1971) *Managing Large Systems: Organizations for the Future*. New York, NY: Harper & Row.

15. Charnes, A. and Cooper, W. W. (eds) (1984) *Creative and Innovative Management*. Cambridge, MA: Ballinger/Harper & Row ... Leonard-Barton, D. (1995) *Wellsprings of Knowledge—Building and Sustaining the Source of Innovation*. Boston, MA: Harvard Business School Press.

16. Davenport, T. H. (1993) *Process Innovation—Reengineering Work through Information Technology* ... Nohria, N. and Eccles, R. G. (1994) Networks and Organizations—Structure, Form, Action. Boston, MA: Harvard Business School Press.

17. Lubos, P. (1994) "Management of Outer Space," *Space Policy*, August, Vol. 10:3, pp. 189–198.

18. Schein, E. H. (1985) *Organizational Culture and Leadership* ... Also see Becker, F.

and Steele, F. (1995) *Workplace by Design—Mapping the High Performance Workscape.* San Francisco, CA: Jossey-Bass Publishers.

19. Padron, J. (1995) "Reinventing NASA," *Aerospace America's AIAA Bulletin,* June, Insiders News, pp. B9/18.

20. The proceedings were entitled *Space Resources* (NASA SP-509) publihsed in 1992 by the U.S. Government Printing Office. Material on space management is in volume 4, *Social Concerns.*

21. Advisory Committee (Norman Augustine, Chairman) (1990) *Report of the Advisory Committee on the Future of the US Space Program.* Washington, DC: U.S. Government Printing Office.

22. Robinson, G. S. and White, H. M. (1986) *Envoys of Mankind.* Washington, DC: Smithsonian Institution Press.

23. Elashmawi, F. and Harris, P. R. (1993) *Multicultural Management* ... . Simons, G. F., Vazquez, C. and Harris, P. R. (1993) *Transcultural Leadership* ... Moran, R. T. and Stripp, W. G. (1991) *Dynamics of Successful International Business Negotiations.* Houston, TX: Gulf Publishing.

24. Two volumes of proceedings from the International Lunar Exploration Conferences for 1994–5 are available from International Space Enterprises (4909) Murphy Canyon Rd., Ste. 330, San Diego, CA 92123, U.S.A.).

25. Office of Air/Space Commercialization (1995) *Trends in Commercial Space;* (1992) *Space Business Indicators;* (1988) *Space Commerce: An Industry Assessment.* Washington, Dc. U.S. Department of Commerce (zip code 2230; tel: 202/482-6125) ... Howerton, B. A. (1995) *Free Space—Real Alternatives for Reaching Outer space,* Port Townsend, Washington: Loompanics Unlimited (PO Box 1197, Port Townsend, WA 98368, U.S.A.).

26. Murphy, K. J. (1992) "New Space Management Structure," *Social Concerns* (Vol. 4, pp. 16–113) edited by McKay, M. F. & D. S./Duke, M. B. in *Space Resources.* Washington, DC: U.S. Government Printing Office (NASA SP-509).

27. Davidson, F. P. with Cox, J. S. (1983) *Macro—A Clear Vision of How Science and Technology will Shape Our Future.* New York, N.Y ... .Davison, F. P., Meador, C. L., and Salkeld, R. (eds) (1980) *How Big is Beautiful? Macroengineering Revisited.* Boulder, CO: Westview Press.

28. Carter, N. E. (1985) "The Challenge of Macro-engineerings," *Battelle Today,* reprint no. 24 (Battelle Institute, 505 King Ave., Columbus, OH 43201, U.S.A.)

29. McFarland, D. E. (1977) "Management, Humanism, and Society: The Case for Macromanagement Theory," *Academy of Management Review,* October.

30. For information on conferences and proceedings, contact the World Development Council (40 Technology Park/Atlanta, Ste. 200, Norcross, GA 30092, U.S.A.)

31. See *Technology in Society, an International Journal,* Vol. 6:2, 1984, which had a selection of six articles on the theme, Education for the Management of Large and Complex Systems. (Available from Institute for Technology Management and Policy, Polytechnic Institute, 333 Jay St., Brooklyn, N.Y. 11201, U.S.A.)

32. Kozmetsky, G. (1985) *Transformational Management.* Cambridge, MA: Ballinger/Harper & Row ... . Gibson, D. V. and Rogers, E. M. (1994) *R&D Collaboration on Trial.* Boston, MA: Harvard Business School Press.

33. Murphy, K. J. (1983) *Macroproject Development in the Third World: An Analysis of Transnational Partnerships*. Boulder, CO: Westview Press.

34. Konecci, E. B. and Kuhn, R. L. (eds) (1985) *Technology Venturing: American Innovation and Risk Taking*. New York, NY: Praeger ... Mendell, W. W. (ed.) *Lunar Bases and Space Activities of the 21st Century*. Houston, TX: Lunar and Planetary Institute.

35. Sayles, L. R. and Chandler, M. K. (1971) *Managing Large Systems: Organization for the Future*. New York, NY: Harper & Row.

36. Write for catalogs for both the regular and the summer session programs at International Space University, Communaute Urbaine de Strasbourg, Blvd. Gontheir D'Andernach, 6700 Illkirch, France (tel: 33 88 65 54 46; fax: 33 88 65 54 47); or ISU North American Office: 3400 International Drive, NW, Washington, DC 20008, U.S.A. (tel: 1-800 677-1987; fax 1-202 237-8336).

37. Beckey, I. and Herman, D. (eds) (1986) S*pace Stations and Platforms: Concepts, Designs, Infrastructures, and Use*. Washington, DC: American Institute of Aeronautics and Astronautics ... Woodcook, G. R. (1986) *Space Stations and Platforms*. Melbourne, FL: Orbit Book/Krieger ... O'Leary, B. (1983) *Project Space Station* Harrisburgh, PA: Stackpole Books ... Martin, F. D. and Finn, T. L. (1987) *Space Station: Leadership for the Future*. Washington, DC: NASA Headquarters, Office of Public Information (also request NP-107/10-88 ... Mark, H. (1990) *The Space Station—A Personal Journey*. Durham, NC: Duke University Press ... McCurdy, H. E. (1990) *The Space Station Decision: Incremental Politics and Technological Choice*. Baltimore, MD: The Johns Hopkins University Press ... . Foley, T. (1995) "Space Station: The Next Iteration," *Aerospace America*, Jan., pp. 22–27 ... Harris, P. R. (1995) "Why Not Use the Moon as a space Station," *Earth Space Review*, Dec., Vol. 4:4, pp. 7–10 ... For further current publications on the International Space Station, contact Office of Space Access and Technology, NASA Headquarters (tel: 202/358-4493 or 4677; e-mail: dstone@hg.nasa.gov).

38. Compton, W. D. and Benson, C. D. (1983) *Living and Working in Space: A History of Skylab*. San Diego, CA: UNIVELT INC. (NASA SP-4208).

39. NASA Office of Space Access and Technology (1995) "International Space Station Assembly Schedule," *Space Technology Innovation*, July–August, Vol. 3:4, p. 17. (To obtain complimentary subscription of this publication, write NASA/OSAT, Code X, Washington, DC 20546, U.S.A.) ... Foley, T. M. (1995) "Space Station: The Next Iteration," *Aerospace America*, January, pp. 22–27 ... The Public Service Broadcasting also has a video on *Space Station: It's About Life on Earth* (call PBS Adult Learning Satellite Service, 1-800. 257-2578).

40. Meyerson, H. and Simonelli, D. K. (1995) *Launchpad for the 21st Century: Yearbook of International Space Year*. San Diego, CA: UNIVELT INC.

CHAPTER 9

1. Robinson, G. S. (1988) "Bicentennial of the Bill of Rights and Rethinking Space," unpublished paper presented at Air/Space America Legal Forum, National University, San Diego, California ... . Refer to (1995) "Natural Law and a Declaration of

Interdependence," Part I—June/Part II—December, Vol. 2:1/2; (1996) "Draft Competition for Space Metanation Constitution," June, Vol. 3:1, in *Space Governance,* journal of United Societies in Space, Inc. (6841 South Yosemite, Englewood, CO 80112, U.S.A.).

2. Robinson, G. S. and White, H. M. (1986) *Envoys of Mankind: A Declaration of First Principles for the Governance of Space Societies.* Washington, DC: Smithsonian Institution Press ... Goldman, N. C. (1996) *American Space Law: International and Domestic.* San Diego, CA: UNIVELT INC ... Kinsley, A. P. (ed.) (1989) *Space—a New Era.* San Diego, CA: UNIVELT INC.

3. Rycroft, M. (ed.) (1991) *The Cambridge Encyclopedia of Space.* Cambridge, UK: Cambridge University Press ... Hayes, W. C. (ed.) (1980) *Space—New Opportunities for International Ventures.* San Diego, CA: UNIVELT INC/American Astronautical Society, Vol. 49.

4. McKay, M. F., McKay, D. S., and Duke, M. B. (eds) (1992) *Space Resources.* Washington, DC: U.S. Government Printing Office, NASA SP-509, 5 volumes.

5. Harvey, B. (1995) *The New Russian Space Programme: From Competition to Collaboration.* Chichester, UK: Wiley–Praxis.

6. Fulda, M. (1995) "US Russian Cooperation—Space Station, A Social Science Perspective," *Spaceflight,* March, Vol. 37, pp. 84–85.

7. Bonnet, R. M. and Manno, V. (1995) *International Cooperation in Space: The Example of the European Space Agency.* Cambridge, MA: Harvard University Press; London, UK: Fitzroy House (11 Chenies St., zip code WC1E 7ET) ... Lafferanderie, G. and Tuinder, P. H. (1994). "The Role of ESA in the Evolution of Space Law," *Journal of Space Law,* Vol. 22:1/2, pp. 97–113.

8. European Space Agency (1995) "ESA Council Ministerial Meeting in Toulouse—European Space: A New Momentum," *ESA News Release,* October 20, No.44-95 ... . The following publications are available from the ESA Public Relations Division (fax: +33 -1-5369-7690 or in U.S.A., 956 L'Enfant Plaza, Ste. 7800, Washington, DC:20024): *All About the European Space Agency* (1995) ... *How to Do Business with ESA* (1994, ESA BR68) ... *Annual Report—European Space Agency.*

9. Taylor, D. C. (1995) "Germany Strives to Build Solid Aerospace Industry," *Aerospace America,* October, pp. 6–7; 36–37.

10. Greenberg, J. S. and Hertzfeld, R. (eds) (1992) *Space Economics.* Washington, DC: American Institute of Aeronautics and Astronautics.

11. Congressional Budget Office (1988) *The NASA Program in the 1990's and Beyond.* Washington, DC: U.S. Congressional Budget Office, May.

12. National Commission on Space (1986) *Pioneering the Space Frontier.* New York, NY: Bantam Books ... AIAA (1987) *The Civil Space Program: An Investment in America,* Washington, DC: American Institute of Aeronautics and Astronautics ... Ride, S. K. (1987) *Leadership and America's Future in Space.* Washington, DC: U.S. Government Printing Office, NASA Report.

13. Ladwig, A. (1995) "A Roadmap for Future NASA Exploration Missions," Nov. 13, *Second Annual International Lunar Exploration Conference,* National Space Soci-

ety. Proceedings available in 1996 from International Space Enterprises (4909 Canyon Rd., San Diego, CA 93123, U.S.A.; fax: 619/637-5776).

14. Advisory Committee on the Future of the US Space Program (1990) *Report*, December. Washington, DC: U.S. Government Printing Office.

15. Will, G. F. "Government is Failing by Dam Site," *Washington Post Writers Group,* August 14, 1995.

16. When as NASA consultant, your author was instrumental in encouraging NASA administration originally to publish an annual *Spinoff*, a detailed report of how space technology has been successfully transferred to commercial use. Past and current copies of *Spinoff* are available from the Center for AeroSpace Information, National Technical Information Service (5285 Port Royal Road, Springfield, Virginia 22161, U.S.A.; tel: 410/859-5300, ext. 241).

17. Harris, P. R. (1989) "Transforming Space Dreams into Realities," *Space Policy*, Vol. 5:4, pp. 273–278 ... . (1987) "Human Futures on the High Frontier," *Futures Research Quarterly,* Summer,Vol. 3:2, pp. 53–81.

18. Simon, M. E. (1994) "International Space Enterprise Now," *Space Governance,* June, Vol. 1:1, pp. 28–31 (journal of United Societies in Space Inc., 6841 So. Yosemite, Ste 3-C, Englewood, CO 80112, U.S.A.) ... Howerton, B. A. (1995) "A New Era of Lunar Exploration," *Ad Astra*, May/June, Vol. 7:3, pp. 40–44 (magazine of the National Space Society, 922 Pennsylvania Ave. SE, Washington, DC 20003, U.S.A.).

19. MacGregor, D. and Slovic, P. (1995) "The Planetary Exploration Survey," *The Planetary Report*, March/April, Vol. XV:2. pp. 4–6. Published by The Planetary Society (65 North Catalina Ave., Pasadena, CA 91106, U.S.A.).

20. Gump, D. P. (1990) *Space Enterprise Beyond NASA*. New York, NY: Praeger ... Harr, M. and Kohil, R. (1990) *Commercial Utilization of Space*. San Diego, CA: UNIVELT INC./Battelle Press ... Kraselsky, E. H. (ed.) (1988) *Space Commerce—An Industry Assessment.* Washington,DC: U.S. Department of Commerce ... Logsdon, J. M. (1988) *$pace Inc.—Your Guide to Investing in Space Exploration.* New York, NY: Crown Books ... Goldman, N. C.(1985) *Space Commerce: Free Enterprise on the High Frontier.* Cambridge, MA: Ballinger ... O'Leary, B. (1982) *Space Industrialization.* Boca Raton , FL: CRC Press, 2 volumes.

21. Goldman, N. C. (1992) "Space Law and Space Resources," *Social Concerns,* Vol. 2, pp. 143–153 in *Space Resources* edited by M. F./D. S. McKay and M. B. Duke. Washington, DC: U.S. Government Printing Office, NASA SP-509 ... . Anaejionu, P., Goldman, N. C., and Meeks, P. J. (eds) (1985) *Space and Society: Challenges and Choices.* San Diego, CA: UNIVELT NC/American Astronautical Society, Vol. 59.

22. Shahrokhi, F., *et al.* (eds) (1988; 1990) *Commercial Opportunities in Space; Space Commercialization Series—Launch Vehicles and Programs*, Vol. 1; *Platforms and Processing*, Vol. 2; *Satellite Technology*, Vol. 3. Washington, DC: American Institute of Aeronautics and Astronautics ... Information on the published volumes of *Space Manufacturing* proceedings may be obtained from the Space Studies Institute (PO Box 82, Princeton, NJ 08542, USA); distributed also by UNIVELT INC. (PO Box 28130, San Diego, CA 92198). The latter also publish *International Space Development Conference Proceedings* for the National Space Society.

23. Wilson, A. (ed.) (1995) *Interavia in Space Directory*. Surrey, U.K.: Jane's Information Group) ... Pardoe, G. K. (ed.) (1995) *Space Industry International—A Directory*. Essex, U.K.: Westgate House (The High, Harlow) ... Euroconsult (1988) *World Space Industry—A Ten Year Outlook*. Paris, FR: Euroconsult (17, Bd. Richard Leonoir, zip code 75011) ... *1995–96 U.S. Space Directory*. Bethesda, MD: Space Publications (PO Box 5782, zip code 20824, U.S.A.; tel: 301/718-9603). All publications are updated annually.

24. Gould, C. L. and Priest, C. C. (eds) (1978) *Space Industrialization: Final Briefings*. Los Angeles, CA: Rockwell International Space Division, 23 March (Contract NASA 8-32198, 4 volumes).

25. Kloman, E. (1984) *Encouraging Business Ventures in Space Technology*. Washington, DC: National Academy of Public Administration.

26. Goldin, D. S. (1994) *Agenda for Change—NASA Commercial Technology*, July ... *NASA Commercial Technology Directory ... How to Hook Up to NASA's Commercial Technology Network ... Partnership Options for NASA and Industry ... NASA Tech Briefs ... Space Technology Innovation ... Technology Today—A Resource for Technology, Science & Math Teachers*. All publications available from NASA Commercial Technology Networks, Office of Advance Concepts, NASA Headquarters (Code XC, Washington, DC 20546, U.S.A.; tel: 202/358-0701 or NCTN 415/528-2555; fax: 202/358-3878; e-mail: http://nctn.hq.nasa.gov).

27. Office of Commercial Programs (1990) *Accessing Space: A Catalogue of Process Equipment and Resources for Commercial Users*. Washington, DC: NASA Headquarters, December (NP-133) ... . McKay, M. F./D. S. and Duke, M. B. (eds) (1992) *Space Resources*. Washington, DC: U.S. Government Printing Office (NASA SP-509).

28. Wheelon, A. D. *et al.* (eds) (1993) *The Corona Papers—The Early Role of Spy Satellites*. Washington, DC: Central Intelligence Agency Press, 360 pp.

29. Graham, D. O. (1982) *High Frontier—a New Strategy*. Also see his numerous editorials in *The Journal of Practical Applications in Space*. Both published by High Frontier Inc. (2800 Shirlington Rd., Ste. 405, Arlington, VA 22206, U.S.A.; tel:1/703/671-4111; fax: 931-6432).

.30. Nozette, S. (ed.) (1987) *Commercializing Defense Technologies*. New York, NY: Praeger ... For the ultimate in space data bases used by the U.S. Space Command, as well as NASA, contact SAR (PO Box 49446, Colorado Springs, CO 80949, U.S.A.).

31. Office of Commercial Space Transportation (1995) *Strategic Plan*, February ... (1995) *Report of the COMSTAC Technology & Innovation Working Group*, May. Washington, DC: Department of Transportation, OCST S-50 (400 7th St. SW, Rm. 5415, zip code 20590).

32. U.S. Department of Commerce (1990) *Commercial Space Ventures—Financial Perspective*, April. Washington, DC: U. S. Department of Commerce (for current reports, contact Office of Air & Space Commerce, 1/202/482-6125).

33. Two other companies as an illustration: Future Port Inc. (326 S. Bundy Dr., Los Angeles, CA 90049, U.S.A.; tel.: 310/472-0846; fax: 472-3286) ... Lunar Corp (4350 N. Fairfax Dr., Ste 900, Arlington, VI 22203, U.S.A.; tel:703/841-9500; fax: 841-9503).

34   Simon, M. C. (ed.) (l994/l995) *International Lunar Exploration Conference Pro-
      ceedings*. San Diego, CA: International Space Enterprises (4909 Murphy Canyon
      Rd., Ste. 330, zip code 92123, U.S.A.; fax: 619/637-1234) ... Lauer, C. J., *et al.*
      (eds) (1995) *Briefing Book—Low Earth Orbit Business Park Concept.* Ann Arbor,
      MI: Peregrine Properties, Ltd. (540 Avis Dr., Ste. E, zip code 48108, U.S.A.).

35.   Howerton, B. A. (1995) *Free Space! Real Alternatives to Reaching Outer Space*.
      Port Townsend, Washington: Loompanics Unlimited (PO Box 1197, zip code
      98368, U.S.A.) ... (1995) "The Cold Hard Realities of Space Business," *Ad Astra*,
      March/April, pp. 44–45 ... See columns on Space Enterprise in *Space Governance*
      (6841 South Yosemite, Englewood, CO 80112); Space Available, an investors'
      newsletter published in *Countdown* (CSpace Press, PO Box 9331, Grand Rapids, MI
      49509).

36.   For further information, contact: California Spaceport Authority (3865-A Constella-
      tion Road, Lompoc, CA 93436, U.S.A.; tel: 805/733-5200; fax: 805/733-0717) ... .
      Pacific Aloha Spaceport Study (75-5751 Kuakini Highway, Kailua-Kona, Hawaii
      96740, U.S.A.; tel: 808/326-2014; fax: 996-2125) ... Lunar Enterprise Corp. (PO
      Box 4457, Burlingame, CA 94011; tel:408/996-9210; fax: 408/996-2125) ... Gen-
      eral Astronautics Corp. (802–810 West Broadway, Vancouver, B.C. V5Z 4C9,
      Canada; tel: 604/876-7640).

37.   Koelle, H. H. (1993) "A Frame of Reference for Extraterrestrial Enterprises," *Acta
      Astronautica*, Vol. 29: 10/11, 735–741.

38.   Goldman, N. C. (1988) "Commercial and Legal Space Frontiers" and "Political Di-
      mensions of Space Enterprise." Unpublished papers. The author may be contacted
      about his publications at 2328 Dryden Rd., Houston, TX 77254, U.S.A. ... (1996),
      "A Lawyers Perspective on the USIS Strategies for Metanation and Lunar Economic
      Development," *Space Governance*, June, Vol. 3:1.

39.   NASA Life Sciences Strategic Planning Committee (1988) *Exploring the Living
      Universe—A Strategy for Space Life Sciences*. Washington, DC: NASA Headquar-
      ters, June.

40.   Glaser, P. E., Davidson, F., and Csigi, K. (eds) (1992) *Solar Power Satellites—The
      Emerging Energy Option*. Chichester, UK: Ellis Horwood.

41.   Gump, D. P. (1990) *Space Enterprise Beyond NASA*. New York, NY:Praeger.

42.   Sheffield, C. and Rosen, C. (1984) *Space Careers*. New York, NY: Quill Publishing.

43.   For information on American Society of Civil Engineers conferences and publica-
      tions on space, contact ASCE Aerospace Division (345 East 47th St., New York,
      NY 10017, U.S.A.), or Dr Stewart W. Johnson (tel/fax: 505/298-2124 or e-mail:
      StWJohnson@aol.com).

44.   Gibson, R. (1990) "Commercial Space Activities: An Overview," *Space Commerce,*
      Vol. 1:1, pp.3–6.

45.   Miller, R. (1993) *The Dream Machines: An Illustrated History of the Spaceship in
      Art, Science, and Literature*. Malbar, FL: Krieger Publishing ... Yenne, B. (1985)
      The Encyclopedia of US Spacecraft. New York, NY: Exeter/Bison
      Books/Bookthrift Marketing.

46.   Yenne, B. (1986) *The Space Shuttle*. New York, NY: Gallery Books/ W. H. Smith
      Publishers (112 Madison Ave., zip code 10016) ... . Torres, G. (1989) *Space Shut-
      tle—The Quest Continues*. Novato, CA: Presidio Press (PO Box 1764SS, zip code

94948) ... Steinberg, F. S. (1980) *Aboard the Space Shuttle*. Washington, DC: NASA Division of Public Affairs.

47. Mansfield, J. E. (1995) Luncheon address at the 31st AIAA Joint Propulsion Conference, San Diego, California, July 11 ... Benson, J. "Conversations with John Mansfield," AIAA's *Aerospace America*, Nov. 1995, pp. 12–13. Conference proceedings and magazine available from American Institute of Aeronautics and Astronautics (370 L'Enfant Promenade SW, Washington, DC 20024, U.S.A.).

48. Throckmorton, D. (1995) "Today's Technologies for Tomorrow's Launch Vehicles," *Aerospace America*, June, pp. 20–24 ... Office of Technology Assessment (1990) *Access to Space: The Future of theU.S. Space Transportation Systems*. Washington, DC: U.S. Government Printing Office.

49. Keith, E. L., *et al.* (1995) "Scorpius: Getting Their at Very Low Cost," *Second Annual International Lunar Conference Proceedings*, Nov. 14. For copies, contact manufacturer, Microcosm, Inc. (2601 Airport Dr., Ste 230, Torrance, CA 90505; tel: 310/539-9444; fax: 539-7268).

50. The Space Transportation Association (2800 Shirlington Road, Ste. 405A, Arlington, VA 22206, U.S.A.) publishes a newsletter, *Space Trans* and co-sponsors *The Journal of Practical Applications in Space*. *JPAS* devoted its entire Summer 1993 issue (Vol. IV:4) to the SSTO theme and regularly carries articles on space transport issues.

51. Sloan, J. H. (1995) "A Two Launch Vehicle Architecture to Reduce Space Transportation Cost", AIAA 95-3089, July 10–12 (paper available from author at Information Universe, 17701 S. Avalon Blvd. #407, Carson, CA 90746, U.S.A.).

52. The Interstellar Propulsion Society (PO Box 1292 LaJolla, CA. 92038, U.S.A. (e-mail: hup://www.digimark.net). The Society produces a quarterly newsletter and a journal, maintains an advanced digital library, while sponsoring research grants and conferences.

53. Goldman, N. C. (1996) *American Space Law: International and Domestic*, 2nd Ed. San Diego, CA: UNIVELT INC.

54. O'Donnell, D. J. and Harris, P. R. (1994) "Is it Time to Replace the Moon Agreement?" *The Air & Space Lawyer* (American Bar Association), Vol. 9:2, Fall, pp. 3–11 ... . "Space-Based Energy Needs a Consortium and a Revision of teh Moon Treaty," *Space Power*, Vol. 13: 1/2, pp. 121–134. Reprints available from the World Bar Association (6841 So. Yosemite, Englewood, CO 80002; tel: 1-800/632-2828; fax: 303/721-1182).

55. Benko, M. and Schrogl, K-W (eds) (1994) *International Space Law in the Making: Current Issues on the Peaceful Uses of Outer Space*. New York, NY or Austria, Vienna: United Nations Publications or Editions Frontier, Gif-sur-Yvette ... Jasentuliyana, N. (ed.) (1992) *Space Law—Development and Scope*. Westport, CT: Praeger ... Meredith, P. L. and Robinson, G. S. (1992) *Space Law: A Case Study for the Practitioner*. Dordrecht, the Netherlands: Martinus Nijhoff Publishers ... Christol, C. (1982) *The Modern International Law of Outer Space*. New York, NY: Pergamon Press.

56. Gabrynowicz, J. I. (1994) "A Beginners Guide to Space Law," *Ad Astra*, March/April, Vol. 6:2, pp. 43–45.

57. Hargrove, E. C. (ed.) (1986) *Beyond Spaceship Earth—Environmental Ethics and the Solar System*. San Francisco, CA: Sierra Club Books.

58. Goldman, N. C. (1986) "The Strategic Defense Initiative: Star Wars and Star Laws," *Houston Journal of International Law*, Vol. III-9.

59. World Space Congress (1993) *Proceedings of the Thirty-Fifth Colloquium on Law in Outer Space*. Washington, DC: American Institute of Aeronautics and Astronautics ... Christol, C. Q. (1991) *Space Law—Past, Present, and Future*. Deventer, the Netherlands/Boston, MA: Kluwer Publishing ... Wassenberg, H. A. (1991) *Principles of Outer Space Law in Hindsight*. Dordrecht, the Netherlands/Boston, MA: Martinus Nijoff.

60. Lafferranderie, G. and Tuinder, P. H. (1994) "The Role of the Evolution of Space Law," *Journal of Space Law*. Vol. 22: 1/2, pp. 97–113.

61. Finch, E. R. and More, A. L. (1984) *Astrobusiness: A Guide to Commerce and Law*. New York, NY: Praeger.

62. Heppenheimer, T. A. (1977) Colonies in Space; (1979) *Toward Distant Suns—A Bold, New Prospectus for Human Living in Space*. Harrisburg, PA: Stackpole Books (PO Box 1831, zip code 17105, U.S.A.) ... . Savage, M. T. (1994) *The Millennial Project—Colonizing the Galaxy in 8 Easy Steps*. New York, NY: Little Brown & Co.

63. Robinson, G. S. and White, H. M. (1986) *Envoys of Mankind: A Declaration of First Principles for the Government of Space Societies*. Washington, DC: Smithsonian Institution Press.

64. Robinson, G. S. (1995) "Natural Law and a Declaration of Humankind Interdependence," *Space Governance*, June, Part 1; December, Part 2, Vol. 2: 1/2. This journal is published by the United Societies in Space Inc. (6841 So. Yosemite, Ste 3-c, Englewood, CO 80112, U.S.A.).

65. O'Donnell, D. J. (1994) "Overcoming the Barriers to Space Travel," *Space Policy*, November, Vol. 10:4, pp. 252–55 ... (1994) "Founding a Space National Utilizing Living System Theory: A New Research Challenge," *Behavioral Science*, April, Vol. 39:2, pp. 93–116 ... with P. R. Harris (1995) "Strategies for Lunar Development and Port Authority."

66. Logsdon, J. M. (1970) *The Decision to Go to the Moon: Project Apollo and the National Interest*. Cambridge, MA: MIT Press ... (1986) "The Space Shuttle Program—A Policy Failure?" *Science*. May 30.

67. Heis, K. P. (1989) "Enterprise and the Strategic Position of Space," *Journal of Practical Applications in Space*, Fall, Vol. 1:1 ... Macaulay, M. (ed.) (1987) Economics and Technology in U.S. Space Policy. Washington, DC: Resources for the Future (1616 P St., NW, zip code 20036) ... McDougal, W. A. (1986) *A Political History of the Space Age*. New York, NY: Basic Books.

68. Congressional Budget Office (1988) *The NASA Program in the 1990s and Beyond*. Washington, DC: U.S. Congressional Budget Office, May ... National Committee on Space (1995) *Space for America—Meeting the Needs in the 21st Century*. Springfield, VA: American Astronautical Society.

69. Bormanis, A. and Logsdon, J. M. (eds) (1992) *Emerging Policy Issues for Long-Duration Human Exploration*. Washington, DC: Space Policy Institute, The George Washington University, December Workshop Report.

70. Goldman, N. C. (1992) *Space Policy: An Introduction.* Ames, Iowa: Iowa State University Press.

71. Wilford, J. N. (1990) *Mars Beckons: The Mysteries, the Challenges, the Expectations of Our Next Great Adventure in Space.* New York, NY: Alfred Knopf.

72. deSelding, P. B. (1995) "European Union Takes Charge of Space Policy," *Space News*, November, Vol. 6:44, p. 1/30.

73. McKurdy, H. E. (1990) *The Space Station Decisions: Incremental Politics and Technological Change.* Baltimore, MD: The Johns Hopkins University Press.

74. Logsdon, J. (1987) "The Space Shuttle Program—A Policy Failure?" *Science*, May 30.

75. Byerly, R. (ed.) (1989) Space Policy Revisited. Boulder, CO: Westview Press.

76. Steinbronn, O. and Cordell, B. (1992) "International Human Expeditions to Mars—Suggestions and Mechanisms," *The Case for Mars IV.* San Diego, CA: UNIVELT INC.

77. Foffert, M. I., Potter, S. D., Kadiramangalam, M. N., and Tubielle, F. (1992) "Solar Power Satellites: Energy Sources for the Greenhouse Century?" in L. Deschamps (ed.) SPS91: *Power From Space, Second International Symposium*, Paris, FR: Societé des Electriciens et des Electroniciens (SSE) and Societé des Ingenieurs et Scientifiques de France (ISF), August, pp. 82–90.

78. ESA Lunar Study Steering Group (1994) *International Lunar Workshop—Towards a World Strategy for the Exploration and Utilization of our Natural Satellite,* November (ESA SP-1170); (1992) *Mission to the Moon—Europe's Priorities for the Scientific Exploration and Utilization of the Moon,* June (ESA SP-1150). Paris, FR: European Space Agency.

79. These American national space studies have all been referenced in Chapters 1 and 2 with the exception of this one: National Space Council (1990) *Report to the President.* Washington, DC: U.S. Government Printing Office.

80. Marshall, A. (1995) "Development and Imperialism in Space," *Space Policy*, Vol. 11:1 ... . (1993) "Ethics and Extraterrestrial Environment," *Journal of Applied Philosophy*, Vol. 10:2.

81. Simon, M. C. (1987) *Keeping the Dream Alive* San Diego, CA: Orbit Books/UNIVELT INC.

CHAPTER 10

1. Refer to special issue, Return to the Moon and Stay, *Ad Astra*, May/June 1995, Vol. 7:3 (National Space Society, 922 Pennsylvania Ave., S.E., Washington, D.C. 20003, USA; tel: 1-800/543-1280).

2. Harrison, A. A., Clearwater, Y. A., and McKay, C. P., (eds) *From Antarctica to Outer Space—Life in Isolation and Confinement.* New York, NY: Springer-Verlag, 1991 ... . Bell, L. *et al.* (eds) *IDEEA ONE Conference Proceedings.* Houston, TX: University of Houston/College of Architecture, 1991. (NOTE: For information on these and other proceedings—*IDEEA TWO*, 1993, University of Montreal, Canada, and *IDEEA THREE*, 1996, University of Oxford, U.K., fax 1-713/743-2358).(cancelled conference)

3. McKay, M. F., McKay, D. S., and Duke, M. B. (eds) *Space Resources.* Washington, DC: U.S. Government Printing Office, 1992, 6 vols. (NASA SP-509) ... Mendell, W. W. (ed.) *Lunar Bases and Space Activities of the 21st Century.* Houston, TX: Lunar and Planetary Institute, 1985.

4. European Space Agency, Paris, France: Press Release on "International Lunar Workshop Declaration" (Nr.12-94, June 6, 1994) ... ESA Lunar Study Steering Group Report, *Mission to the Moon* (SP-1150, 1992) ... Kyodo News Service/News Net: "Moon Station Blueprint Drawn Up by Lunar Society," Tokyo, Japan, May 31, 1994 ... "Joint Power Projects with the Russians," Vol. 1:1, p. 32; "The Need for a Global Space Based Power Transmission Consortium" by Peter Glaser, Vol. 1:2, pp. 32–33, in *Space Governance,* journal of United Societies in Space, Inc., Englewood, Colorado, U.S.A .... Criswell, D. R. "No Growth in the Two Planet Economy," IAF proceedings (8.1.704) of the 45th Congress, Jerusalem, Israel, October 9, 1994.

5. Simon, M. C. "International Space Enterprise Now," *Space Governance*, journal of United Societies in Space, Englewood, Colorado, Vol. 1:1, June 1994, pp. 28–31 ... . Simon, M. C. (ed.) *Proceedings of the International Lunar Exploration Conference,* Nov. 14–16, 1994 (available from International Space Enterprises, 4909 Murphy Canyon Rd., Ste. 330, San Diego, CA 92123, U.S.A.; fax: 1-619/637-5776) ... See also Howerton, B. A., "International Space Enterprise: A New Era of Lunar Exploration," *Ad Astra,* May/June 1995, pp. 40–44; Goldstein, E. "Luna Corp's Enchanted Pathway Back to the Moon," pp. 37–39.

6. Krauthammer, C. "Clinton is Cutting NASA to Pieces," Washington Post Writers Group in *San Diego Union,* April 9, 1995, G-2.

7. Beckey, I. "Prospects for SSTO," *The Journal of Practical Applications in Space,* Fall 1944, Vol. VI:1, pp. 71–84 ... Wasser, A., "The Most Valuable Real Estate Off Earth," *Ad Astra,* Vol. 7:3, May/June 1995, p. 29 ... Lawler, A. "Agencies Eye Clementine 2 Mission to Moon," *Space News,* August 15, 1994 ... Dasch, P. "Lunar Prospector," *Ad Astra*, May/June 1995, pp. 31–33.

8. Reynolds, G. H., "Return of the Moon Treaty," *Ad Astra*, May/June 1994, pp. 27–29 (National Space Society, Washington, DC) ... O'Donnell, D. J. and Harris, P. R., "Is It Time to Amend or Replace the Moon Treaty," *Air & Space Lawyer*, Vol. 9:3, Summer 1994 (American Bar Association, Chicago, IL ... O'Donnell, D. J. "Founding a Space Nation Utilizing Living Systems Theory: A New Research Challenge," *Behavioral Science* Vol. 39:2, April 1994, pp. 93–116 (Journal of International Society for the Systems Sciences, Box 26519, San Diego, CA 92196, U.S.A.). "Survey of Top Ten Space Policy Problems at 1995," *Space Governance* Vol. 2:2, December 1995, pp.40–43 ... Reijnen, B. C. M. (1992) *The United Nations Space Treaties Analyzed*; plus Benko, M. and Schrogi, K-U (eds) (1993) *International Space Law in the Making—Current Issues in the UN Committee on Peaceful Uses of Outer Space, Vol.1*. Paris, France: Editions Frontières.

9. Benaroya, H. "Lunar Industrialization," *The Journal of Practical Applications in Space,* Fall 1994, Vol. Vl:1, pp. 85–94 ... Perek, L. "Management of Outer Space," *Space Policy*, August 1994, Vol. 10:3, pp. 189–198.

10. Robinson, G. S., "Natural Law and a Declaration of Humankind Interdependence," *Space Governance*, the journal of United Societies in Space, Inc., June 1995, Vol.

2:1 Part I; Dec. 1995, Vol. 2.2, Part II ... See also Robinson, G. S. and White, H. M., *Envoys of Mankind—A Declaration of First Principles for the Governance of Space Societies*. Washington, DC: Smithsonian Institution Press, 1986 (See Chapter 23, "The Moon Treaty and Res Communis").

11. O'Donnell, D. J. and Harris, P. R. "Space-Based Energy Needs a Consortium and a Revision of the Moon Treaty," *Space Power,* 1994, Vol. 13:1/2, pp. 121–134 ... "Overcoming Barriers to Space Travel," *Space Policy*, Nov. 1994, Vol. 10:4 pp. 252–54 ... . Harris, P. R., "A Case for Lunar Development and Investment," *Space Policy*, Aug. 1994, Vol. 10:4, pp. 187–88 ... See also Heiken, G. H., Vaniman, D. T., and French, B. M. (eds) *The Lunar Sourcebook*. New York, NY: The Cambridge University Press, 1991.

12. O'Donnell, D. J., "Metaspace—A Design for Governance in Outspace," *Space Governance*, June 1994, Vol. 1:1, pp. 8–17 ... O'Donnell, D. J. and Harris, P. R. "Space-Based Energy Needs a Consortium and Revision of the Moon Treaty," *Space Power—Resources, Manufacturing and Development*. Vol. 13: 1/2 1994, pp. 121–134 ... Fasan, E. (1994) "Human Settlements on Planets: New Stations or New Nations," *Journal of Space Law*, Vol. 22:1/2, pp. 47–56.

13. Interagency Arctic Research Policy Committee (1987) *United States Research Policy Committee*. Washington, DC: National Science Foundation ... *Antarctic Journal of the United States*, published quarterly by the National Science Foundation and available from the U.S. Government Printing Office, Washington, DC 20402, U.S.A ... Re "Conservation and the Tennessee Valley Authority," refer to the *United States Code*, Vol. 16, Chapter 12-A, Section 831-N-4-b.

14. Jasentuliyana, N. *Space Law—Development and Scope*. Westport, CT: Praeger Publishing, 1992, pp. 98–9 ... Meredith, P. L. and Robinson, G. S. *Space Law: A Case for the Practitioner*. Dordrecht, the Netherlands: Martinus Nijhoff Publishers, 1992 ... Goldman, N. C. (1996) *American Space Law*. San Diego, CA: UNIVELT Inc ... Benko, M. and Schrogl, K. H. (eds) (1994) *International Space Law in the Making*. Paris, France: Editions Frontières.

15. Cleveland, H. *Birth of a New World Order: An Open Moment for International Leadership*. San Francisco, CA: Jossey-Bass Publishers, 1993 ... Davidson, F. P. and Meador, C. L. (eds) *Macro-engineering Solutions*. Chichester, UK: Ellis Horwood, 1992.

16. For further information on proposals for Metanation and the Lunar Port Authority, contact United Societies in Space, inc. (6841 South Yosemite, Englewood, CO 80112, U.S.A.; tel: 1-800/895-META; fax: 1-303/721-1117; net: WWW.Tagsys.Com). USIS Countdown Conference, Aug. 2–4, 1995 and 1996, Colorado, were on the subject of lunar development; the proceedings were published in its journal, *Space Governance*, Dec. 1995, Vol. 2.2: Dec. 1996, Vol. 3:2.

17. Koelle, H. H., Chairman, IAA Subcommittee on Lunar Development's draft report on *Recommended Lunar Development Scenario*, January 1995. (Available from Prof. Koelle at Aerospace Institute, Technische Universität, Berlin, Marchstr. 14, D-10587, Berlin, Germany; fax: 030/ 314-22866.)

18. Stuhlinger, E and Ordway III, F. I., *Wernher von Braun: Crusader for Space*, Vol. 1, *A Biographical Memoir*; Vol. 2, *An Illustrated Memoir*. Malabar, FL: Krieger

Publishing, 1994 ... . O'Neill, G. K. *The High Frontier—Human Colonies in Space*. Princeton, NJ: The Space Studies Institute, 1989 (PO Box 82, zip code 08542) ... Savage, M. T. *The Millennial Project—Colonizing the Galaxy in 8 Easy Steps*. Denver, CO: Empyrean Publishing, 1992 (l6l6 Glenarm, Ste. 101-B, zip code 80202).

19. Lauer, C., *et al.* (1995) "A Reference Design of a Near Term Low Earth Orbit Commercial Business Park;" "Legal & Regulatory Aspects of Low Earth Orbit Business Park Development," published by the American Institute of Aeronautics and Astronautics (see Appendix C) ... *Briefing Book: Low Earth Orbit Business Park Concept*, published by Pererine Properties, LTD. (540 Avis Dr., Ste. E, Ann Arbor, MI 48108, U.S.A.)

20. Harris, P. R. "Viewpoint—Space Policy for the New U.S. Administration," *Space Policy*, May 1993, Vol. 9:2, pp. 82–84.

21. Howerton, B. A. *Free Space! Real Alternatives for Reaching Outer Space*. Port Townsend, WA: Loompanics Unlimited, 1995 (PO Box 1197, zip code 98368).

EPILOGUE

Bormanis, A. and Logsdown, J. M. (1992) *Emerging Policy Issues for Long Duration Human Space Exploration*. Washington, DC: The Space Policy Institute/The George Washington University.

Butler, G. V. (ed.) (1990) *The 21st Century in Space*. San Diego, CA: UNIVELT Inc., AAS Volume 70.

Hargrove, E. C. (ed.) (1986) *Spaceship Earth—Environmental Ethics and the Solar System*. San Francisco, CA: Sierra Club Books, Ch. 1 by David Brin.

Harris, P. R. (1996) *New Work Culture and HRD Transformational Management;* (1994) *High Performance Leadership*. Amherst, MA: Human Resource Development Press.

Lewis, R. S. (1990) *Space in the 21st Century*. New York, NY: Columbia University Press.

McKay, M. F., McKay, D. S., and Duke, M. B. (eds) (1992) *Space Resources*. Washington, DC: U.S. Government Printing Office, NASA SP-509, 5 volumes (see volume 4, *Social Concerns*).

National Commission on Space (1986) *Pioneering the Space Frontier*. New York, NY: Bantam Books.

O'Donnell, D. J. (1994) "Founding a Space Nation Utilizing Living Systems Theory," *Behavioral Science*, April, Vol. 39:2, pp. 93–116.

O'Neill, G. K. (1981). *2081 A Hopeful View of the Human Future*. New York: Simon & Schuster/Touchstone Books.

Robinson, G. S. and White, H. M. *Envoys of Mankind*. Washington, DC: Smithsonian Institution Press, 1986.

Simon, M. C. (1987) *Keeping the Dream Alive*. San Diego, CA: UNIVELT/Earth Space Operations.

Steinbronn, O. and Cordell, B. (1990) "International Human Expeditions to Mars—Suggestions and Mechanisms" in *The Case for Mars IV*. San Diego, CA: UNIVELT, Inc.

Tichy, N. M. and DeVanna, M. A. (1986) *The Transformational Leader*. New York: John Wiley & Sons.

APPENDIX B—CASE STUDY

Angelo, J. and Easterwood, G. W. (1989) *Lunar Base Concepts, Technologies and Applications*. Melbourne, FL: Orbit Books/Krieger Publishing.

Battrick, B. and Ockels, W. J. (eds) (1994) *A Moon Programme: The European View*. Paris, France: European Space Agency, May (ESA BR-101).

Bodie, S. (ed.) (1990) *Lunar Power System Planning Workshop Proceedings*, San Diego, CA: Lunar Power System Coalition, July 9–11, 1990. (See LPSC address in Appendix C.)

Criswell, D. R. (1995) "Lunar Solar Power System: Scale and Cost Versus Technology Level, Boot-strapping, and Cost of Earth-to-orbit Transport." 46th International Astronautical Federation Congress, October 2–6, Oslo, Norway (IAA-95-R.2.02—see address of either IAA or IAF in Appendix C.)

Criswell, D. R. (1994) "Net Growth in a Two Planet Economy," 45th International Federation Congress," October 9–14, Jerusalem, Israel (IAA-94-IAA.8.1.704).

Criswell, D. R. (1993a) "World Energy Requirements in the 21st Century," Houston, TX (77204): University of Houston/Institute of Space Systems Operations.

Criswell, D. R. (1993b) "Lunar Solar Power System and World Economic Development," World Solar Summit, *Solar Energy and Space Report*, July 5–9 (Chapter 2.5.2). Paris, France: United Nations/UNESCO Publications.

Criswell, D. R. (1992) "Solar Power System Based on the Moon," in *Solar Power Satellites—The Emerging Energy Option* by Glaser, P. E., Davidson, F., and Csigi, K. (eds) Chichester, UK: Ellis Horwood, Part 4.8.

Criswell, D. R. (1989) "Lunar Power System—Summary of Studies" in *Report of the NASA Lunar Energy Enterprise Case Study Task Force*, NASA Technical Memorandum # 101652, July, pp. 84–96.

Criswell, D. R. (1980) *Lunar Power System Preliminary Financial Analysis,* 18 pp., prepared for NASA Code Z Lunar Enterprise Committee.

Criswell, D. R. (1979) "The Initial Lunar Supply Base," *Space Resource and Space Settlements*, NASA SP-428, pp. 207–224. Washington, DC: NASA Headquarters.

Criswell, D. R. and Waldron, R. D. (1993a) "Lunar Solar Power System: Options and Beaming Characteristics," 44th International Astronautical Federation Congress, October 16–22, Graz, Austria. Paris, France: International Astronautics Federation (IAF-93-R.2.430).

Criswell, D. R. and Waldron, R. D. (1993b) "International Lunar Base and Lunar-Based Power System to Supply Earth with Electric Power," *Acta Astronautica*, Vol. 29:6, pp. 469–480. Paris, France: International Astronautical Federation.

Criswell, D. R. and Waldron, R. D. (1991), "Results of Analysis of Lunar-based Power System to supply Earth with 20,000 GW of Electric Power," paper presented at SPS '91, Paris, France, August.

Criswell, D. R. and Waldron, R. D. (1990a) *A Power Collection and Transmission System*, United States Patent—Granted, 1990.

Criswell, D. R. and Waldron, R. D. (1990b) "Lunar System to Supply Solar Electric Power to Earth," paper #900279—area #243, 25th Intersociety Energy Conversion Engineering Conference, Reno, Nevada, August 11–17.

Criswell, D. R., Waldron, R. D., and Erstfeld, T. (1980) *Extraterrestrial Materials Processing & Construction,* 500 pp. (Microfiche from National Technical Information Service).

General Dynamics (1969) *Operations and Logistics Study of Lunar Exploration Systems for Apollo*; (1979) *Lunar Resources Utilization for Space Construction.* (NASA Contract No. NAS9-155560; GDC-ASP79-001). San Diego, CA: General Dynamics Convair Division (now Lockheed-Martin).

Glaser, P. E. (1977) "Solar Power from Satellites," *Physics Today,* pp. 30–38, February ... . "The Solar Power Satellite (SPS)—Progress So Far," Appendix B-3, 68-83, Report of NASALunar Energy Enterprise Case Study Task Force, July 1989.

Glaser, P. E. (1989) "The Solar Power Satellite—Progress So Far" in *Report of NASA Lunar Energy Enterprise Task Force Case Study,* July , NASA Tech. Memo. 101653, pp. 68–83.

Glaser, P. E. (1994) "The Power Relay Satellite," *Space Power,* Vol. 13:1/2, pp. 1–23. Published by the Council for Economic and Social Studies (1133 13th St. NW, Ste 2-C, Washington, DC 20005, U.S.A.).

Glaser, P. E., Davidson, F., and Csigi, K. (eds) (1992) *Solar Power Satellites—The Emerging Energy Option.* Chichester, UK: Ellis Horwood.

Harris, P. R. (1991) "Promoting the Coalition for a Lunar Power System," *Space Policy,* May (Butterworth Scientific Ltd., Oxford, U.K.).

Harrison, A. A., Clearwater, Y. A., and McKay, C. P. (eds) (1991) *From Antarctica to Outer Space: Life in Isolation and Confinement.* New York: Springer-Verlag, 1991. (See Ch. 7 by P. R. Harris, "Personnel Deployment Systems—Managing People in Polar and Outer Space Settings", pp. 156–188).

International Astronautic Federation)/IISL (Institute of Space Law). *Colloquium on the Law of Outer Space.* Washington, DC: American Institute of Aeronautics and Astronautics/IAF/IISL,   3 vols., 1969. 1988, 1987.

Johnson-Freese, J. (1990) *Changing Patterns of International Cooperation in Space.* Melbourne, FL: Orbit Books/Krieger.

McFarland, D. E. *The Managerial Imperative—the Age of Macromanagement.* Ballinger/Harper & Row, 1986.

McKay, M. F., McKay, D. S., and Duke, M. B. (eds) (1992). *Space Resources.* Washington, DC: U.S. Government Printing Office (NASA SP-509, 5 vols).

Mendell, W. W. (ed.) (1985) *Lunar Bases and Space Activities of the 21st Century.* Houston, TX: Lunar and Planetary Institute.

NASA Technical Memorandum #101652, *Report of NASA Lunar Energy Enterprise Case Study, July 1989*, NASA Office of Exploration, Washington, DC (Refer to "Executive Summary," pp. 1–6.)

O'Donnell, D. J. and Harris, P. R. (1994) "Space-Based Energy Needs a Consortium and a Revision of the Moon Treaty," *Space Power,* Vol. 13:1/2, pp. 121–134.

Peterson, M. N. A., Criswell, D. R., and Greenwood, D. (1990) "Clean, Sustainable Solar Power from the Moon." (Paper presented at the Earth Observation Conference, October 24, at the National Press Club in Washington, DC)[*]

Prisnyakov, V. "SPS Interest and Studies in the USSR," *Space Power*, Vol. 13:1/2, pp. 25–38.

Sarkisyan, S. A., *et al.* (1985) "Socio-Economic Benefits connected with the Use of Space Power and Energy Systems." (Paper presented at the International Astronomical Federation Congress; proceedings subsequently published by Pergamon Press, IAF-85-188.)

Simon, M. S. (1994/1995) *Proceedings of the International Lunar Exploration Conference,* vols. 1 & 2. Published by International Space Enterprises (4909 Murphy Canyon Rd., Ste. 330, San Diego, CA 92123, U.S.A.).

---

[*] Lunar Power System Coalition (LPSC)—contact for proceedings and papers: Dan Greenwood, LPSC Chairman; President, NETROLOGIC Inc., Suite 201, 5080 Shoreham Pl, San Diego, CA 92122, U.S.A. (Tel. 619/587-0970 or Fax: 619/ 458-1624) ... Dr David R. Criswell, Director, Institute for Space Systems Operations, College of Science and Mathematics, University of Houston, Houston, Texas 77204-5502 (Tel. 713/749-9135 or Fax: 619/713/486-5019) ... Dr Robert D. Waldron, LPSC President, 15339 Regaldo St., Hacienda Heights, CA 91745, U.S.A.

Index